D1546187

FORCE CONTROL OF ROBOTICS SYSTEMS

DIMITRY M. GORINEVSKY
Honeywell-Measurex
Vancouver, Canada

ALEXANDER M. FORMALSKY
Moscow University, Russia

ANATOLY YU. SCHNEIDER
Fraunhofer Institute for Industrial Automation
Magdeburg, Germany

CRC Press
Boca Raton   New York

Acquiring Editor: *Norm Stanton*
Senior Project Editor: *Susan Fox*
Cover Design: *Dawn Boyd*
Prepress: *Kevin Luong*
Translator: *Alla M. Gorinevsky*

Library of Congress Cataloging-in-Publication Data

Catalog record is available from the Library of Congress.

Developed by Nauka Publishers, Fizmatlit of the Academy of Sciences of Russia, Moscow.

Contents

Preface to English Edition

In preparation of this edition, the survey of the subject in the first chapter was re-worked to reflect the current state of force control research. About 150 new references were added to the bibliography. On the other hand, some Russian references of the original edition were removed as inacessible to the mainstream reader. Readers familiar with technical Russian can still access this material through the Russian edition.

Though several years have passed since the original version of the book was written, recent developments in the field have not changed the fundamental issues of force control discussed in the book. Therefore, the modifications made in Chapters 2–11 are mostly confined to the update of the references and selected topics, without changing the structure of the book.

This edition would be impossible without the dedication and patience of my wife Alla who translated the book from Russian.

Dimitry Gorinevsky   Vancouver, December 1996

Preface

Many important early developments in the theory and practice of Automatic Control are related to problems of controlling moving objects, such as airplanes, spacecraft, missiles, ships, and submarines. Control of such vehicles is usually based on the information on their position, orientation, linear and angular velocities. Control of free motion of manipulators is also based on the information on coordinates and velocities of their links. At the same time, to control a manipulator performing assembly or machining operations it is important to know deformations and stresses in the manipulator structure. Such information allows one to make judgements on the interaction between the manipulator and the environment and correct manipulator motions. The information on the stresses in the structure can be obtained from force and torque sensors.

By installing force sensors in the manipulator arm, it becomes possible to monitor and control micro-displacements of the manipulator end-effector with respect to external objects. Compliance of the manipulator structure does not spoil the accuracy of such displacement measurement. Thus, in many problems rigidity requirements of the manipulator structure can be relaxed. Note that information on internal stresses and deformations is also important in control of aerospace vehicles with nonrigid structure.

In flexible automated manufacturing and other applications of Robotics, robotic manipulators perform an ever expanding number of tasks requiring to use information about the environment for robot control. Force information, in particular, is critical in assembly, packing, surface machining (grinding, fettling), and other tasks. By using force information, reliability and performance can be improved for almost all tasks in which a manipulator keeps contact with external objects. A high performance can be achieved with simpler and cheaper manipulator hardware, since less precision is required. Information from force sensors or load cells is very important for many other types of automated industrial devices as well.

The scientific problems of manipulator force control are complex and remain current; the literature on these topics is extensive and growing, which shows the vitality of the field. A few books on Robotics touch upon the related problems. However, to the authors' knowledge, the literature lacks a monograph concentrating on the fundamentals of manipulator force control.

Much of the available literature on the subject of manipulator force control considers various advanced force control algorithms that assume knowledge of detailed models of manipulator dynamics and environment. At the same time, more fundamental issues pertaining to the practice of robotics control are often overlooked. There are a few fundamental issues related to force control that students and engineers venturing into this area need to be aware of. The first issue is force sensor design, or integration. Since a force sensor is installed in a manipulator arm structure and bears structure loading, its elasticity influences the system dynamics. Second, a force sensor signal actually gives the information on the manipulator micro-displacements with respect to contacted objects. Hence, a feedback of the force sensor signal closely interacts with the position control loop. This interaction of force and position control loops needs to be thoroughly understood. Third, force control of manipulator motion in contact with external objects should be theoretically treated as a motion with imposed mechanical constraints.

This book consistently exposes and analyses the problems related to force control. The theoretical analysis in the book is based on the methods of Analytical Dynamics and Automatic Control Theory. Control algorithms and force sensor design methods developed in the book are experimentally verified and the emphasis is on the practically important problems.

The book is based on the research the authors and their colleagues have performed since the beginning of the 1980s. The group in which the authors worked was one of the first to engage in force control research in the former Soviet Union. Because of the limitations on information exchange between Soviet and Western research communities at that time, only some of the results presented in the book were published in international conference proceedings and international technical journals. A part of the results was published in major Russian academic journals, which - though translated into English - remain less accessible to Western readers.

The book considers the following fundamental problems related to force control: design of force sensors, force feedback synthesis and stability analysis (with an emphasis on one- and two-dimensional motion), and design of simple and efficient control algorithms for performing certain applied tasks with a robot. In the book, problem statements are defined mainly by mechanical and physical features of manipulator force control and are fundamental. The solutions of these problems do not depend on the ever improving level of computer technology. At the same time, computational problems related to manipulator kinematics and dynamics were left outside of the scope of the book. A large number of published papers and books consider these problems in much detail.

The book is intended for students, engineers, and scientists working or interested in the fields of automation and robotics systems. One can use it as a textbook; different parts of the book contents were used in several graduate courses.

The book has three parts corresponding to the three main stated problems of force control. The material is organized in eleven chapters.

The readers who are interested just in the issues of force sensor design need to read only Chapter 1 (Section 1.2) and Part II of the book. Conversely, the readers who are primarily interested in force control of manipulator systems can omit Part II from their reading. In order to understand the algorithms for performing applied tasks described in Part III, it is sufficient to read Sections 1.3, 6.1, 6.2, and 7.1–7.3 before reading this Part.

The first chapter is introductory, it provides a brief overview of the subject. The main objective is to survey the state of the art in the use of force-related information for manipulator system control.

The second and the third chapters concentrate on force sensors and principles of their design. The second chapter presents an analysis of force sensor design, stressing two major approaches. One approach is to compose a sensor from separate modules, each being mechanically sensitive to one or two components of force or torque. Another method is to design a multi-component sensor without decoupled channels. In such a sensor, individual components of force and torque can be obtained by a linear transformation of the raw data after careful calibration of the sensor.

Usually, force sensors are mounted in the manipulator wrist or built into the gripper fingers. Forces are determined by measuring deformations or displacements of loaded compliant elements. The most widespread and promising are strain-gauge force transducers, in which deformations of a compliant element are converted into a change of the electric resistance of strain gauges attached to the element. Such sensors are the main subject of Chapters 2 and 3. Chapter 3 concentrates on force sensor theory and design methods. The sensors are characterized by such parameters as sensitivity, stiffness, cross coupling and linearity. The basic relations between these parameters and the sensor design features are discussed. To provide for an adequate sensitivity, a highly flexible compliant element should be used. On the other hand, this element should be stiff enough to ensure a high natural frequency of the manipulator with a built-in sensor. The book material on force sensor design includes many insights based on the extensive practical experience of the authors and could be valuable for design of any automation system using force or torque measurement. The presentation is supported by an extensive survey of the international patent literature on force sensor design, including a few patents by the authors.

Chapter 4 describes a research manipulator system comprising a manipulator linked to a minicomputer and equipped with force sensors of various types. The manipulator has special force sensing tools for various operations such as assembly, parts machining, parts retrieval, and gripping. Most of the experimental research described in the book was carried out with this system, and the mathematical models presented are primarily applicable to it.

In Chapters 5 to 8, mathematical models for force controlled manipulator systems are developed and control methods are designed.

Chapter 5 addresses the problem of keeping a manipulator in contact with an object. A mathematical model for one-dimensional manipulator motion is derived, linear and variable structure control approaches are discussed. The stability of the contact maintenance is studied in the presence of a delay in the force feedback loop and a lumped compliance of the manipulator structure. Chapter 6 studies a two-dimensional motion of a manipulator making contact with an object. The motion is considered to be a superposition of the motions along the normal and the tangent to the object contour. A contouring motion with an optical proximity sensor is also considered.

Chapter 7 presents a mathematical model for force control of a general type manipulator. The manipulator links are assumed to be absolutely rigid. In many tasks, the manipulator should move the gripped object to comply with the imposed mechanical constraints. This is the case where use of force information becomes imperative for the control. A general approach to force control design for manipulator systems is developed. The approach assumes the commanded velocity vector to be the sum of vectors directed along normals and tangents to constraints. Chapter 8 studies discrete-time control of a manipulator motion with a digital computer. The problem of keeping contact in contour tracking is solved using a dynamic model description of the contour.

Chapters 9 to 11 discuss control algorithms for a manipulator performing certain applied tasks and describe experimental studies of the algorithms. Force control algorithms are designed with the methods developed in the book. The control of complex motions is designed by integrating together control laws for simple motions.

Chapter 9 discusses control algorithms for tasks of surface machining (for example, fettling) with a manipulator. The motion of a tool (a grinding wheel) along the surface of a fettled part is treated as a motion with an imposed mechanical constraint. Two most widespread assembly tasks are considered in Chapter 10: insertion of a cylindrical peg into a hole and assembly of a screw connection. The most essential and difficult step in these tasks is the initial alignment (engagement) of the mating parts. At this step, the type of imposed constraint changes in a complicated manner, as mating parts come into contact with each other. The last, eleventh, chapter of the monograph discusses problems of searching and grasping workpieces with unknown positions. Ferromagnetic workpieces can be extracted out of a bin with a weight-sensitive electromagnetic gripper. In other methods, workpieces are retrieved with a force sensing probe and force sensing gripper fingers. The performance of all the algorithms considered in Chapters 9 to 11 is illustrated by experimental results. Most of the other control algorithms described in the book are also experimentally verified.

Acknowledgments

The book is based on the results obtained by the joint research group of the Russian Academy of Sciences (Institute for Problems of Information Transmission) and Moscow State University (Institute of Mechanics), where the authors of this book worked. The authors are deeply thankful to A.V. Lensky, whose participation was instrumental in obtaining many of the results presented herein, to S.A. Mozhzhevelov, and A.B. Lizunov, who greatly contributed to the research described in the book. We are pleased to acknowledge results of the doctoral students: S.A. Anishenko, D.G. Marinova, S.N. Osipov included in this book. V.S. Gurfinkel and E.A. Devjanin, who headed the divisions where the authors worked, created a most supportive research environment and contributed to many of the published results. We have also enjoyed encouragement and help of our colleagues E.V. Gurfinkel, A.A. Grishin, and L.G. Shtilman.

Many results of this book were presented in the research seminars chaired by D.E. Okhotsimsky at Moscow University; the authors benefited from the useful discussions there.

Chapter 1

Force Information and its Use in Robotics Systems

The objective of this chapter is to describe the state of the art in the use of force-related information for manipulator control. The chapter presents a brief survey of the subject and introduces some basic notions and ideas that are considered in subsequent chapters.

1.1 Use of force information: objectives and problems

Today, most industrial robots execute fixed programs by repeating a pre-set sequence of motions. However, such robots may be inadequate for a variety of manufacturing tasks, where part sizes and positions may vary. Some tasks require programming and reproducing motions with a precision that cannot be easily achieved. These tasks can be alternatively performed by using sensor information to adjust robot motions.

In many cases, sensor-based robotic systems are preferable to the rigidly programmed. The use of sensor systems can relax precision requirements and allow the use of less expensive hardware in a robotic system. Regardless of the sensor type used, design of a sensor-based robot system involves solution of the following two basic problems. The first problem is to have in place the necessary sensor technology. This includes both *primary transducers* performing initial conversion of measured physical quantities electrical signal and real-time *processing of sensor signals,* so as to extract the information needed for manipulator control in real-time. The second problem is to design *control algorithms* for robots using this information.

External sensors used in Robotics belong to two large groups: contact and noncontact. The robotic vision system is a prominent example of a non-contact sensor. *Force sensors,* which provide the information on forces and torques applied to the manipulator end effector, are the most widespread and useful contact sensors.

Use of force information provides optimal solutions for many different problems. One of the simplest uses of force sensors is monitoring the task execution without modifying the robot motion. In this way, safe operation of a robot can be ensured (e.g., by stopping the system if a collision with an obstacle is detected).

Force sensors in Robotics can be used for *weighing*, for instance, to distinguish between different types of workpieces. Section 11.2 describes a problem where a force sensor is applied to weighing workpieces.

Force sensor information can be used for handling randomly oriented workpieces, e.g., for recognizing a workpiece and determining its center of mass. A force sensor enables a robot to *follow* a previously unknown or unspecified *object contour*. This feature is employed in welding, grinding, and other manufacturing tasks requiring motion along a surface, contour, or a joint of workpieces. Probing and contour tracking are used in tasks involving *pick up of parts randomly stored in a bin*. Chapter 11 of the book considers algorithms using force sensing for handling randomly positioned workpieces.

Force information is essential for a large group of applications where *constrained objects* are manipulated. Among these applications are mating and *assembly* tasks, where parts need to be moved in contact with each other. Many *surface machining tasks* performed by robots, such as milling, grinding, deburring, and polishing may be treated as constrained motion problems. In these tasks, a tool (for example, a grinding wheel) should follow the surface of a workpiece. Chapters 9 and 10 discuss how surface machining and assembly tasks can be performed using force information.

A problem of simultaneous motion control for *cooperating robot arms* belongs to constrained motion problems as well [41, 99, 104, 159, 180, 233]. In the absence of force control, internal deformations in the system can build up to levels where manipulators or a manipulated object may be damaged. Problems of controlling *multifingered articulated hands* closely resemble those of cooperating robot control. An object is held and moved relative to the hand by a coordinated motion of the fingers [19, 130, 165, 176, 224]. Similar problems arise in control of *legged robots*. These statically underdetermined systems require control of foot force distribution [82, 133, 190, 193, 278].

As mentioned above, design of sensor-based robotic systems using force sensors requires solution of three problems: force sensor design, processing of force sensor signals, and robot control using force information. In turn, force control includes at least two levels: lower level, i.e., force feedback for controlling basic manipulator motions, and higher level, or task level, i.e., control algorithms for a specific task based on these motions.

Following the above classification of the major problems of force control, the book material is divided into three parts. The first part, which includes Chapters 2 and 3, is devoted to conceptual design, basic theory, and design computations for force sensors. The second part (Chapters 4–8) develops *mathematical models* for robotic systems with force sensors. These models are used in design of low-level control algorithms, in particular, force feedback synthesis. Motion control of a manipulator with imposed mechanical constraints is developed and analyzed. Finally, the third part (Chapters 9–11) presents a number of *applications*, for which force sensing control algorithms have been designed and experimentally tested.

The issues of initial processing of force sensor signals and their transformations into different coordinate systems are mostly left out of the scope of this book, as they are treated in sufficient detail elsewhere, e.g., see [36, 197]. The rapidly growing level of digital computer technology makes the problems of real-time force information pre-processing easily solvable.

The problems of force information use in robotic control are discussed, to some extent, in the books [11, 33, 34, 36, 37, 68, 134, 165, 176, 177, 197, 208, 214, 215, 247, 276].

1.2 Force measurement

This section briefly considers the main principles for force sensor design and their placement in robotic systems. We formulate a mathematical model of a force sensor used later in the book. The material of this section is used in the analysis of force sensor design in Chapter 2 and design computations in Chapter 3. The reading of this section (without Chapters 2 and 3) will generally suffice to understand Chapters 5 - 11 devoted to control.

1.2.1 Methods for converting force into electrical output

In all force sensors, forces and torques are measured through *deformations or displacements in an elastic element of a sensor.* Various physical principles are employed to convert these quantities into a sensor output [14].

In the simplest design, displacements in an elastic element are measured by *potentiometric displacement transducers.* Such transducers possess a comparatively low sensitivity, so deformation of the elastic element caused by the measured force are large enough, and this sensor is highly compliant. *Photoelectric displacement transducers* are more accurate and sensitive. They measure a displacement through a change of light flow from a light source to a photoreceiver.

Displacement can be related to as a change in magnetic flux or field density. *Inductive displacement transducers* employ this principle. They measure a change of mutual inductivity of two coils that either have a moving iron core inside, or move relative to each other. *Integral semiconductor Hall effect devices* also seem promising for measurement of magnetic field.

Capacitive displacement transducers that measure a change in capacitance caused by moving one capacitor plate with respect to the other are also used in some sensors [341].

Strain can be transformed into electrical output with the aid of *piezoelectric transducers*, which achieve a high measurement accuracy as well as a large dynamic range and high accuracy of conversion. However, piezoelectric transducers cannot measure DC force components, and they must be equipped with special charge amplifiers for measuring low-frequency signals.

There exist force sensors based on *electro-conductive polymers*, which change electrical resistance under strain.

Strain can be converted into electrical signal by using *magnetostrictive effect*, i.e., a change of magnetic permeability under strain. This effect is weak, so this type of transducers can only be used with sufficiently sensitive amplifying equipment [173].

Strain gauge transducers are most widely used for force measurement [24]. In these transducers, strain causes a change in electrical resistance of strain gauges mounted to an elastic element. There are several relatively simple techniques for mounting strain gauges to the elastic element depending on the type of gauge (metal-foil, film, semiconductor). Strain gauge force sensors have high sensitivity and measurement accuracy, and require relatively simple amplifiers. The main drawback of strain gauges is sensitivity to temperature, which has to be counteracted by special methods. This book mainly considers strain gauge force sensors.

1.2.2 Placement of force sensors on a robot

A manipulator arm is a set of links connected in a kinematic chain. Configuration-dependent gravity moments as well as motion-induced inertia forces inevitably influence the output of force sensors installed anywhere in the arm. Sensors mounted in the base of a manipulator are most affected by dynamical effects and gravity. They can only be used to monitor collisions of the manipulator arm with obstacles within its operational space. The dynamical errors are smaller for sensors installed in a robot wrist, immediately preceding the end-effector. *Wrist force sensors* are the most widespread type of sensors in Robotics. Yet their output is likewise influenced by weight and inertia of a manipulator end-effector and a load. If it is necessary to accurately measure small values of force acting on a manipulator, sensors are built into its end-effector as *force-sensing fingers* or a *sensory probe.*

Forces acting on the end-effector of a manipulator are sometimes determined indirectly, by measuring a current supplied to an electric motor, pressure in a hydraulic actuator, or even a tracking error in a position servo system. The end-effector forces can be measured by installing torque sensors in the manipulator joints [158, 180, 271] or measuring tension of pulleys that transmit forces from an actuator to a joint [149].

Instrumented platforms (pedestal sensors) present a special group of sensors. These sensors are external to the robot manipulator structure. They can be used in control of assembly tasks, where one mating part is mounted to the sensor and the other is held by the robot gripper, as well as for determining coordinates of a workpiece placed on the platform.

Chapter 2 of the book discusses comparative features of force sensors depending on their placement on a robot.

1.2.3   Initial transformation of sensor output

When designing a multicomponent sensor, it is not imperative to demand selective sensitivity of its channels to individual components of force and torque. Instead, a six-component sensor measuring all three components of the force vector and three torque vector components may use an additional transformation of primary sensor outputs to determine force and torque components [288]. Let n be the number of channels of such a sensor ($n \geq 6$). Usually, $n = 6$ or $n = 8$. Denote by u an n-component column vector with output values of the sensor channels as its components. Denote by Φ a six-component column vector composed of force and torque components that are applied to the sensor in the chosen coordinate system. We shall further jointly refer to force and torque vectors acting on a sensor simply as a (generalized) force vector. Assuming that each channel output linearly depends on the components of the force applied to the sensor, we can write

$$u = S\Phi + u_0, \tag{1.2.1}$$

where S is an $n \times 6$ matrix called a *sensitivity matrix*, u_0 is an n-component offset vector for the sensor, i.e., the sensor output in the absence of an external load.

If the sensor design is correct, then the matrix S has rank 6, and the matrix is well-conditioned. If the matrix S is known, one can find components of the applied force from the sensor output. The least mean-square estimate of the force vector can be obtained from (1.2.1) as

$$\Phi = R(u - u_0), \tag{1.2.2}$$

where $R = (S^T S)^{-1} S^T$ is a $6 \times n$ pseudoinverse matrix for S, and the superscript T denotes the matrix transpose. The matrix R is usually called a decoupling matrix. Force vector components are generally computed according to (1.2.2) by an embedded microprocessor of the sensor. If we assume that the sensor outputs (after the transformation) are the components of the vector Φ, then there is no coupling between the components, provided that the sensor characteristics are linear and the sensitivity matrix S is known accurately enough. One can experimentally determine the sensitivity matrix S by *calibrating* the sensor. To this end, known values of force vector components or their combinations are applied to the sensor, and the corresponding outputs are logged.

Force information at the output of the sensor is represented in the *coordinate system attached to the sensor* (usually, with the origin at the center of the sensor). Yet, for many control problems, we need to find force and torque components in the *coordinate system attached to the manipulated object*. We denote by $D = [D_x, D_y, D_z]^T$ a vector connecting the origins of the two coordinate systems and by G, a 3×3 matrix of cosine projections for the object coordinate axes in the coordinate system of the sensor. Let $\Phi = [F^T, M^T]^T$ be a force component column vector in the sensor coordinate system. Then a force component vector Φ_0 in the object coordinate system can be written as

$$\Phi_0 = G_0 D_0 \Phi, \tag{1.2.3}$$

where the matrix D_0 defines transformation of force and torque components for the coordinate system origin offset

$$D_0 = \begin{bmatrix} 1 & 0 & 0 & 0 & 0 & 0 \\ 0 & 1 & 0 & 0 & 0 & 0 \\ 0 & 0 & 1 & 0 & 0 & 0 \\ 0 & D_z & -D_y & 1 & 0 & 0 \\ -D_z & 0 & -D_x & 0 & 1 & 0 \\ D_y & D_x & 0 & 0 & 0 & 1 \end{bmatrix}$$

Block-matrix G_0 of size 6×6 defines transformation of the force and torque components in the coordinate system rotation,

$$G_0 = \begin{bmatrix} G^T & 0 \\ 0 & G^T \end{bmatrix}$$

For a six-component sensor, calculations in accordance with (1.2.3) are often combined with the transformation of the sensor output according to (1.2.2). To this end, the matrix $R_0 = G_0 D_0 R$ is substituted for the matrix R in (1.2.2), which allows one to obtain force and torque components directly in the coordinate system attached to the manipulated object.

1.2.4 Force sensors as displacement sensors

In this book, we consider force sensors as sensors measuring deformations in their elastic element. For a sensor built into the manipulator structure, such deformations reflect a displacement of links adjacent to the manipulated object and located "after" the sensor with respect to the links that are closer to the manipulator base. Gain of such displacement measurement tends to be rather high. By way of example let us consider a force sensor with a measuring range ±100 N and a dynamic range of 2000. Such a sensor can detect a force change of 0.1 N. If the sensor stiffness is 10 N/mm (which is not too much for this sensor), then it can measure a displacement of 0.1 μ. Thus, force sensors, in fact, offer advanced capability for monitoring microdisplacements.

We now proceed to compose a *mathematical model* of a force sensor. We start with a sensor measuring *a single force component*. The sensor usually has two *rigid flanges*, between which a measured force is applied. The flanges are connected through an elastic element. Since the mass of the elastic element is generally much smaller than that of the manipulator parts attached to the sensor, we consider the elastic element to be a lumped one. The action of an external force Φ produces strain in the elastic element that results in a relative displacement ξ of the sensor flanges in the direction of the applied force.

Assuming the elastic element has damping, we can write the relationship between the force and strain

$$\Phi = k\xi + b\dot{\xi}, \tag{1.2.4}$$

where k is the sensor stiffness, and b is its damping.

The displacement ξ in the sensor elastic element, or the deformation proportional to it, is converted into electrical signal. The sensor output u and the displacement in the elastic element are related by the formula

$$u = s_1\xi + u_0, \tag{1.2.5}$$

where s_1 is a coefficient of sensitivity, and u_0 is a "zero offset" for the sensor signal (an output in the absence of force). The force acting on the sensor can be calculated from the output u by neglecting dynamic effects:

$$f = s^{-1}(u - u_0) \tag{1.2.6}$$

Here $s = s_1/k$ is the *sensor gain*. According to (1.2.5), under the static loading (i.e., for $\dot{\xi} = 0$), f (1.2.6) is equal to the applied force Φ (1.2.4). The sensor gain s is usually determined experimentally. The formula (1.2.6) can also be used for a dynamic loading, though it is approximate in this case. Unlike the force Φ (1.2.4), according to (1.2.5) and (1.2.6), f is related to the deformation by the formula

$$f = k\xi \tag{1.2.7}$$

Let us now consider *a six-component force sensor*. Again, we consider a sensor with two rigid flanges connected by an elastic element. We introduce the coordinate system attached to one of the sensor flanges. The load applied to a flange of the sensor is defined by three components of the principal force vector $\boldsymbol{F}$ and three components of the principal torque $\boldsymbol{M}$ applied at the origin of the coordinate system. They compose the vector $\Phi = [\boldsymbol{F}^T, \boldsymbol{M}^T]^T$. The strain in the sensor elastic element results in a small linear displacement δ and angular displacement φ of the second flange relative to the first one. We define these displacements by a six-component vector $\xi = [\delta^T, \varphi^T]^T$. The relationship between the displacement ξ and Φ is given by the formula analogous to (1.2.4):

$$\Phi = K\xi + B\dot{\xi}, \tag{1.2.8}$$

where K is a 6×6 *stiffness matrix*, B is a 6×6 damping matrix. Similar to the single-component sensor, an n-component output vector u of the sensor is related to a small displacement vector ξ of the elastic element. When a static force is applied, for instance during calibration, the output vector is linearly related to the applied load by (1.2.1). Then for the load F, which is determined from the sensor output by (1.2.1) and (1.2.2), we can write by analogy with (1.2.7)

$$F = K\xi \tag{1.2.9}$$

The expressions (1.2.8) and (1.2.9) represent the simplest mathematical model of a six-component force sensor.

1.3   Use of force information in control

This section surveys low-level *force feedback* algorithms. Task-level control algorithms are based on the low-level algorithms and depend on specific applications. A survey of force control applications in specific practical tasks is included in Chapters 9–11, which consider such applications. The survey of this section develops from force control principles, to analysis of one-dimensional motion with force feedback, to control of multidimensional motion. A special emphasis is placed on hybrid control of constrained motion, which is a central theme of this book.

1.3.1 Main approaches and principles of force control

Robot manipulators are used to move parts or tools in space. Manipulator motions can generally be subdivided into *gross* motions, where reaching a given final manipulator position is a primary goal, and *fine* motions, where manipulator position relative to the environment is accurately adjusted [291]. Gross motions are usually fast and have large amplitude, while fine motions are slower and smaller. Force sensor information is primarily used in control of fine motions. Since these motions are slow, in many problems of force control nonlinear dynamics of the manipulator are of little importance.

Manipulator control approaches using force information can be subdivided into two major groups. The first group of approaches uses *logic branching* of control when the measured force satisfies certain conditions. The second group introduces continuous *feedback* of force sensor measurements in the control loop. Often, force feedback is combined with logic branching.

The most common use of the logic branching is in a *guarded motion* [154, 300], where the manipulator stops after the measured force exceeds a given threshold value. A guarded motion allows one to establish manipulator contact with an object where position of the object is *a priori* unknown. For a spatial motion, logic branching can involve different thresholds on individual components of the measured force, a force magnitude threshold, etc. [142, 197, 245, 258]. When a threshold is exceeded, the manipulator may only stop corresponding component of the motion, or stop completely. As a sequence of guarded motions, it is possible to perform precision assembly, such as peg-in-hole insertion [86, 119], or tasks of packing and placing parts [219, 327].

A sophisticated force controller based on logic branching [241] is designed as a stochastic automaton which inputs the sensor information and outputs manipulator control signals. The controller has a learning capability and was successfully used for control of peg-into-hole insertion.

The logic branching algorithms are often used in practice jointly with continuous force feedback, in particular, to control multi-stage strategies of task execution. Examples of such multi-stage strategies for some applied tasks are presented in Chapters 9–11.

Many early algorithms of force control were formulated using logic branching [199, 292]. The task execution, however, might be slow, since the velocity of each motion is low, and the manipulator may stop between the motions. The control approaches based on continuous force feedback are more attractive. With the feedback, the manipulator motion at each time instant depends on current force sensor readings. The force feedback can be used in two groups of tasks [240]. In the tasks of the first group such as fettling, grinding, etc., the force itself is an important process parameter. In the second group of tasks, a force sensor is used to obtain information on mutual position of parts, such as in assembly or other part mating tasks. Microdisplacements of the mated parts are closely related to force sensor deformations, and, hence, to force measurements. An ability to *regulate an interaction force between the manipulator and environment* is a key requirement for all force control tasks.

A general way of maintaining a given interaction force between a manipulator and environment is by introducing an *additional compliance* into the system. If this compliance is large, then a position error causes a small change in the interaction force. A special compliant mechanical element can be placed in the manipulator wrist to help control forces exerted onto the environment. This approach is used in automated assembly [183, 286, 297] and grinding tasks [2, 8]. Mechanical compliance can also be concentrated in the manipulator joints, as in SCARA manipulators [161]. However, it is usually impossible to adjust additional mechanical compliance to different tasks. Compliance in the manipulator structure can be undesirable, as it may result in oscillations of the manipulator load if the manipulator moves fast.

It is possible to introduce compliance into the system in a different way - by reducing feedback gains of a manipulator positional servo system [72, 111, 197, 198, 270]. In this case, a manipulator displacement will cause a small variation of the control torque. Unlike mechanical, or *natural* compliance, such *artificial* compliance is software-adjustable depending on a specific task [72, 270]. In the limit, the servo system of an actuator can be disabled and the actuator commanded to generate a desired end-effector force. Controlling end-effector forces through the actuator commands is known as *implicit force control*.

In reality, the torque produced by the motor and the contact force between the manipulator and the environment are not always directly related. To control the force more accurately, one has to take into account gravity forces, servo system dynamics, and a Coulomb friction in the gear train [54, 197, 236].

The implicit force control methods can only be used for lightweight direct-drive manipulator arms, or if the friction in the gear train of the drives is small (the gear train is "back-drivable"). It is common knowledge, however, that for most industrial manipulator systems, joint friction by far exceeds torques generated in contact tasks. It is possible to eliminate the influence of friction in the gear train, i.e., to render it back-drivable, by installing a torque sensor at the output of the manipulator joint and using the output torque feedback in the drive control system [74, 158, 205, 271, 272, 301]. The experiments show that in this way effective friction in the joint can be reduced by a factor of 20 or more. Conceptually, introducing torque sensor feedback in the control loop is similar to introducing a wrist force sensor feedback and should be considered as an *explicit force control* or *active force feedback* method.

Having established the importance of the compliant behavior in manipulator force control, let us now define such behavior in more specific terms. Such specification can then be used for the controller design, as will be discussed further on. Obviously, a desired compliant behavior of a manipulator arm can be specified as an elastic spring behavior of the form

$$x - x_d = c(F - F_d), \tag{1.3.1}$$

where x is a coordinate of the arm, x_d is a reference coordinate, F is an applied force, F_d is a reference force, and c is a desired compliance. Since the behavior (1.3.1) is achieved by means of active feedback, it is referred to in the literature as *artificial* or *active compliance* [133, 191, 222].

The desired manipulator behavior is also often specified as a damper behavior of the form

$$\dot{x} - v_d = g(F - F_d), \tag{1.3.2}$$

where $\dot{x}$ is an arm velocity, v_d is a reference velocity, F_d is a reference force, and g is a damping factor. This behavior corresponds to that of a particle in a viscous environment under the action of an external force and is referred in the literature as *active accommodation* [291] or *generalized damping* [163].

In a more general way, the relation between the external force and the manipulator motion can be specified as a desired *impedance* of the manipulator [107, 108, 128]. The impedance can be defined as a transfer function between the external force acting on the manipulator and its displacement. The notion of impedance generalizes elastic compliance and viscous damping behaviors. A desired impedance of the manipulator end-effector can be specified as the first stage of force control design. Usually only one to three lower order terms of the transfer function are taken into account. The lower order impedance terms describe a steady-state motion under the action of a constant force, elasticity, damping, inertia in the system, etc.

The concept of impedance control is related to modeling of biological motor control systems where a stiffness or impedance of a human arm joint can be regulated by co-activating two antagonist muscles powering the same joint [106].

The specified impedance can be achieved by different means - by using implicit or explicit force control. This book is primarily devoted to explict force control. The explicit force control includes an active feedback on force sensor measurements. Main issues of the explicit force control are briefly overviewed in the next subsection.

When designing force feedback based on impedance specifications it is usually assumed that the external force is given. In fact, this force is defined by the manipulator interaction with the environment and is coupled with the manipulator moves. Therefore, as a part of the control design, the closed-loop dynamics of the force controlled manipulator in contact with the environment need to be considered. Transient processes in the control of a manipulator with a force sensor and their stability are studied in Chapters 5–8. The issue of transient dynamics of force control is briefly surveyed in the next subsection.

1.3.2   Design and analysis of low-level force feedback

In this subsection, we shall mainly consider force control of a single degree of freedom manipulator motion.

Fine manipulator motions, where force control is required, are usually slow and are well controlled by a positional servo system. In practical robotic systems, force sensor feedback is typically implemented as an external loop, which modifies reference input of a positional (or velocity) servo system. By specifying commanded position or velocity as a function of force, the desired behavior (impedance) of the manipulator is defined.

A force sensor placed in the manipulator end-effector provides a noncollocated feedback for the manipulator joint drives. It is a common practical experience that force feedback is prone to instability problems. Therefore, accurate analysis and design of the force control loop is of practical importance. A detailed study of the force feedback loop dynamics is given in Chapter 5 of the book. Herein we present a brief overview of the subject in order to introduce main ideas and issues of such analysis and survey some related literature.

Methods of force feedback implementation are discussed in [47, 102, 138, 155, 164, 240, 292], among other papers. A force sensor signal is informative when the manipulator is in contact with the environment. The measured contact force is a product of the system stiffness and its deformation in the contact. This deformation includes the deformations of force sensor, manipulator structure, and, possibly, of the environment. A simple model for the force feedback dynamics can be obtained by considering a rigid manipulator and a compliant force sensor. For sensor measurement, a model of the form (1.2.7) can be used. In this model, the force feedback in the contact task is equivalent to an extra position feedback. If the force sensor is stiff, this positional feedback has very high gain.

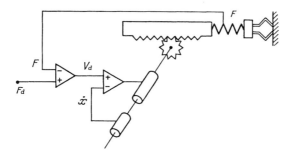

FIGURE 1.3.1
Manipulator contacting environment with force and velocity feedback

The simplest design is a linear force feedback [211]. Such feedback is equivalent to a high-gain position feedback. For such feedback, the naturally present damping in the system (such as back e.m.f. in the electric drive or viscous friction in the hydraulics) can be insufficient resulting in a highly oscillatory system. To increase the damping, a velocity feedback should be introduced in the system.

Let the system have force feedback and a feedback of the manipulator velocity $\dot{x}$ [90, 91, 254]. In the cascade scheme of Figure 1.3.1, the commanded velocity is proportional to the force error. Thus, an external force acting on the sensor will lead to a steady state motion with the velocity proportional to the force. Such behavior fits into the generalized damping paradigm (1.3.1). The scheme of Figure 1.3.1 provides a good design of the force controller. Such design is used in most of the experiments described in this book and studied in most of the theoretical analysis.

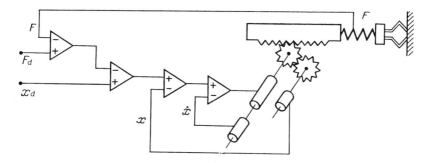

FIGURE 1.3.2
Manipulator contacting environment with force, position, and velocity feedback

It is not always possible to implement the scheme of Figure 1.3.1, because force feedback often has to be added to an existing manipulator position servosystem. A force feedback cascaded with a position servosystem, as shown in Figure 1.3.2, modifies the commanded position proportionally to the force error. For external force acting on the system, the steady-state behavior can be described by the active compliance paradigm (1.3.2). By tuning feedback gains, one can achieve a high quality of transient processes for the manipulator in contact with the environment. Yet the above feedback architecture yields a large static error in maintaining a commanded force if the commanded coordinate differs significantly from the object contact coordinate.

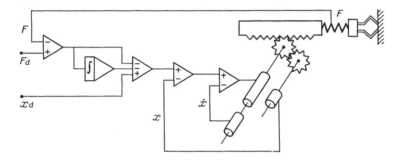

FIGURE 1.3.3
Manipulator contacting environment with force and PID position feedback

In order to get rid of the static error in maintaining a commanded force, an integrator can be introduced into the force feedback loop [240, 291] as shown in Figure 1.3.3. If an external force acts on the manipulator, it will move with the velocity proportional to the difference between this force and the commanded force. Such manipulator motion fits into the generalized damping paradigm (1.3.1). However, an integrator in the feedback loop increases the order of the system, which can become unstable for large integral gains [61, 299].

Analysis of closed-loop dynamics and stability for one degree-of-freedom manipulator motion with the above described schemes can be found in [47, 75, 90, 91, 291]. One of the issues in the force control design and analysis is that the positional gain of the force feedback depends on the joint compliance of the force sensor, manipulator structure, and the environment. While force sensor compliance is constant, manipulator structure compliance is configuration-dependent, and environment compliance may vary. An adaptive control approach to force control that estimates the desired feedback parameters on-line is proposed in [39, 303].

An important related issue is analysis of the contact transition for a manipulator with a contact sensor. When establishing contact, a manipulator usually impacts a still object with a certain velocity and, despite the force sensor feedback, the maximal contact force in the transient process may be high. Also, in contact transition control, the system structure changes at contact time, and the stability analysis needs to be extended to take this transition into account. Force control of the contact transition has recently received an increased attention in the literature [172, 257, 274, 310, 312]. This problem is studied in Sections 5.5 and 8.5 of the book.

The simplest lumped model described above overlooks some practically important effects in manipulator force control. In particular, rigorous analysis shows that the feedback schemes of Figures 1.3.1 and 1.3.2 are stable for any high gains as long as the gains have a correct sign. In practice, however, force feedback loops become unstable for a high gain of the force feedback. Several possible causes of such instability were discussed and analyzed in the literature. One possible reason is *feedback loop delay* caused by sampling in the control computer or current dynamics in the electric motors [77, 91, 210, 291]. The influence of delays on stability is studied in Sections 5.6 and 8.2 of the book.

Another possible factor limiting the gain and closed-loop bandwidth of the force feedback is *influence of structural compliance of the manipulator.* The dynamics of a manipulator with structure compliance and a force sensor is considered in [55, 213, 237, 249, 275, 310]. Effects of the structural compliance on the force feedback stability were studied in [55, 56] using numerical analysis of the examples for simple lumped models of the compliance. A rigorous analytical study for similar lumped models is presented in [78]. Sections 5.7 and 5.8 study this problem. Sections 8.4 to 8.6 present design and experimental study of dynamical discrete-time controllers compensating for the manipulator structural compliance.

An alternative approach to force feedback design and robustness analysis is by considering an unstructured uncertainty model for the dynamics of the manipulator and applying the Small Gain Theorem conditions of stability [128, 277]. This approach allows one to consider more generic types of uncertainty, yet it yields overly conservative stability results for structured uncertainties such as feedback loop delay and structural compliance of the manipulator. The unstructured uncertainty description can be applied to the dynamics of the environment as well.

The models and analysis of the force feedback for the moving contact point with an object are considered in [46, 47]. This and related issues are discussed in Sections 5.9 and 8.6 to 8.8 of the book.

1.3.3   Force control of multidimensional motion

Force control design for multidimensional, spatial manipulator motion is based on the approaches to force control of one-dimensional motion discussed in the previous subsection. The main difficulty in designing multidimensional force controllers is that an abundance of parameters should be selected for each executed task. In a practically usable robotic system, the controller design process needs to be automated.

In the pioneering force control work by Whitney [184, 291], empirical approaches to controller design were proposed based on shaping of the end effector impedance (active accommodation matrix) for each task. In particular, these approaches were targeting robotic systems performing assembly and part mating tasks. In the task of peg-in-hole insertion, the feedback gains are chosen to be large for the force and torque components orthogonal to the hole axis, and small with respect to the force and torque acting along this axis. The force feedback gains can be chosen in such way when using the active accommodation algorithm [269, 291], active compliance [184, 222, 240], or more generic impedance control methods [239].

In the above mentioned papers as well as in many others, the desired impedance matrix is shaped to be diagonal in the coordinate systems with the origin at the peg lower end. After it was demonstrated that such impedance greatly assists the assembly, passive *Remote Center Compliance* devices have been developed which provide similarly shaped passive compliance matrix thanks to their mechanical design [297]. Some of the designed passive compliance devices may even be adjustable for different tasks [38].

The major approach to automated design of force controllers is by having a compact formal description for force control tasks and using this description for the design. An adequate and very attractive description of force control tasks can be obtained using the notion of the *mechanical constraint*. If the constraint is ideal (frictionless), the constraint force is directed at a normal to it. A manipulated object sliding on the top of a smooth table surface presents a simple example of an ideal constraint. If the positional control of the manipulator is "stiff", then errors in positioning the object with respect to the table may result in very large constraint forces. These forces can be damaging and, thus, need to be controlled. As discussed in Subsection 1.3.1, the main idea for controlling the forces is to introduce an active or passive compliance into the system. Control of constrained motion is the main paradigm used in this book for design of multi-dimensional force controllers.

The simplest and historically earliest systematic method for control of the constraint manipulator motion is the *compliant joint method* [197, 199]. The designated compliant joints of the manipulator are controlled through torque sensor feedback or just have positional servo control switched off.

The remaining joints are kept under position control. As the compliant joints, the joints that are most involved in the motion orthogonal to the constraint are chosen. The number of such compliant joints should coincide with the dimensionality of the constraint. Sections 8.6–8.8 of the book consider control algorithms for the motion of a gantry manipulator along an object surface using force control of a designated degree of freedom. The control algorithm of Section 7.6 works in the same way.

A more advanced representation of constrained motion tasks was proposed in the influential paper by Mason [163]. He suggested describing *natural constraints* imposed in a manipulation task by selecting axes in a task frame, where the end effector cannot move translationally or rotationally. A controller should complement these natural constraints by *artificial constraints* on the position of the manipulator, which define positionally controlled axes. At the same time, generalized force can be controlled along the axes of the natural constraints. A further development of Mason's specification method can be found in [47].

Mason's task description was used by Raibert and Craig in their formulation of the *hybrid control* approach to constraint manipulator motion [211]. This formulation is based on the concept of a *selection matrix*, which is a diagonal 6×6 matrix with zeros and ones on the diagonal. The selection matrix defines which components of forces and torques in the task-related frame should be controlled. The complementary directions are controlled positionally. The initial formulation of the *hybrid control* was modified to be more consistent and extended to include compensation of dynamical effects in a number of subsequent papers [5, 131, 289, 290, 306].

The hybrid control formulation based on the selection matrix concept was demonstrated to be non-invariant to coordinate system transformation. More consistent, but also more mathematically involved, descriptions of the hybrid control task were suggested in [116, 150, 243] based on the screw theory. These formulations use reciprocity condition for force wrenches and position twists instead of orthogonality of the compound 6-D vectors of force-and-torque and infinitesimal displacement-and-rotation.

A generic and conceptually clear specification of the force control task can be alternatively obtained by using an analytical description of a mechanical constraint imposed on the generalized coordinates of the system. The imposed constraint is considered as an analytical relationship satisfied by the Cartesian and angular coordinates of either the manipulator end effector or the manipulated object. Such analytical description was used in different forms and for different types of control formulations in [120, 168, 171, 194, 306]. The analytical description of the constraint is closely related to that used in the hybrid control formulation [45]. Many of the above formulations ignore closed-loop dynamics of the force control and concentrate on the dynamics of the motion "along" the constraint assuming the constraint and the nonlinear dynamics of the manipulator to be perfectly known.

Chapter 7 of this book presents a generic practical formulation of the constrained motion control based on the analytical description of the constraint. The manipulator motion in the directions normal to the constraint is controlled using force sensor feedback. The motion "along the constraint", i.e., at a tangent to the constraint as a generalized hypersurface remains under positional control. Force feedback for the motion along the (generalized) normals is implemented using designs developed for one-dimensional motion of the manipulator. As demonstrated in the book, such formulation provides a good mix of a practically convenient task description and a consistent control scheme and can be used in different practical tasks. The formulation of Chapter 7 is based on the specification of the desired velocity in a constraint tangent direction and desired force for each of the constraint normal directions. These desired velocity and force are then tracked using sensor feedback. The definition of the velocities (virtual displacements) and the constraint forces compatible with the constraint naturally follows from the analytical description of the mechanical constraint.

So far, it was assumed that *the constraint is known exactly*. This can be assumed for such tasks as steering wheel rotation (Section 6.6), hatch lid opening (Section 7.6), coordinating the motion of cooperating robot arms, abrasive grinding (Chapter 9), and in some assembly tasks, such as peg-in-hole insertion and driving in a screw (Chapter 10). To design control of the manipulator motion, one needs to describe the geometry of the constraint and identify the motions *tangential* and normal to the constraint. This can be achieved by geometrical reasoning [53, 163] or other means.

In some tasks, directions of the constraint normals and tangents can be identified on-line from the sensory data. A typical example is the task of following an object surface or contour. In the absence of friction, contact force measured by the sensor is directed along the surface normal and can be used to identify constraint geometry. An alternative approach is to identify a constraint tangent by observing manipulator velocity or incremental change of position during the constrained motion. The described approaches are discussed in [25, 47, 73, 90, 125, 262, 309]. Sections 6.1–6.5 of the book study the contour-following algorithms in detail.

If the information about a constraint is insufficient, the force control of manipulator motion can be extended to all directions. This is similar to the impedance shaping approaches discussed in the beginning of this subsection. Such approach is most useful where even the dimensionality of the imposed constraint is not known exactly, for instance, in exploring the ribs, surfaces, and convex angles of an unknown object [40].

A common situation where a constraint is unknown is part mating in assembly tasks. For instance, a peg inserted into a hole consecutively establishes a one-point contact of the peg edge with the hole chamfer, a one-

or two-point contact with the hole edge, and a three-point contact with the hole wall. Since the peg position relative to the hole is known with an error, it is hardly possible, even after a detailed advance analysis of the problem, to determine the character of a currently imposed constraint. Yet control algorithms for the peg motion during the peg-in-hole insertion can be heuristically designed using the concept of motion along the imposed constraint. In so doing, the feedback gains can be heuristically chosen so as to ensure the motion along the constraint imposed on the peg that is already inserted in the hole. Sections 10.1 and 10.2 give some examples of such assembly algorithms.

1.3.4 Dynamics and stability of constrained motion control

Stability of manipulator force control in a generic multi-dimensional setting was studied in a number of papers, e.g., [31, 32, 200, 290, 302]. These studies assume a rigid-body model for the manipulator while the force sensor is modeled as a spring element between the manipulator end-effector and constraint. The force feedback is then equivalent to an additional positional feedback. Provided that in a chosen coordinate system the overall linear positional and velocity feedback can be described by positively definite matrices, the action of feedback torques is equivalent to that of an additional natural spring and dashpot. By extending the mechanical energy of the system to include these additional spring and dashpot elements one obtains a Lyapunov function for the stability proof.

In some force feedback implementations, linear servo control can be combined with feedforward compensation of the nonlinear dynamic effects (computed torque control). The compensation can also be extended to ensure that the manipulator follows a reference dynamics, such as in the resolved acceleration control approach [131, 158]. The feedforward compensation of nonlinear dynamics effects, however, is not always practical. One reason is that it is difficult to obtain a detailed mathematical model of a manipulator. Another reason is that in real life, even for direct drive arms, the influence of unmodeled friction is often much larger than that of nonlinear dynamics.

Efficient compensation of the unmodeled dynamics and friction can be achieved by using a high-gain servo feedback, while force feedback is used in an external loop, cascade to the positional servo system. For limited actuator effort, the high-gain feedback can be implemented in the form of switching or variable structure system (VSS) control. Switching control implementations of force feedback are considered in [153, 157]. Force feedback using a special version of VSS control, which is called robust control in robotics literature, is considered in [153]. Switching control of manipulator with force sensor is discussed in Section 5.4 of the book.

The force feedback stability analysis of the above referenced papers focuses on stability of a manipulator equilibrium under force control. At the same time, the stability of a task-execution motion, where the constrained task frame keeps changing, is also very important. Such stability analysis is presented in [146, 147] for the task of contour following where the normal to the contour is estimated from the measured force. Sections 6.1–6.4 of the book present an analytical study of stability for the contour-following task. Section 7.5 presents such study for the motion along a screw constraint.

Control of constraint force is subject to disturbances caused by nonlinear dynamics and friction forces. By using prior information on the constraint, one can reduce the influence of these disturbances. For this purpose, a feedforward signal can be added to the force feedback signal. The signal required to compensate for the perturbations can be either computed using a detailed model of the constraint and manipulator dynamics [305], determined by "learning" (an iterative procedure that involves repetition of the required motion [117, 124]), or computed on-line [132]. The disturbances acting on the process of maintaining given interaction forces can also be represented by a dynamical model [160, 216, 249, 295]. Algorithms for disturbance compensation using an internal dynamical model are well known in control theory. Chapter 8 considers some dynamical models of disturbances that affect the manipulator moving along an object contour and algorithms for their compensation.

Interesting extensions of the force control of constraint manipulator motion are related to control of flexible manipulators. Structural deformations of such manipulator are related to the constraint forces and can be controlled, provided an accurate description of the flexible dynamics is available. Such control can use structure deformation sensors in addition to or instead of the force sensor. Though related to this book topic, flexible manipulator control issues are outside of its scope. Some initial references on force control of flexible manipulators are [202, 212, 307].

A number of papers on the subject of force control stability contain theoretical studies of dynamics and stability for implicit force control schemes [4, 26, 43, 64, 126, 246]. Such studies are mostly outside of the scope of this book, which focuses on issues of explicit force control.

Part I.

Force Sensors

This part of the book considers the fundamentals of force sensor design. Efficient sensor designs satisfy a number of constraints, which are derived from practical experience. The second chapter surveys the most typical structures of force sensors used in practice. This survey is aimed at giving the reader an understanding of practically acceptable designs. Chapter 3 treats the issues of design computations and optimization of parameters and characteristics for the major types of sensors outlined in Chapter 2.

Chapter 2

Conceptual Designs of Force Sensors

This chapter contains an analytical survey of the major known conceptual designs for force sensors. The chapter is organized into sections and subsections according to sensor categories. For each category, we discuss a few major design ideas. The chapter contains a survey of patent and other literature on the force sensor design.

We describe in more detail several force sensors developed by a joint research group of the Russian Academy of Sciences (Institute for Problems of Information Transmission) and Moscow State University (Institute of Mechanics), where the authors of this book worked. We shall further refer to these sensors as IPIT–IM sensors. Some of these force sensors are used in experiments described in the following chapters of the book. Though the conceptual designs of the developed sensors are generally based on one of the major discussed schemes, they also employ some original technical ideas. The authors gratefully acknowledge an important contribution by Dr. A.V. Lensky of Institute of Mechanics, Moscow State University, who conceived many design ideas described in the book.

2.1 Fundamentals of Sensor Design

A force sensor design depends on a task it is intended for, sensor placement on a robot, and a primary transducer used to convert force or torque into electrical signal (see Section 1.2). Further design constraints include overall sensor dimensions and a range of loadings to be measured. For instance, bending elastic elements are typically used for measuring small forces, while compression-tension and/or shear elements are used for measuring large forces (exceeding 10^3 N). Intellectual property (patent) considerations can also influence sensor design.

Though multicomponent sensors come in a variety of configurations, they tend to follow one of the two main conceptual design schemes. The first scheme, mentioned in Section 1.2, is based on computation of individual force and moment components for a six-component sensor through a linear transformation (1.2.2) of primary transducer signals, with transformation weights determined by calibration. The transformation is performed in real time by an embedded or external microprocessor unit.

A transformation consisting in summation and subtraction of individual strain gauge signals can be performed in a Wheatstone strain gauge bridge. In this way, one can obtain a sensor with channels that are selectively sensitive to individual force components. However, for such design, tolerances of sensor manufacturing, inaccuracies in placement of strain gauges, or variations of their parameters will influence the accuracy. Because of these issues, cross-coupling of the sensor channels cannot always be made sufficiently small.

Sensors measuring less than six or sometimes even six force components may be designed in a different way, i.e., by combining individual elements that are sensitive to one or two force or moment components each into an integrated structural unit. Channels of such composite sensor are decoupled with respect to the measured components due to mechanical selectivity of the individual elements. Such modular elements can be independent units that are assembled together into a multicomponent sensor. An assembled sensor design has a number of advantages. First, this design facilitates manufacturing, because individual modules are generally simpler to design and machine than a multicomponent sensor. Second, one can obtain a sensor capable of measuring as many force components as needed through an arrangement of standard modules. If the measurement range requirements for one of the components change, or the element fails, only a single module needs to be replaced. The drawbacks of an assembled sensor are that it is more bulky, and that the inaccuracies of manufacturing and assembly of individual modules may accumulate.

In some cases, individual parts of a monolithic multicomponent sensor can be treated as modular elements. Arrangement of an assembled or monolithic sensor structure from modular elements facilitates conceptual design.

Force sensors arranged from modular elements are most common among strain-gauge force sensors, which are the focus of attention of this book. One- and two-component modular elements are widely used in measurement systems, for example, in weighing devices, as aerodynamic balance sensors, or as parts of more complex multicomponent sensors. Let us consider some most typical and widespread configurations of the modular elements that are used in or can be of interest for Robotics.

2.1.1 Structural elements of force sensors

One of the simplest bending elastic elements is an *elastic beam*. Such a beam is selectively sensitive, because its strain due to an axial force is much less than that caused by a bending moment. An elastic beam is often used to measure a lateral force normal to its axis. The point of the force application should be fixed and sufficiently distant from strain gauges. It is possible to make a beam mechanically selective to only one component of the lateral bending moment by making its cross-section elongated in one direction. A modular element using such a beam is described in [280].

A simple elastic element for force measurement is a *double bending beam* [314] (Figure 2.1.1). Strain gauges *1* are mounted near the opposite ends of the beam *2*. The drawback of this configuration is a weak protection from the moments caused by the change of the application point of the force F. Such an element is usually employed when a high measurement accuracy is not required.

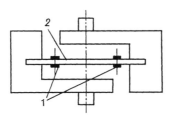

FIGURE 2.1.1
Double bending beam element

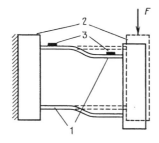

FIGURE 2.1.2
Parallelogram elastic element

A *parallelogram elastic element* [318] has a more advanced design (Figure 2.1.2). It consists of two parallel elastic beams *1* connecting two rigid flanges *2* and measures a force acting in the direction from one beam to

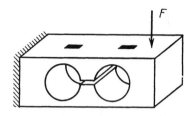

FIGURE 2.1.3
Easy-to-manufacture parallelogram elastic element

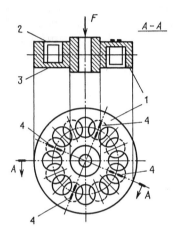

FIGURE 2.1.4
Spatial parallelogram mechanism with uniform protection from bending moment

the other. Strain gauges *3* are attached to the base of the beam to measure strain. The strain in each beam caused by the measured force F is analogous to that of a double bending beam.

Parallelogram elastic elements are widely used in force sensors. Many variations of this element exist. For example, a parallelogram elastic element shown in Figure 2.1.3 is easy to manufacture. It is made of a parallelepiped with two drilled through holes and a slot between them [317]. The holes create concentration of stress at places where strain gauges are attached.

A sensor with a spatial parallelogram mechanism for a uniform protection from bending moment is shown in Figure 2.1.4 [313]. It consists of a cylinder *1*, in which blind holes are drilled along the circumference. The holes in the end *2* are shifted by half a step relative to the holes in the end *3*. Strain gauges *4* are mounted to the elastic struts left between the holes. The

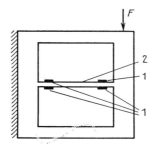

FIGURE 2.1.5
Parallelogram element with enhanced protection from bending
moment

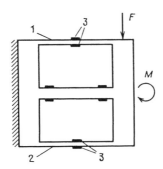

FIGURE 2.1.6
Module measuring the lateral force and the lateral moment

structure is easily machinable and has a good protection from the influence
of bending moments. A large cross-sectional size, however, may hinder the
use of this design.

A bending moment acting on a parallelogram elastic element causes tension of one elastic beam and compression of the other. In the element shown
in Figure 2.1.5, strain gauges *1* are mounted on the third middle beam *2*
to reduce the influence of the lateral moment on the output of the element
[332].

A module for measuring both the lateral force *F* and lateral moment
M, illustrated in Figure 2.1.6, has a similar design. When measuring the
force, the module works as a parallelogram elastic element with a middle
beam. The lateral moment causes tension of one thin outer beam *1* and
compression of the other beam *2*. The strain gauges *3* are attached to the
middle of the outer beams. They do not respond to strain caused by the
lateral force. Such a module is used in aerodynamic balance [319].

Let us now consider elements used for torque measurements. The sim-

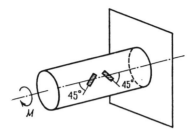

FIGURE 2.1.7
Measuring torque by torsional strain

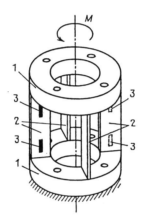

FIGURE 2.1.8
Hollow cruciform torque sensor

plest way to measure torque is through the shear strain of a cantilever beam subjected to torsion. To measure such strain, strain gauges are mounted to the beam normally to each other and connected as adjacent arms of the strain gauge bridge (see Figure 2.1.7). The angle between the sensitivity axes of the gauges and the axis of the beam should be 45° [320]. A 90° strain gauge rosette can also be used to measure shear strain. However, this configuration does not ensure mechanical selectivity to the measured torque M. Thus, the coupling influence of unmeasured components on the measurement can be significant.

A torque sensor of a hollow cruciform type shown in Figure 2.1.8 [316] is a variation of a parallelogram elastic element. The sensor has two rigid round flanges *1* connected by four elastic beams *2*. The beams have elongated rectangular cross-sections and are spaced at 90° intervals around the sensor at the same distance from the sensor axis. The beam cross-sections are extended along the radius of the sensor. The strain gauges *3* are attached in the middle of the wide lateral side of the beams, near the top and bottom flanges. Each beam bends tangentially and twists due to the axial torque *M*. If the radius of the sensor is much larger than the dimension of the beam cross-section, then torsion can be neglected.

The modular elements described in this section may be used both individually and as a part of various multicomponent sensors.

If a highly selective measurement is not required, a single-component force sensor with a double bending beam (see Figure 2.1.1) can be employed. Such a module is used in the sensory probe described in Section 2.2 and as a wrist sensor for a robot with an electromagnetic gripper (see Section 2.3).

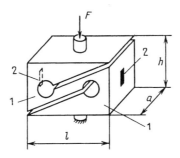

FIGURE 2.1.9
Compact element for axial force measurement

In a number of tasks, a force applied along a mechanical link should be measured. In this case, it is important to keep the cross-sectional dimension of a sensor small. An original design of a force-sensing module with a small cross-sectional dimension was proposed for use as a foot force sensor for a legged robot [48]. The sensor shown in Figure 2.1.9 has elastic struts *1*, which are symmetrical with respect to the sensor center. The struts are equally strained by the measured force *F*. The strains caused by the bending moment in the struts have opposite signs. The strain gauges *2* mounted to the struts should only measure the axial force, hence they are connected as opposite arms of the strain gauge bridge. The struts are formed by the holes and slots made in the sensor body. The drawback of the design is its lack of mechanical selectivity to the measured force.

Force sensing modules with parallelogram elastic elements (see Figures 2.1.2, 2.1.3) and a module for measuring axial torque (see Figure 2.1.8) are employed in a number of multicomponent modular force sensors developed in IPIT-IM. These sensors are further described in Section 2.3 of the book. Figure 2.1.10 shows a variation of the parallelogram elastic element that is used in these sensors and is simple to manufacture. The sensor is made

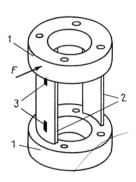

FIGURE 2.1.10
Parallelogram element made of cylindrical shell

of a hollow cylindrical shell with two rigid flanges *1*. Wide openings are
machined in the walls of the cylinder, so that the remaining parts of the
walls form two diametrically opposite beams *2*. Strain gauges *3* are mounted
at the center line of the beam outer surface, which is easily accessible.
Design computations for such modules are presented in Section 3.2 of the
next chapter.

2.2   Gripper Sensors

In this section, we consider three distinct categories of force sensors
placed in the manipulator gripper. These are *gripping force sensors, force-
sensing fingers*, and *sensory probes*.

2.2.1   Applications and design features

Grasping (gripping) force needs to be measured when grasping a fragile,
easily deformable object, as well as to make sure that an object is reliably
held. Single-component gripping force sensors have a simple design.

Sensory probes and finger sensors measure forces and torques directly,
without intermediate links. Inertia and weight of other manipulator links
do not influence the work of such sensors. Hence one may relax the requirements to sensor stiffness in order to ensure high sensitivity and measurement accuracy of sensors. The two mentioned sensor types are used in the
tasks that involve probing, determining a shape, size, and orientation of an
object, or controlling a manipulator motion in contact with an object. For

example, probes may be used for seam tracking in welding.

Unlike probes that can only be used for measurement, sensing fingers combine both sensory and motor functions. Of particular interest is using sensing fingers as a force sensor to measure external forces (and torques) acting on the object grasped in the fingers. In this case, two or more force sensing fingers can be considered as a single sensor measuring only external forces applied to the grasped object and insensitive to the gripping force. Finger sensors are used in various assembly operations. These sensors are often difficult to calibrate and suffer from the inevitable gripping force influence on measurement of external forces and torques (a gripping force can be an order of magnitude greater than external forces).

Sensors of the above-mentioned categories also have some features in common. Sensing fingers may be used to measure the gripping force as well. In many cases, sensory probes and sensing fingers have similar designs.

Sensors placed in the manipulator gripper tend to use strain gauge transducers. Displacement transducers are rarely used, in part, due to constraints on dimensions.

Let us now consider the major known conceptual designs for sensors of the above categories. Sensors placed in the gripper are not designed as six-component sensors requiring an additional linear transformation. Multicomponent sensors mounted to the gripper are typically arranged from individual modules measuring one or two components of force or torque. To be used in sensory probes and sensing fingers, such modules should satisfy strict constraints on dimensions. At the same time, they are subjected to smaller lateral moments as compared to wrist sensors, so less mechanically selective modular elements can be used.

2.2.2 Gripping force sensors

To determine a gripping force, it is sufficient to measure elastic deformation in the gripper jaws. As an example, one can refer to the strain gauge force sensors described in [57, 166, 280].

In some cases, it is impossible to measure the deformation in the gripper jaws directly. Then the gripping force may be determined, for instance, through the strain in elements linking the jaws with the drive or by measuring pressure in the pneumatic drive of the gripper [201]. Gripping force sensor for a prosthetic hand [23] exemplifies a typical design where a gripping force is measured through the strain in intermediate links. Various pressure sensors based on electro-conductive polymers [96, 123] can also measure the gripping force. A variety of gripper designs results in a great variety of gripping force sensors. Yet the sensor designs are usually relatively simple. The most tedious part, generally, is to built a sensor into an existing gripper.

In some cases, in addition to the gripping force, one needs to measure

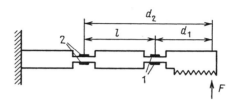

FIGURE 2.2.1
Measuring bending moments at two points in a gripper jaw

a distance from the jaw edge to the grasped object. It may be useful if
the position of the grasped object is unknown. To find this distance, it
is sufficient to measure bending moments at two points of the gripper jaw
[280]. One can obtain signals proportional to the bending moments by
mounting two strain gauge pairs *1* and *2* (see Figure 2.2.1) on the two sides
of the elastic beam aligned along the gripper jaw. The strain gauges are
connected into adjacent arms of the bridge.

Let the strain gauges *1* and *2* be placed at unknown distances d_1 and d_2
from the application line of a point force F. Then the bending moments
$M_1 = d_1 F$ and $M_2 = d_2 F$ are measured. Hence, we may easily find the
force $F = (M_1 - M_2)/l$ and the distance from the point l to the force
application point $d_1 = M_1 l/(M_2 - M_1)$, where $l = d_2 - d_1$ is the known
distance between the points *1* and *2*.

2.2.3   Fingers force sensors

By using multicomponent force sensing finger in a manipulator gripper,
one can measure up to six components of force and torque vectors applied
to a grasped object. Such sensors, in addition to the measurement of external forces applied to the object, inevitably respond to internal forces and,
sometimes, torques, arising due to gripping of the object. These forces and
torques are not external with respect to the manipulator, they are, in effect,
the forces of interaction between the fingers applied through the grasped
object.

Commonly, components of force and torque vectors acting on each finger
are measured. Then the sum of the two force sensor outputs represents
the external force applied to the object, and the half of the difference, the
respective gripping force component. A number of known designs for force-
sensing fingers are based on this approach. Another approach implies that
each finger measures a part of force and torque components. Then external
forces acting on the object may be determined by additional calculations
[165, 333]. For this approach, the contact between the finger and the object
should satisfy certain assumptions, in particular, a point contact is assumed.

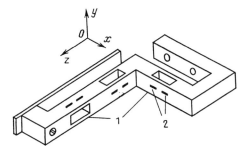

FIGURE 2.2.2
Three-component finger sensor with parallelogram elastic elements

A simple design of force-sensing fingers includes a spring element and a displacement transducer [95]. However, most designs employ strain gauge transducers.

Let us start with a finger sensor configuration designed to measure three force components. A standard three component force sensor can be used for each finger [251]. The arrangement that seems to be the best of its kind and is used in a few different sensors consists of three parallelogram elastic elements connected in series. The elements are typically machined from a solid billet of metal and transform force components acting along three mutually orthogonal axes. Figure 2.2.2 shows a design [113, 340] where parallelogram elastic elements *1* are formed by holes in an L-shaped beam. The strains are measured by strain gauges *2*.

Some force-sensing finger configurations, in addition to three force components, allow one to measure two lateral moments (normal to the finger axis). Such a sensor consists of a rod with two bending elastic elements which measure two orthogonal components of a lateral bending moment. By measuring bending moments at two points, one can determine the force component acting normal to the sensor axis and the force application point, as described above (see Figure 2.2.1). A separate elastic element, to which the rod is attached, measures the force component acting along the sensor axis.

Figure 2.2.3 illustrates this finger sensor design. Four pairs of strain gauges *2* which respond to bending moments are mounted to the rod *1*. Strain gauges *3* attached to the elastic beam fixed at both ends measure the force acting along the sensor axis. The rod is mounted to the middle of the beam. If it is not necessary to know the point of force application [187], then the strain gauge signals needed to calculate the lateral force components can be added up in the strain gauge bridges.

The drawback of the last two configurations is that they are not mechan-

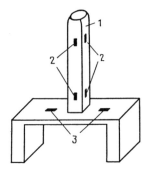

FIGURE 2.2.3
Finger sensor measuring three force components and two lateral moments

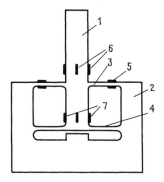

FIGURE 2.2.4
Finger sensor with double parallelogram elastic element

ically selective to individual components. This may cause relatively large coupling. Figure 2.2.4 [209] displays a similar finger sensor design, yet with greater stiffness and selectivity. In this configuration, a double parallelogram elastic element is used to measure a longitudinal force component. The rod *1* serves as a common base for the two parallelogram elements. The rod is connected to the base *2* through four parallel elastic struts *3* and *4*. The strain gauges *5* mounted on the struts *3* measure the force acting along the rod. Four pairs of strain gauges *6* and *7*, attached to the rod *1*, are used to determine two lateral components of the force acting on the rod and two lateral moments defining the application point of this force. The drawbacks of this design are that it is difficult to manufacture and that under the action of lateral force components, the deformation of the rod *1* is passed to the struts *3* and distorts the output of the strain gauges *5*.

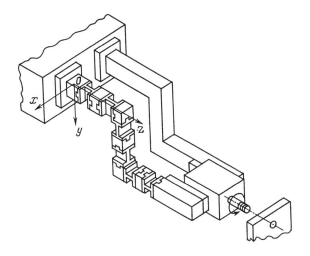

FIGURE 2.2.5
Six-component finger sensor composed of modules

In principle, it is possible to design a finger sensor which allows one to determine all six force and torque components acting on it. The finger sensor shown in Figure 2.2.5 [280, 326] has six uniform modules measuring a bending moment with a beam elastic element (see Section 2.1). Components of force and torque are computed through the corresponding algebraic operations on the modular signals. Since the module sensitivity axes are orthogonal, the computation is comparatively simple. The drawbacks of such a configuration are that it is complicated to manufacture and has a relatively low stiffness.

One more approach to force measurement relates to the development of multifingered articulated hands. Each articulated finger of such a hand has a few degrees of freedom. The fingers are typically driven via cables (tendons) by motors located outside the hand. For many designs, tendon tension is determined by measuring the pressure exerted on a pulley axis. Thus, by measuring joint torques for the fingers, one can determine force and torque components acting on the object grasped by the fingers [40, 114, 165]. Alternatively, a load cell can be embedded into each finger tip [40].

2.2.4   Sensory probes

In some tasks, grippers instrumented with sensory probes are used. Such probes can be made retractable. This design is justified if the main part of the gripper either needs retooling in the course of the manipulation process

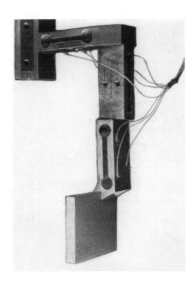

FIGURE 2.2.6
Finger sensor measuring three force components

or is unsuitable for force sensor installation. Light-weight probes can be used to provide information on the contact with external objects that might be easily moved out of their position, or for avoiding the gripper damage.

In many cases, it is not advisable to use sensory probes, because the system productivity may suffer if the sensing and action are separated. The use of probes is appropriate if probing may be performed simultaneously with the manufacturing task, or when there is enough time for the task. For instance, probes are used for seam tracking in welding [186] and in fettling of welding seams [252].

A probe typically consists of a long elastic beam with attached strain gauges and a separate module for measuring the axial force. The gauges sense bending caused by the action of lateral force components in two orthogonal directions [350]. Such probe design resembles a commonly used force-sensing finger design (see, for example, Figure 2.2.3). Probe sensors have been used in measuring machines for many years.

2.2.5   IPIT–IM sensors

Let us describe force sensors employed in the end-effectors of the experimental robotic system. These sensors are used to solve the problems considered in Chapters 9, 10, and 11. The descriptions of the robotic system and the end-effectors are given further in Chapter 4.

Figure 2.2.6 illustrates force-sensing fingers measuring three force com-

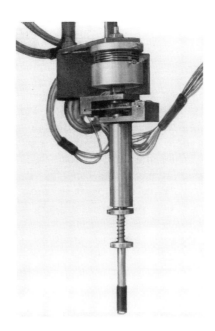

FIGURE 2.2.7
Sensory probe measuring three force components

ponents simultaneously. The finger sensor is designed by analogy with the configuration shown in Figure 2.2.2. Each finger sensor has three parallelogram elastic elements similar to those shown in Figure 2.1.3 and connected in series. The elements are aligned so as to measure forces acting in each of the three orthogonal directions. The sensor dimensions are 90 × 10 × 20 mm, the weight is 70 g, the force measurement range is 0.1–12 N, the coupling between the channels does not exceed 3.5%, and the sensor stiffness is $3 \cdot 10^4$ N/m.

A sensory probe shown in Figure 2.2.7 can measure three force components. The probe has a standard design. It consists of a module with a double bending beam elastic element (see Figure 2.1.1), measuring a longitudinal force component, and a rod. Two pairs of strain gauges mounted along the rod axis at 90° intervals measure a bending deformation of the rod. In the photo, the part of the rod with the strain gauges is covered by the protective sheath. The probe dimensions are 90 × 40 × 18 mm, the force measurement range is 0.1–8 N, the coupling does not exceed 2.5%.

2.3 Wrist Sensors

The closer a force sensor is located to a gripper or an end-effector, the less it is affected by the manipulator dynamics. Most commonly, a sensor is placed in the manipulator wrist, immediately preceding the gripper.

The gripping force does not influence wrist sensor readings as it does for sensing fingers. At the same time, a wrist sensor has a massive gripper attached to it. Therefore, dynamical factors affect its output more than the output of a finger sensor. Wrist sensors are the most common force sensors employed in Robotics. They are used in all tasks where force information is required, except gripping force measurement.

Designing a wrist sensor, one has to satisfy contradictory requirements for its sensitivity and stiffness. An elastic sensor with an attached massive gripper is an oscillatory system with six degrees of freedom. For an optimal performance, the lowest eigenfrequency of this system should be beyond the bandwidth of the manipulator motion. The center of mass of a gripper with a load tends to be located at a significant distance from the wrist sensor center, typically, close to the sensor axis. Hence in the oscillations with the lowest eigenfrequency, the sensor experiences angular displacements. Therefore, the sensor should possess a high stiffness with respect to the lateral bending moment. To reduce the magnitude of lateral bending moments, sensor dimensions in the axial direction should not be too large.

This section is subdivided into subsections in accordance with a primary transducer type used in the sensor and a range of loadings to be measured. The first two subsections consider sensors with strain gauge transducers and a bending elastic element, which are used for moderate loadings, and strain gauge sensors with a compressive-tensile and shear elastic element, which measure relatively large loadings. The third subsection considers sensors with various displacement transducers.

Even though a large variety of different force sensors are known, it is possible to single out one or two principal conceptual designs in each of the above categories. For instance, for strain gauge sensors with a bending elastic element, which enjoy a particular variety of designs, a Maltese cross type elastic element is a commonly used design.

2.3.1 Strain gauge sensors with bending elastic elements

For many problems, *a small number of force and torque components* (one to three) should be measured. A sensor measuring a small number of components can be arranged from the required number of simple elements connected in series. For example, for a two-component sensor, two elastic beams connected at a right angle may be used.

A sensor arranged from one to three single-component force sensing modules with parallelogram elastic elements is proposed in [361]. Each module is analogous in design to that shown in Figure 2.1.7. The advantage of this configuration is that it is possible to rearrange standard modules. Also, each module has good metrological characteristics and is relatively easy to manufacture. The drawback of this configuration is its comparatively large length, all the module lengths add up.

Another sensor arranged from individual modules is mentioned in [191]. Three modules with parallelogram elastic elements are used to measure three orthogonal force components. In addition, the sensor measures two lateral moments by using a fourth module.

A spatial structure of a six-component sensor considered in [308] includes six parallelogram elements built into the three-dimensional cross-shape structure. The advantage of the design is that the elements are mechanically selective. The drawback is that the structure is bulky and difficult to manufacture.

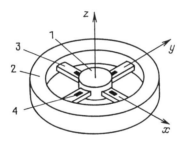

FIGURE 2.3.1
Maltese cross elastic element

A different approach to the multicomponent sensor design is to use an elastic element that responds to several force and torque components simultaneously. One of the most common and widely used elastic element configurations for a three-component strain gauge sensor is *the Maltese cross elastic element* shown in Figure 2.3.1. The sensor has two rigid flanges, the inner flange *1* and the outer flange *2*, connected by four elastic beams *3*. The beam cross-sections are elongated in the plane of the sensor. Four pairs of strain gauges *4* attached to the opposite surfaces of the beams are connected into four half bridges. The sensor measures the axial force F_z and two lateral moments M_x and M_y acting in the plane of the sensor.

The lateral force components F_x and F_y applied to the center of the sensor cause compressive-tensile deformations of the beams, but the sensor does not respond to these components. The axial moment M_z also causes little error in the measurements of F_z, M_x and M_y, since the beam cross-sections are elongated. A hardware or software solution can be used for computing the three measured components from the four sensor signals. Generally, one can assume that the moment acting along one of the beams is only measured by the strain gauges attached to the other beam. The corresponding addition or subtraction of the strain gauge signals can be performed by strain gauge bridges as well. Sensors of this type are described in [84, 245].

A Maltese cross elastic element is often used under the assumption that the point of the measured force application does not change, point torques are not applied, and, therefore, lateral moments yield the data on lateral components of the force vector. The advantage of such a sensor is the simplicity of manufacturing and design. Its drawback is an insufficient stiffness to the action of lateral moments. If the beams in the elastic element are sufficiently thin, then the sensor output may be nonlinear for large loadings. The nonlinearity can be caused by the bending-related tension in the beams fixed at both ends. Sensors of this type often have a significant hysteresis, if beams are attached to the flange (for instance, by screws) rather than machined as a single unit with it. In some arrangements, a Maltese cross elastic element (see Figure 2.3.1) is connected in series with an element measuring axial torque [285, 342, 356]. This configuration yields a four-component sensor.

Let us now consider designs for *six-component sensors with bending elastic elements*.

Several attempts to design a six-component sensor with mechanically selective (i.e., sensitive to individual components of force and torque) channels are known. To mechanically resolve force and torque vectors into individual components, a moment-free force transmission to each beam is arranged through a rod [100], pulleys [66], or ball joints. Such sensors are very complicated to manufacture, highly compliant, and tend to have a significant coupling in their channels. Besides, an assembled sensor structure may exhibit a substantial hysteresis, especially if sliding joints are used.

Let us now look into designs that do not try to achieve mechanical decoupling. As already noted, a Maltese cross elastic element (see Figure 2.3.1) is one of the most widely used elastic element configurations. In many six-component sensors, this element is modified to allow for the measurement of lateral forces. For example, the outer beam ends may be supported as balls freely sliding in the radial direction in bashings of the outer flange [184, 334]. Strain gauges attached to the four surfaces of each elastic beam measure strains. The drawback of this design is the hysteresis caused by the Coulomb friction in the sliding ball bearings. Roller bearings may be used instead of sliding bearings [356], but they do not do away with hysteresis or backlash, either.

A more sophisticated design is to support the outer beam ends of the Maltese cross elastic element through compliant support elements such as

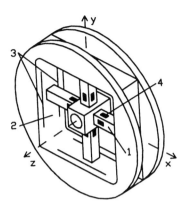

FIGURE 2.3.2
Maltese cross elastic element with elastic beam support

flexible plates. Such an element, highly compliant in the radial and two rotational directions, allows one to avoid hysteresis and increase sensor accuracy, and behaves similarly to a sliding bearing. Designs of this sort are described in [228, 323, 339, 344].

Figure 2.3.2 shows a six-component wrist sensor designed as described above [238]. The sensor is machined from a monolithic billet of metal. The ends of elastic beams *1* are attached to thin plates *2*, compliant in the radial directions. The plates are separated from the rest of the outer flange by slots *3*. Strain gauges *4*, attached in the middle of the beam surfaces and close to the inner flange, are connected into eight bridges. The manipulator arm is mounted to the outer sensor flange, the gripper, to the inner flange. This sensor configuration seems to be one of the best known designs in terms of accuracy, linearity, and measuring range. Various modifications of this design are used [17, 268]. Instead of four radial beams, only three may be used in some design variations. Such sensors are commercially available from Assurance Technologies Inc., Garner, North Carolina (formerly Lord) [109, 151].

The papers [49, 230, 335] propose a six-component sensor, where a Maltese cross elastic element sensing an axial force and two lateral moments is connected in series with a parallelogram elastic element transforming lateral forces and an axial torque. The view of the sensor is shown in Figure 2.3.3. The sensor and flanges are machined as a single unit. The sensor annular base flange *1* is connected by four radial beams *2* with a hub *3*. The hub, in turn, is connected with a loading flange *5* through four supporting beams *4*. Strain gauges *6*, are wired into eight bridges.

A six-component force sensor with two Maltese cross elements is shown in Figure 2.3.4 [104]. Beams with square cross-section are clamped in the two flanges *1* and *2*, and the hub *3*. Strain gauges *4* are attached to the upper and lower surface of each beam and to the lateral surfaces of the two diametrically opposite beams of the upper Maltese cross element. One can determine the lateral forces through the difference in the lateral moments relative to the centers of the two Maltese cross elastic elements. An additional transformation is needed for converting the signals from the sensor ten pairs of strain gauges, connected into seven bridges, into force and torque components.

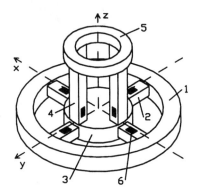

FIGURE 2.3.3
Maltese cross elastic element connected in series with spatial parallelogram elastic element

One more six-component force sensor design is a monolithic structure consisting of two rigid flanges connected by elastic rings fitted with strain gauges [238]. A commercially available JR3 sensor has a structure with strain rings and is produced in different sizes and measurement ranges [115].

In some cases, it is necessary to fit a sensor inside the walls of a hollow cylinder. For example, if a gripper motor is to be located inside the sensor, then a Maltese cross elastic element cannot be used. Figure 2.3.5 [220, 35] illustrates a six-component wrist sensor that conforms to such constraints. The sensor consists of two three-component blocks machined from hollow cylinders. Eight pairs of strain gauges *1* measure bending strain in the elastic beams. The block *2* measures lateral forces and an axial torque, the block *3*, an axial force and lateral moments. When assembled, one block is embedded into another. The forces and torques are applied between the flanges *A* and *B*. The arrangement is easy to manufacture. The sensor drawbacks are that it is bulky and the flange stiffness is low.

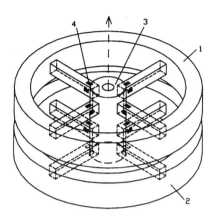

FIGURE 2.3.4
Double Maltese cross elastic element

2.3.2 Strain gauge sensors with compressive-tensile and shear elastic elements

To measure large loadings using strain gauge sensors, it is advisable to employ compressive-tensile and shear elastic elements. Such elements have a high stiffness.

The sensor design proposed in [288] and in patent [329] is now well known (see Figure 2.3.6). The sensor is machined from a solid billet of metal and consists of two annular flanges connected by three short struts located along the circumference at 120° intervals. Each strut of a rectangular cross-section has strain gauges and shear strain gauge rosettes attached to it. The strain gauges measure longitudinal compression or tension in the struts, and the rosettes measure shear strain related to the tangential forces acting on the struts. Force and torque components are derived from the output of the six strain gauge bridges, for example, using specialized hardware [288, 298]. It should be emphasized that this sensor configuration can hardly be used to measure small forces and torques, because the minimum size of the struts is bounded by the size of strain gauges attached to it as well as by mechanical strength requirements. A modification of the above-described arrangement with four elastic struts instead of three is also used [235, 268]. By combining strain gauge signals in the bridges, one can obtain a sensor compliance matrix close to a diagonal one.

A common design idea is to convert applied forces and torques into compression and tension of six single-component load cells. Such a sensor design is shown in Figure 2.3.7 [336, 351]. Two rigid flanges *1, 2* are connected via

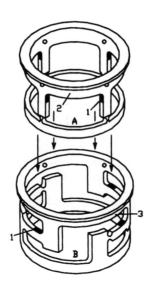

FIGURE 2.3.5
Sensor placed in the walls of a hollow cylinder

six linear single-component load cells *3, 4* placed at an angle to the sensor axis. Since each load cell has a high axial stiffness and low lateral stiffness, the system is statically determinate, which improves the sensor metrological properties, in the opinion of some authors [69]. To determine force and torque components, one should compute appropriate linear combinations of the load cell output signals.

In one prototype sensor of this configuration, load cells have been connected with the flanges via sliding joints, which causes a significant hysteresis. The introduction of elastic joints allows noticeable reduction of hysteresis [235]. A similar sensor arrangement with a slightly different

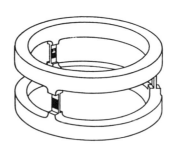

FIGURE 2.3.6
Sensor with compressive-tensile and shear elastic element

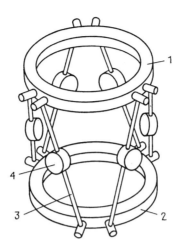

FIGURE 2.3.7
Sensor with six one-component load cells

placement of load cells is described in [351].

2.3.3 Wrist sensors with displacement transducers

Some force sensors operate by measuring displacements in the elastic element, rather than deformations. In the simplest devices, a spring may serve as an elastic element, and potentiometric or photoelectric transducers measure displacements. Such sensors were the first to find industrial applications. Wrist sensors with spring elastic elements and similar displacement transducers were installed in the first industrial assembly manipulators by DEA (Italy), Olivetti (Italy), Hitachi (Japan) [42, 185, 255]. A sophisticated design of a four-component sensor of this sort is described in the patent [327].

Let us now consider designs of six-component force sensors with displacement transducers. Most sensor of this type have similar design. Two rigid annular flanges are connected through three or four elastic elements, which are typically placed at regular intervals along a circumference. Between the elastic elements, three or four pairs of displacement transducers, respectively, are located along the same circumference. Each transducer pair (or a single two-component transducer) measures relative flange displacements in tangential or/and axial directions. Forces and moments are computed as respective linear combinations of transducer output signals.

The sensor designs differ by the number of elastic elements and transducer pairs (three or four), by elastic element structure, and transducer

types used. A six-component force sensor with potentiometric transducers
is described in [219, 324]. Inductive transducers [112, 204, 325, 328, 337] and
photoelectric transducers [101, 324] yield a higher measurement accuracy.
Capacitive displacement transducers may also be used [341, 355]. A variety
of elastic element types is used in sensors measuring displacements. Bending elastic elements are used to ensure high sensitivity [325, 328, 331, 337].
If high sensitivity is not required (the measurement range is large), then
compressive-tensile and shear elastic elements may be employed [204]. In
this case, elastic elements analogous to those of the strain gauge force sensors can be used. For example, various modifications of a Maltese cross
elastic element (see Figure 2.3.2) [325, 328, 354] or an elastic element with
three struts shown in Figure 2.3.6 [204] are used. Some of the elastic elements used in sensors with displacement measurement, such as rubber
spacers [324, 355] or specially shaped spring elements [341], cannot be used
for strain gauge sensors.

The main challenge in the design of such sensors is to accommodate
displacement transducers in the sensor and transfer relative displacements
in the sensor flanges to them. Many authors propose to change only elastic
elements in sensors designed for various measurement ranges, while leaving
the rest of the structure the same [204, 331].

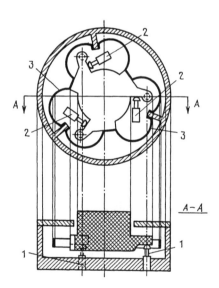

FIGURE 2.3.8
Sensor with inductive displacement transducers

Consider the following typical configuration of a sensor measuring displacements. Figure 2.3.8 shows a sensor with inductive displacement trans-

ducers [331, 337]. The sensor hub has three radial arms attached to it and six displacement transducers to measure their displacements. The three transducers *1* measure displacements along the sensor axis, three other transducers *2* measure tangential displacements. The sensor elastic element consists of three arc-shaped springs *3* which are attached to the body and hub of the sensor. The sensor sensitivity may vary with the type of spring.

2.3.4 IPIT–IM wrist sensor designs

We shall now describe some typical wrist sensor configurations designed by the IPIT–IM research team. The majority of these sensors were used in the tasks considered in Chapters 9–11. The three-component sensor with a Maltese cross elastic element shown in Figure 2.3.1 measures the axial force and two lateral moments. The sensor is machined from a solid billet of metal. Metal-foil strain gauges attached to both surfaces of the elastic beam are connected into four half bridges, and the bridge signals are independently amplified. The four output signals of the half bridges are linearly transformed by the analog hardware into the three sensor output signals. The measurement range is 100 N for force and 10 N·m for lateral moments, the coupling after the analog transformation is less than 2%. The sensor is used as a wrist sensor for a PUMA-560 manipulator, as a part of an electromagnetic gripper for a ZIM-10 manipulator, and in the three-component instrumented joystick for supervisory control of a walking robot.

As mentioned above, one of the promising approaches to multicomponent sensor design is to compose a sensor of one- or two-component modular elements. Original designs for such modular sensors have been suggested by the IPIT–IM group [358, 359, 360]. These sensors are assembled by embedding individual modules into each other so that the overall sensor dimensions are not much larger than the dimensions of a single module.

Several types of four-component modular sensors for various measurement ranges have been developed. Each sensor includes three parallelogram modules (see Figures 2.1.2– 2.1.10) for measuring three force components and a hollow cruciform module (see Figure 2.1.8) to measure axial torque.

The sensor shown in Figure 2.3.9 uses silicon strain gauges. The sensor measurement range is ±50 N for the force and ±2.5 N·m for the axial torque. The cross-coupling does not exceed 2% for each channel under the action of lateral moments of up to 8 N·m, the sensor eigenfrequencies with a mass of 5 kg attached to its flange are no less than 25 Hz, the mass of the sensor is 450 g. The sensor is 90 mm in diameter, and its height is 63 mm. The parameters and characteristics of the sensor individual modules and design computations for them are presented in Section 3.2. The main advantages of the sensor are that it measures individual force and torque components directly and has a high eigenfrequency and measurement accuracy. In case of a structural failure, it is possible to replace only a single module without abandoning the whole sensor.

A version of the modular sensor designed for smaller forces and moments (5 N and 0.05 N·m, respectively) is used in the end-effector described in Section 4.3 (see Figure 4.3.3). The wrist has been used in assembly tasks considered in Chapter 10.

FIGURE 2.3.9
IPIT–IM four-component modular sensor

It follows from the analysis of the present section that one of the most promising designs for a six-component strain gauge force sensor is based on a Maltese cross elastic element with membrane elastic joints. Figure 2.3.10 shows a sensor of this type designed by the IPIT-IM group. The sensor hub with diameter of 60 mm is connected with four elastic beams of 5.2 × 5.2 mm cross-section and 20 mm length each. The outer ends of the beams are connected to elastic membranes 40 mm long, 1.5 mm thick, and 20 mm high. The sensor is machined from a single block of D16T aluminum alloy. Silicon strain gauges attached to all the beam surfaces close to the hub are connected into eight strain gauge bridges. The sensor has a measurement range of ±12.5 N for the lateral forces, ±25 N for the axial force, ±1.25 N·m for the lateral moments, and ±2.5 N·m for the axial torque. With a mass of 1 kg attached at the 130 mm arm of force, the sensor eigenfrequency exceeds 80 Hz. The cross-coupling after the transformation is about 1%, the dynamical range is as high as 2000. The sensor is 110 mm in diameter, the height is 40 mm.

A sensor analogous to that shown in Figure 2.3.10 was designed for the industrial robot PUMA-560. The two sensors differ in dimensions, measurement range and mechanical interface. The sensor for PUMA-560 is made from D16T alloy and has an external diameter of 90 mm. The elastic beams are 20 mm in length and 5×5 mm in cross-section, the elastic membrane dimensions are $38 \times 15 \times 1.5$ mm. The sensor is designed for a measurement range of ± 5 N for forces, ± 0.5 N·m for moments.

A force sensor with a compressive-tensile and shear elastic element is shown in Figure 2.3.11. The sensor is designed for a manipulator with a payload capacity of 100 kg.

The sensor design is similar to that displayed in Figure 2.3.6, differing only in that it has four elastic struts rather than three. The long side of each strut of 6×13 mm cross-section is normal to the radius. The strut length is 20 mm. Shear strain rosettes are attached to the external strut surfaces and linear strain gauges, to the lateral surfaces. In addition, linear strain gauges are attached to the sensor flanges for temperature compensation. The sensor is machined from a single block of D16T alloy. The sensor conceptual design allows one to measure all six force and torque components as described in Section 3.4. In the described sensor, which only measures forces, all linear strain gauges are connected into a single bridge measuring the axial force. The strain gauge rosettes are connected into two bridges measuring two lateral force components. The sensor has a force measurement range of $\pm 2 \cdot 10^3$ N, the cross-coupling in the channels does not exceed 7% for forces applied at an arm of force up to 0.3 m, the dynamical range is greater than 10^3. The sensor is 150 mm in diameter, the height with the flanges is 120 mm, the mass is 2.8 kg.

2.4 Other Types of Force Sensor

Force sensors can be installed on a robot in other ways than those presented in Sections 2.2 and 2.3. Let us consider two more sensor types.

2.4.1 Instrumented platforms

In some cases, sensors measuring forces between a manipulator and a manipulated object are located in the base where the manipulated object is placed. Instrumented platforms (pedestal sensors) have several special features. An instrumented platform is a separate sensing device, exterior to a manipulator. The coordinate system for force and torque components at the output of such sensor does not depend on the manipulator arm configuration at a given time. On the other hand, when an instrumented platform is used, the tasks requiring force information may only be performed on the platform, rather than anywhere within the manipulator workspace.

Instrumented platforms typically consist of a base and a support plate connected by several force sensing supports. Depending on the platform configuration, each sensor measures one to three force components. Since the support plate has considerable linear dimensions, it may experience significant bending deformations under loading, and this may lead to non-linear errors in the sensor readings. To increase its stiffness, the plate can be strengthened by stiffening ribs. Another way to reduce influence of the support plate deformation on the measurement is by using special modifications of force transmitting elements, such as ball supports. However, a ball support can only transmit one, vertical force component. Elastic joints, produced by thinning down structural elements transmitting force to sensor supports, significantly reduce the influence of the lateral moments caused by the plate deformation [338].

Three-component instrumented platforms measuring a vertical force and two lateral moments are frequently used. These measurements suffice to determine coordinates for the point of vertical force application such as the center of mass for an object placed on the platform [223]. Such platforms typically comprise three or four single-component force sensing supports to measure the vertical force [321, 338, 343]. The single-component force sensors vary in design and use different elastic elements. In particular, any force measuring module described in Section 2.1 may be used for this purpose.

Instrumented platforms do not allow one to determine the orientation of an object placed on them. However, it is conceptually possible to define an object orientation by measuring bending moments acting on the force sensing supports provided the upper platform plate is elastic. This idea is considered in [362]. It proposes to specifically strengthen the elastic upper plate by stiffening ribs, which allows determination of an object orientation by the readings of sensing supports.

Considerable experience with instrumented platform design is accumulated in medical, biomechanical, and sports applications [322, 357].

For a platform of small dimensions, it is sometimes possible to get by with a single multicomponent sensor mounted in the middle of the platform, or design a platform by neglecting the deformations of the support plate, i.e., by assuming that the plate is rigid. Such instrumented platform configurations, described, for example, in [121, 288], are analogous to the multicomponent wrist sensor designs considered in Section 2.3.

One of the instrumented platforms developed by the IPIT–IM research group measures a vertical force component and two lateral moments, which allows one to determine the weight and coordinates of the center of mass for a part placed on it. The platform consists of two plates connected by three force measuring modules using double bending beams as elastic elements (see Figure 2.1.1). To reduce the influence of the upper plate deformations on the module outputs, the forces are transmitted to the modules through steel balls.

2.4.2 Foot force sensors

Force sensors are also used to measure foot force distribution, i.e., forces acting in the legs of walking robots [48, 82, 133, 231, 278]. Output accuracy and measurement range requirements are lower for the foot force sensors than for robotic system force sensors. This is due to the fact that small forces arising from microdisplacements should be measured in robotic systems, while large force loadings are measured for walking robots. Sensors for legged robots should have a higher stiffness.

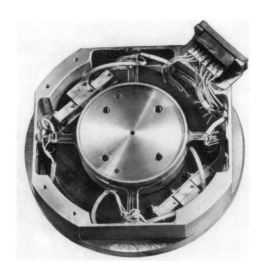

FIGURE 2.3.10
IPIT–IM six-component Maltese cross sensor

FIGURE 2.3.11
IPIT–IM six-component force sensor with compressive-tensile
and shear elastic element

Chapter 3

Basic Theory and Design Computation of Force Sensors

As the analysis of the previous chapter suggests, design of a r ulticomponent sensor can follow one of two main schemes. A multicom ponent sensor can be composed of individual modular elements, each measuring one or two components of force or torque with a high accuracy. These modular elements are usually designed to be mechanically decoupled, i.e., to have selective mechanical response to the measured components.

Another approach is to design a six-component force sensor employing a general linear transformation of primary output signals. With this approach, there is no need for mechanical decoupling; the sensor just needs to be accurately calibrated. Sensors of this type are most commonly placed in the manipulator wrist.

Strain gauge force sensors designed using the two above schemes are the subject of this chapter. Section 3.2 studies decoupled modular elements of force sensors, while Sections 3.3 and 3.4 deal with six-component sensors without mechanical decoupling.

3.1 Main Characteristics of Force Sensors

Let us consider the main force sensor characteristics that need to be calculated in the design process. We start with the fundamental question: what is the sensor output under the action of an external force?

3.1.1 Conversion of force into electrical output

The force sensor is designed for a certain measuring range. This range can be defined through a nominal measured force F_0 or nominal measured torque M_0, i.e., through the loads, for which the sensor output should have a specified magnitude. For multicomponent sensors, the nominal force and nominal torque tend to be the same for different measured components. The relationship between the nominal force F_0 and nominal torque M_0 can be defined via the characteristic force arm D, where $M_0 = F_0 D$. Since torques acting on a sensor are generally produced by forces offset from its center, D often is the characteristic distance between the sensor center and the point of force application. For wrist sensors, the distance between the sensor center and the gripper jaws or the center of mass of the end-effector can be taken as the characteristic force arm.

In robotic systems, a force sensor is usually subjected to complex loading. Several components of force and torque are applied to it simultaneously. Hence a possible influence of auxiliary load components must be taken into account even for a single-component sensor. The sensor measurement error must be small for auxiliary load components that do not exceed the nominal force (torque).

We shall consider force sensors using strain gauges. A strain gauge is a sensing element that experiences a change in electrical resistance under strain, namely, tension or compression in the direction of its sensitivity axis. Denote the axial strain in the strain gauge by ε. Let $\varepsilon = \Delta l / l$, where l is the strain gauge length, and Δl is the length variation. Denote by R the strain gauge electric resistance, and by ΔR, the resistance change caused by the strain ε. Then

$$\Delta R / R = \varepsilon \chi, \qquad (3.1.1)$$

where the constant χ is a gauge factor. The gauge factor χ is about two for metal strain gauges, and $50 \leq |\chi| \leq 400$ for semiconductor ones. Also, $\chi > 0$ for n-type semiconductors, and $\chi < 0$ for p-type semiconductors [14, 24, 50].

It is necessary to pass an excitation current through the strain gauge in order to register a change in its resistance. The maximum excitation current is limited by the allowed power of heat dissipation and acceptable change in the strain gauge temperature during the measurement. Note that the change in resistance caused by a change in temperature may be commensurable with that produced by measured strain. Therefore, the strain gauges are normally connected into a bridge circuit (see Figure 3.1.1). In so doing, the useful signal across the output diagonal of the bridge depends on the difference in the change of resistances, connected as adjacent arms of the bridge. Thus, the bridge circuit with the same heat dissipation conditions and equal thermal sensitivity of the strain gauges in its arms is not sensitive to the temperature change.

Consider a balanced strain gauge bridge with an excitation voltage $\pm u_0$ shown in Figure 3.1.1. Let the resistances of the arms of the bridge be

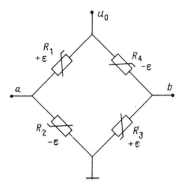

FIGURE 3.1.1
Balanced strain gauge bridge

$R_i = R + \Delta R_i$ $(i = 1, \ldots, 4)$, where $|\Delta R_i| \ll R$. Then the voltage across the output diagonal of the bridge [24] is given by

$$u = \frac{u_0}{2} \left(\frac{\Delta R_1}{R} - \frac{\Delta R_2}{R} + \frac{\Delta R_3}{R} - \frac{\Delta R_4}{R} \right) \tag{3.1.2}$$

Provided that the changes in resistance are produced by strains in the strain gauges with the same gauge factor χ, we obtain from (3.1.2) by using (3.1.1)

$$u = \frac{\chi u_0}{2} (\varepsilon_1 - \varepsilon_2 + \varepsilon_3 - \varepsilon_4) = k u_0, \tag{3.1.3}$$

where ε_j, $j = 1, \ldots, 4$ are strains in the respective strain gauges. The factor k is called a transfer factor, and it depends on the strains ε_j. This factor is proportional to the gauge factor χ. To obviate the dependence of the sensor elastic element design on the type of strain gauge, we shall use average strain e in the strain gauge bridge instead of the transfer factor, where

$$e = \frac{1}{2} (\varepsilon_1 - \varepsilon_2 + \varepsilon_3 - \varepsilon_4) \tag{3.1.4}$$

In view of (3.1.4), we rewrite (3.1.3) in the form

$$u = \chi u_0 e \tag{3.1.5}$$

In many cases, a half bridge of two active strain gauges connected as adjacent arms is used for measurement instead of a full bridge. The two remaining arms of the bridge are passive. In this case, we take $\varepsilon_3 = \varepsilon_4 = 0$ in (3.1.4).

In some instances, unequal gauge factors of strain gauges need to be taken into consideration, in particular, when defining the influence of unmeasured force and torque components on the sensor output. Denote by χ_i the gauge factor of the i-th arm of the bridge, whereas χ is the nominal (average) gauge factor. Then the bridge output can be found according to (3.1.5), where e will now denote the *apparent average strain*

$$e = \frac{1}{2}\left(\varepsilon_1\chi_1/\chi - \varepsilon_2\chi_2/\chi + \varepsilon_3\chi_3/\chi - \varepsilon_4\chi_4/\chi\right)$$

$$= \frac{1}{2}(\varepsilon_1 - \varepsilon_2 + \varepsilon_3 - \varepsilon_4) + \frac{1}{2}(\varepsilon_1\eta_1 - \varepsilon_2\eta_2 + \varepsilon_3\eta_3 - \varepsilon_4\eta_4), \quad (3.1.6)$$

where $\eta_i = (\chi_i - \chi)/\chi$ $(i = 1, \ldots, 4)$ are relative gauge factor deviations from the nominal value. By design, for an unmeasured component of force or torque, the average strain $(\varepsilon_1 - \varepsilon_2 + \varepsilon_3 - \varepsilon_4)/2$ in the bridge is usually zero. Therefore, e is defined by the dispersion of the gauge factors.

Let us give three common examples of connecting strain gauges into a bridge for measuring different types of strain in the elastic element. Consider measuring bending strain for the beam shown in Figure 3.1.2 as the first example.

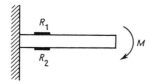

FIGURE 3.1.2
Measuring of bending strain in a beam

Two strain gauges, connected as an active half bridge, are mounted to the opposite faces of the beam with a symmetrical cross-section. When the bending moment is applied to the beam, the strains in the strain gauges are equal in magnitude and opposite in sign. Then the average strain

$$e = \frac{1}{2}[\varepsilon - (-\varepsilon)] = \varepsilon \qquad (3.1.7)$$

in the bridge has the same magnitude as the strain in each strain gauge.

The second example deals with measuring shear strain by means of a strain gauge rosette (Figure 3.1.3).

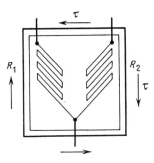

FIGURE 3.1.3
Strain gauge rosette for shear strain measurement

The rosette incorporates two strain gauges R_1 and R_2 with orthogonal sensing axes. The axis of the rosette is directed at 45° to these axes. With shear stress τ acting along the rosette axis, the shear strain of the elastic element material is $\gamma = \tau/G$, where G is the shear modulus. Therefore, the two strain gauges in the rosette experience axial strains $\pm\gamma/2$. If these strain gauges are connected as a two-active-arm bridge, then the average strain in it is given by

$$e = \frac{1}{2}[\gamma/2 - (-\gamma/2)] = \gamma/2 \qquad (3.1.8)$$

If the elastic element undergoes elongation or compression in the direction of the rosette axis, then the axial strains in the strain gauges are equal, and the average strain in the bridge is zero.

Measurement of uniaxial strain of tension-compression is our third example (Figure 3.1.4).

A four-active-arm strain gauge bridge is typically employed for such measurement. Two strain gauges with sensing axes parallel to the direction of strain are connected into opposite arms of the bridge. Only these two strain gauges are active, while the other two provide for the temperature compensation. Let ε be the uniaxial strain, then the average strain in the bridge is given by

$$e = \frac{1}{2}(\varepsilon - 0 + \varepsilon - 0) = \varepsilon \qquad (3.1.9)$$

Average strain in the bridge is defined by an elastic element of the sensor,

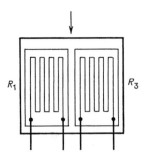

FIGURE 3.1.4
Measuring uniaxial tensile-compressive strain

a method of strain gauge placement, and loading conditions. Regardless of all these conditions, the output signal of the strain gauge bridge at a nominal load must be amplified to a standard level.

The necessary magnitude of average strain in the strain gauge bridge is defined by a practically acceptable voltage amplifier gain for a voltage across the diagonal of the bridge. The required amplifier gain differs depending on the type of strain gauges used (semiconductor or metal). Semiconductor strain gauges have gauge factors approximately 50 times greater than metal strain gauges, and they are smaller in size. However, they have a high temperature coefficient of resistance. Also, nonlinearity of semiconductor strain gauges is higher. Thus, semiconductor strain gauges are mostly employed where it is desirable to use less complicated amplification electronics, or for elastic elements of a small size.

Metal-foil strain gauges seem to be more general-purpose. They are appropriate for sensors with no constraint on dimensions, including force sensors with small nominal strain in the elastic element. Low temperature coefficient and substantially lower level of thermal noise in these strain gauges compensate for their lower gauge factor.

If an appropriate strain gauge amplifier is used, then the maximum practical amplification factor tends to be limited because of the temperature dependence of resistance. For most practical purposes, at the output signal ± 10 V, the voltage gain should not exceed $2 \cdot 10^4$ for metal-foil strain gauges and 500 for semiconductor gauges. For this output signal, these gains correspond to the signals $5 \cdot 10^{-4}$ V and $2 \cdot 10^{-2}$ V across the output diagonal of the bridge. Let us find average strains e_1 and e_2 corresponding to these signals for metal (at $k = 2$) and silicon strain gauges (at $k = 100$). We assume the bridge excitation voltage to be $u_0 = 3$ V. According to (3.1.5), we obtain the required average strain $e_1 = 0.83 \cdot 10^{-4}$ for metal strain gauges and $e_2 = 0.66 \cdot 10^{-4}$ for silicon ones. In the subsequent discussion, we assume that the average strain must be of the order 10^{-4}, regardless of the type of strain gauges.

3.1.2 Design parameters of force sensors

To design a force sensor, one should compute and, if possible, optimize a number of its parameters.

Sensitivity.

We shall define sensitivity of a sensor channel to the measured force or torque component in terms of an average strain in the respective bridge under the action of this component. In so doing, we assume that a loading of the nominal magnitude is applied to the sensor. So defined, sensitivity only depends on the sensor design and strain gauge location, and does not depend on the type of strain gauges and output signal amplifiers used.

For multicomponent sensors, it is necessary to define a sensitivity matrix that establishes a relationship between the components of force and torque vectors applied to the sensor and a vector of average strain values for the strain gauge bridges.

Stiffness.

Linear or angular stiffness is defined as a ratio of an applied loading to the linear or angular elastic displacement produced by this loading. Sensor compliance is an inverse of the stiffness. Single-component modules are typically mechanically decoupled. For these modules, much greater displacements are caused by a measured force or torque component than by an unmeasured one. Therefore, linear or angular compliance is only significant in one direction. For multicomponent sensors, it is necessary to define a stiffness matrix of the form (1.2.8) relating displacements in the elastic element to applied force and torque components. Different entries of this matrix are usually of the same order of magnitude.

Eigenfrequency.

The sensor stiffness defines the eigenfrequency of a sensor with an attached inertial load. The type of the attached load may vary for different sensors. For force sensing fingers, the load is the mass of an object held between the fingers. In the case of a wrist sensor, the mass of the end-effector must also be taken into consideration. For modules with mechanical selectivity to a single component, the lowest eigenfrequency is defined by their compliance to the action of this component. For multicomponent sensors, eigenfrequency of the system is jointly defined by the inertial properties of the attached load and the stiffness matrix of the sensor.

Decoupling.

Decoupling is described by a coupling coefficient, which is a ratio of characteristic magnitudes of the channel output caused by an unmeasured and measured loading. It is assumed that the measured loading is of the nominal magnitude, and the unmeasured one has its characteristic magnitude, which usually coincides with the nominal force or nominal torque. For a multicomponent sensor with decoupled channels, the coupling coefficients are off-diagonal entries of the sensitivity matrix.

Decoupled sensors are designed to have no coupling. The actual coupling they have is a consequence of inaccurate placement of strain gauges and variation of their parameters. We shall compute *the worst coupling coefficient*, that is, the worst possible value of the coupling coefficient for the specified strain gauge placement error and parameter variation. We calculate coupling for each measured force or torque component assuming that spurious signals add up in the bridge, that is, the strain gauge placement and variation of their parameters are the worst possible within the specified error limits. In keeping with this definition, actual coupling coefficients for any particular sensor should not exceed the worst one.

Linearity.

A sensor output signal can be represented as a sum of zero-, first-, and higher order terms in the components of the applied loading. In such expansion, zero-order terms describe offsets of the sensor outputs, while the weights of the first-order terms describe sensitivities and coupling of the sensor channels. Nonlinearity can be defined as a relative magnitude of higher-order terms for loading components within the nominal limits. Quadratic terms tend to be the most significant among the nonlinear terms [148]. To evaluate them experimentally, one has to apply, in a sequence, all possible combinations of two force and torque components to the sensor.

Nonlinearity can be both of mechanical and electrical origin. A beam fixed at both ends and loaded in the middle exemplifies a mechanically nonlinear element commonly used in sensor designs. A large bending deformation would cause a longitudinal tension in the beam, which results in the dependence of strain on the applied force being nonlinear. Correct design of the sensor can generally ensure its mechanical linearity. Strains in the sensor flanges often produce nonlinear effects as well. If the flanges deform, the sensor output may not only depend on the principal force vector and the principal torque vector, but also on spatial distribution of applied loadings, a method of sensor mounting, and a method of mounting strain gauges to it. Nonlinear errors typically cannot be eliminated by an additional signal transformation in real time, as this would require excessively complex calibration of the sensor.

Hysteresis.

Hysteresis is the difference in the sensor output for incrementally increasing and decreasing loading sequences. Strains in the elastic element of the sensor are usually small (less than 10^{-3}) and do not produce plastic deformations. The hysteresis is often caused by friction in the assembled sensor structure or by inadequate mounting of strain gauges. To reduce the hysteresis, it is advisable to manufacture the elastic element and flanges as a monolithic structure. However, this may still leave the hysteresis caused by friction and microdisplacements between the deformable flanges and their mounts.

Other error sources in strain gauge force sensors are related to strain gauges themselves or electronic amplifiers of their signals. The error effects include creep, temperature drift, to name just a few. These errors, similarly to hysteresis, do not generally depend on the sensor design, rather they are related to sensor manufacturing. These and other issues of sensor technology are beyond the scope of this book.

3.2 Design Computations for Sensor Modular Elements

As mentioned above, many multicomponent force sensors are composed of one- or two-component modular elements. Such modules should only be sensitive to the measured component among multiple components of force and torque acting on it. Selective sensitivity of the modular element is not defined by its mechanical design alone. It also depends on such factors as accuracy of strain gauge placement and variation of their parameters. These factors have to be taken into account when designing a force sensor module.

Let us consider three commonly used designs of modular elements described in the survey of Section 2.1.

3.2.1 Elastic beam

A bending moment can be measured by using an elastic beam with a rectangular cross section. Such an element is mechanically selective provided its cross-section has an elongated shape. This element is described in Chapter 2.

Figure 3.2.1 shows a beam subjected to strain. Strain gauges *1* and *2* are connected as adjacent arms of a two-active-arm strain gauge bridge and convert the strain into an electrical signal. They are attached at midpoints of the opposite wide faces of the beam and oriented along its longitudinal axis. Let us denote by a, h, and l the width, thickness, and length of the beam, respectively. Consider a coordinate system with an origin at the center O of the beam. The axis Oz of the system is directed along the axis of the beam, the axis Oy is normal to the wide face of the beam, the axis Ox complements the axes to a right-handed Cartesian frame.

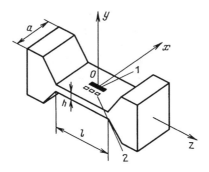

FIGURE 3.2.1
Beam element for bending moment measurement

Let us give an example of design computations for this modular element.

Sensitivity.

The nominal moment M is applied to the beam along the axis Ox and bends the beam. The center line of the bent beam acquires a curvature $\kappa = M/EI$, where $I = ah^3/12$ is the central moment of inertia for the beam cross-section, and E is the Young's modulus of the beam material [260]. The strains measured by the strain gauges *1* and *2* are $\varepsilon_{1,2} = \pm\kappa h/2$. The average strain in the bridge is $e = (\varepsilon_1 - \varepsilon_2)/2 = \kappa h/2$. By using the expression for the curvature, we obtain

$$e = 6M/(Eah^2) \qquad (3.2.1)$$

The expression (3.2.1) defines the sensor sensitivity.

Stiffness.

The bending moment M applied to the beam causes an angular deflection of one end of the beam relative to the other. This deflection can be found as $\varphi = \kappa l$, or

$$\varphi = 12Ml/(Eah^3) = 2el/h \qquad (3.2.2)$$

As can be seen from (3.2.2), if the sensitivity of the beam element is constant (e is given), the angular deflection φ decreases with the decrease of the length l and increase of the thickness h of the beam. In effect, by decreasing the beam length, we reduce the strained part of the beam free of the strain gauges. One cannot, however, make the beam excessively short, as the strain in the site of strain gauge attachment should be sufficiently uniform. Therefore, strain gauges should be mounted at some distance from the beam ends. Besides, there should be enough space on the beam to place a strain gauge with electrical contacts. According to (3.2.1), with the sensitivity held constant, an increase of the thickness h leads to the reduction of the beam width a. This would undermine selective sensitivity of the beam strain to the measured bending moment. Consequently, requirements for stiffness, sensitivity, and coupling should be considered simultaneously.

Coupling.

Consider the influence of unmeasured force and moment components on the sensor output (coupling).

If the strain gauges are attached exactly to the center line of the beam faces, the lateral bending moment M_y does not affect them. In order to find the worst case coupling coefficient ϑ for M_y, we shall assume that the strain gauges are displaced from the center line by a placement error Δ in the opposite directions. Assume that the moment M_y of the same magnitude as the nominal bending moment M is applied to the beam. Subjected to this moment, the beam bends in the direction of the greater stiffness. The curvature of the beam center line is $\kappa_y = M/EI_y$, where $I_y = a^3h/12$ is the moment of inertia of the cross-section relative to the axis Oy. The strains of the strain gauges are $\varepsilon_{1,2} = \pm\kappa_y\Delta$. The average strain in the bridge is $e_1 = (\varepsilon_1 - \varepsilon_2)/2 = \kappa_y\Delta$. The worst case coupling coefficient ϑ is the ratio of magnitudes for this strain and the average strain (3.2.1) created by the nominal moment. By substituting the above expressions for the average strains, we obtain

$$\vartheta_1 = 2h\Delta/a^2 \qquad (3.2.3)$$

The torque M_z causes shear strain in the beam. The principal axes of the shear strain tensor are directed at $45°$ to the beam axis (the axis Oz). If the strain gauges are oriented along the axis Oz, their length does not change. To find the worst coupling coefficient, we assume that the strain gauges are rotated relative to the axis Oz by an angular placement error ψ. When the torque of the same magnitude as the nominal bending moment M is applied along the axis Oz, the strains in the plane of strain gauge placement are: $\sigma_{xz} = \sigma_{zx} = \tau = \sigma_{xx} = \sigma_{zz} = 0$. For the beam with an elongated cross-sectional area ($h \ll a$), we get $\tau = 3M/(ah^2)$ [261]. The strain in the beam is given by $e_{\alpha\beta} = \sigma_{\alpha\beta}(1+\mu)E$, where $\alpha, \beta = x, z$, and μ is the Poisson's ratio for the beam material. The axial strain of the strain gauge is $\varepsilon_r = \sum \varepsilon_{\alpha\beta} n_\alpha n_\beta$, where $n = [n_x, n_z]$ is the orientation vector of the gauge, $n_x = \sin\psi$, $n_z = \cos\psi$. By assuming the angular placement error ψ to be small ($\psi \ll 1$), we obtain $\varepsilon_r \approx \tau 2\psi(1+\mu)E = 6M\psi(1+\mu)/(Eah^2)$. The strain in the second active gauge is of the same magnitude and opposite in sign. Therefore, the average strain in the bridge is ε_r. Now we can find the worst case coupling coefficient ϑ_2 as a ratio of ε_r to the average strain e (3.2.1):

$$\vartheta_2 = 2(1+\mu)\psi \qquad (3.2.4)$$

The effects of lateral forces F_x and F_y are defined by the lateral bending moments M_y and M_z, created by these forces. Hence we shall not discuss these effects in detail. The lateral force F_x also causes shear strain, which can affect the output signal if the strain gauges are shifted from the axis Oz. However, this strain is much less than the shear strain caused by the torque M_z. One can demonstrate that the ratio of the two characteristic strains has the order of h/D, where h is the thickness of the beam, and D is the characteristic arm of force.

The action of the longitudinal force F_z results in the uniform strain of the gauges. The strain gauges are connected as adjacent arms of the bridge, so in the ideal case the output of the bridge is zero. Denote by η a maximum relative variation of strain gauge sensitivity. If the force F_z equals the characteristic force M/D, where M is the nominal measured moment, then the strain in each of the gauges is $\varepsilon = M/(EDah)$. Therefore, according to (3.1.6), an apparent average strain in the bridge is $e = M\eta/(EDah)$, and the worst case coupling coefficient for the longitudinal force is given by

$$\vartheta_3 = \eta h/(6D) \qquad (3.2.5)$$

Linearity.

Nonlinear mechanical coupling might be caused by the simultaneous action of the tensile (compressive) force F and the measured bending moment M on the beam. The force creates an additional bending moment $= F\delta$, where δ is a deflection of the beam end due to the measured moment. The relative nonlinearity can be estimated as

$$\kappa = F\delta/M = \delta/D = 6Ml^2/(DEah^3) \qquad (3.2.6)$$

As an example, let us find parameters of the module used in a number of IBM sensors [280]. The dimensions of the elastic beam in the module are: $l = 6.35$ mm, $h = 1.78$ mm, $a = 12.7$ mm. The module is designed for the nominal moment $M \approx 0.3$ N·m and the characteristic arm of force $D \approx 100$ mm. Assume the placement error of strain gauges is $\Delta = 0.25$ mm, and the angular placement error is $\psi = 0.001$rad ($0.6°$). Then we obtain the following parameters of the module: the average strain (3.2.1) in the strain gauge bridge due to the nominal moment is $e = 2 \cdot 10^{-4}$, the torsional stiffness is $Eah^3/(12l) = 0.6$ N·m (see (3.2.2)). The worst case coupling coefficient for the lateral bending moment (3.2.3) $\vartheta_1 = 2.2\%$, torque coupling coefficient (3.2.4) $\vartheta_2 = 1.3\%$, longitudinal force coupling coefficient (3.2.5) $\vartheta_3 = 0.03\%$; the relative mechanical nonlinearity $\kappa = 0.05\%$ (see (3.2.6)).

3.2.2 Elastic parallelogram element

We have described this widespread force sensor design in Section 2.1. Let us choose a coordinate system with the origin O at the center of the sensor as shown in Figure 3.2.2, i.e., O is the center of the interval between the midpoints of the two elastic beams. The axis Ox is directed from one beam to the other, the axis Oz is parallel to the elastic beams, the axis Oy complements the axes to the right-handed Cartesian frame. Two strain gauges *1* and *2*, connected as adjacent arms of a two-active-arm strain gauge bridge, measure beam deformation. They are mounted close to the ends of the elastic beam.

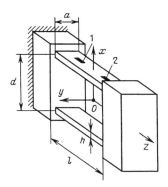

FIGURE 3.2.2
Elastic parallelogram element

Let us compute design parameters for this module. We assume that the beam has an arbitrary cross-section with a fixed length. We consider the axis Oy to be the principal axis of inertia of the cross-section, otherwise its shape is arbitrary.

Let l be the length of the elastic beams, d, the distance between them, E, Young's modulus of the beam material. Denote by I_0 the moment of inertia of the beam cross-section with respect to the axis Oy, by I_1, the moment of inertia with respect to the axis Ox, and by h_d, the distance between the point of strain gauges placement and the neutral axis of the beam cross-section. The strain gauges 1 and 2 are attached at distances $l/4$ from the beam ends.

Sensitivity.

A nominal force F acting on the elastic parallelogram bends the beams in the direction of the axis Ox so that one base remains parallel to the other. The bending moment diagram for one of the two beams is shown in Figure 3.2.3.

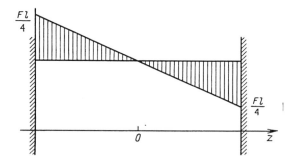

FIGURE 3.2.3
Bending moment diagram for the parallelogram element beams

The gauges are attached at distances $l/4$ from the beam ends, and they acquire the strains $\varepsilon_{1,2} = \pm Flh_d/(8EI_0)$. Hence the average strain in the bridge is given by

$$e = Flh_d/(8EI_0) \tag{3.2.7}$$

Stiffness.

One can find a relative deflection δ of the elastic element flanges subjected to the nominal measured force F by solving the equation for the elastic line of the beam, or by applying the Castigliano theorem [260, 261]. This displacement is given by

$$\delta = Fl^3/(24EI_0) = el^2/(3h_d) \tag{3.2.8}$$

Equation (3.2.8) shows that keeping the sensitivity constant, one can increase stiffness by reducing the length l of the beams and increasing the distance h_d. We mentioned above that the finite size of strain gauges imposes the limitation on the length l. By changing the shape of the cross-section to increase the distance h_d, we would undermine mechanical decoupling of the measurement. It follows from the expression (3.2.7) that at a constant sensitivity, a decrease in the Young's modulus of the material leads to the growth of the cross-sectional dimensions. Consequently, sensors made of materials with a lower Young's modulus, such as aluminum, are stiffer provided their sensitivity and cross-section shape are the same as for sensors made of higher Young's modulus materials.

Eigenfrequency.

The characteristic mass attached to the sensor is $m = F/g$, where F is the nominal force. By finding the sensor stiffness $k = F/\delta$ from (3.2.8), we obtain the eigenfrequency

$$\nu = \sqrt{k/m}/(2\pi) = \sqrt{6EI_0 g/(Fl^3)}/\pi \qquad (3.2.9)$$

Coupling.

We start by finding the worst case coupling coefficient for the lateral bending moment M_x. This moment deflects the beams in the direction of the axis Oy. The strain gauges do not deform, if they are placed on the center line of the beam face. Suppose the strain gauges are attached with an offset Δ on opposite sides from the neutral axis of lateral bending. Then their strains are given by $\varepsilon_{1,2} = \pm\kappa\Delta$, where κ is the curvature of the beam. We assume that the characteristic magnitude of the bending moment M_x is equal to FD, where D is the characteristic arm of the force. From here we obtain $\kappa = FD/EI_1$. By calculating the ratio of the average strain in the bridge $e_1 = (\varepsilon_1 - \varepsilon_2)/2$, caused by the lateral bending moment, to the strain (3.2.7), we obtain the worst case coupling coefficient

$$\vartheta_1 = 8\Delta DI_0/(lh_d I_1) \qquad (3.2.10)$$

One can see from (3.2.10) that a small cross coupling can be obtained if the ratio I_0/I_1 is small, i.e., the cross-section of the beam is elongated along the axis Oy.

The effect of the lateral force F_y is defined by the bending moment M_y, created by this force in the beam. Hence we need not discuss the effect of the force F_y in more detail. Similarly, if the distance d between the beams is sufficiently large, then the effect of the torque M_z can be reduced to a couple of forces applied along the axis Oy and bending the beams in the opposite directions.

The measured force F_x is applied with an offset D from the coordinate origin O and creates the lateral bending moment M_y. This moment, in a first approximation, results in the tension of one beam of the parallelogram and compression of the other. The tensile (compressive) force is FD/d. The strain gauges attached to the same beam are connected as adjacent arms of the bridge, therefore, they are strained equally and should not cause bridge imbalances. However, an imbalance can arise due to a variation of strain gauge sensitivity. The strains of the gauges are given by $\varepsilon_{1,2} = \pm FD/(dEs)$, where s is the area of each beam cross-section. Denote by η the maximum relative variation of strain gauge sensitivity. Then, according to (3.1.6), the maximum apparent average strain in the bridge is $e_2 = \eta FD/(dEs)$. Thus, in view of (3.2.7), we obtain the worst case coupling coefficient for the moment M.

$$\vartheta_2 = 8\eta DI_0/(sldh_d) \qquad (3.2.11)$$

Linearity.

Nonlinear coupling may be caused by the simultaneous action of the tensile (compressive) force F and bending force on the beam. The maximum tensile force FD/d acting on the beam is caused by the bending moment M_y. This force creates an additional bending moment $\delta FD/d$, where δ is a deflection of the beam end (3.2.8) due to the nominal measured loading F_x. By comparing this moment with the bending moment $Fl/2$ caused by the measured force, we obtain the expression for the relative nonlinearity

$$\kappa = 2\delta D/(dl) = Fl^2 D/(12dEI_0) \qquad (3.2.12)$$

Let us also address the issue of the sensor mechanical structure strength. The measured strain in the elastic element does not exceed 10^{-3} and is significantly less than yield limits for aluminum, steel, titanium, etc. However, the lateral moment M_y causes compression of the parallelogram element beam, which may potentially result in the element collapse due to the beam buckling. Let us define buckling safety factor as a ratio of the critical load (the Euler load) applied to the beam with both ends fixed, to the maximum compressive force FD/d.

$$n = 4\pi^2 EId/(FDl^2), \qquad (3.2.13)$$

where I is the least of the moments of inertia I_0 and I_1 (usually, I_0).

3.2.3 Torque sensor

The hollow cruciform module for axial torque measurement is described in Section 2.1. Let us consider such a module shown in Figure 3.2.4. Four strain gauges *1, 2, 3, 4* connected as a bridge circuit are attached to two diametrically opposite beams along the center line of their wide faces near the top and bottom flanges. Let us introduce a coordinate system attached to the sensor with the origin O in its center. The axis Ox is directed along the diameter passing through the opposite beams with strain gauges, the axis Oz is the sensor axis, the axis Oy complements the axes to the right-handed Cartesian frame. Denote by h the thickness of each beam (the smaller side of its cross-section). Denote by a, l, and R, respectively, the beam width, the beam length, and the distance between the center line of the beam and the sensor axis. Assume that the beams are sufficiently distant from the axis Oz, that is, $a \ll l$. This assumption allows us to treat the strain caused by the measured moment M_z in each beam as a pure bending. We can now define the parameters of the sensor, supposing that the strain gauges are attached at a distance $l/4$ from the flanges.

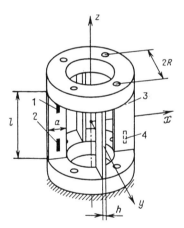

FIGURE 3.2.4
Hollow cruciform module for torque measurement

Sensitivity.

The forces $M/(4R)$, exerted on the flanges by each beam, counteract the action of the nominal torque M directed along the axis Oz. The bending moment diagram for each of the beams is similar to that shown in Figure 3.2.3, with the difference that the bending moment is now $Ml/(8R)$ instead of $Fl/4$. The bridge consists of four active strain gauges, so the average strain in the bridge is given by

$$e = 3Ml/(4REah^2) \tag{3.2.14}$$

Stiffness.

We can calculate the deflection δ of each beam end caused by the nominal measured torque M_z by solving the equation for the elastic line of the beam. Using the expression (3.2.14), we obtain (at $\delta = \varphi R$):

$$\varphi = Ml^3/(4Eah^3R^2) = el^2/(3hR) \tag{3.2.15}$$

The expressions (3.2.14), (3.2.15) demonstrate that for a given sensitivity, the stiffness of the sensor can be increased by reducing the length l and increasing the thickness h of the beams. The length of the beams can only be reduced to a certain limit, defined by the finite size of the strain gauges to be mounted on the beam. According to (3.2.14), greater thickness h leads to a reduced width a, and, therefore, to a poorer decoupling. The radius R of the sensor should be increased to obtain greater stiffness. The maximum dimension of the radius is defined by the limits on the external dimensions of the sensor.

Decoupling.

The lateral force F_x causes the beams with attached strain gauges to bend in the direction of the greatest stiffness. The strain gauges placed exactly in the middle of the beam faces are not deformed in the bending. We find the worst case coupling coefficient for the force F_x in the same way as for the parallelogram elastic element. By taking the nominal lateral force to be $F = M/D$, we obtain for the placement error Δ

$$\vartheta_1 = 4Rh\Delta/[D(a^2 + h^2)] \tag{3.2.16}$$

The strain computation for the beams subjected to the lateral force F_y is similar to that for the force F_x. The only difference is that the beams with strain gauges bend in the direction of the least stiffness. The strains in the two strain gauges are $3F_yl/[4Ea(a^2 + h^2)]$, the strains in two other gauges are opposite in sign. If the gauge factors are equal, strain gauge outputs are balanced in the bridge. Suppose that the maximum relative variation of the gauge factors is η, and the gauge factor in the bridge differs from the average value in the worst possible manner. Then the apparent average strain in the bridge is given by $e_2 = 3\eta F_yl/[2Ea(a^2 + h^2)]$, according to (3.1.6). Assuming the characteristic magnitude of the force $F_y = M/D$

and taking into account (3.2.14), we obtain the following expression for the worst case coupling coefficient ϑ_2 of this force

$$\vartheta_2 = 2\eta h^2 R/[D(a^2 + h^2)] \qquad (3.2.17)$$

The modular element in consideration is often used when the characteristic magnitude M_1 of the lateral moments M_x and M_y is greater than the nominal measured moment M. The action of the lateral bending moment M_1 directed along the axis Oy causes tension of one beam with strain gauges and compression of the other beam. The tensile (compressive) force equals $M_1/2R$. The disbalance of the strain gauge bridge is defined by a gauge factor variation. By using (3.2.14), we find the worst case coupling coefficient

$$\vartheta_3 = 4\eta M_1 h/(3Ml) \qquad (3.2.18)$$

The other lateral moment M_x mostly strains the beams without the strain gauges, so it can be neglected. The longitudinal force F_z applied to the sensor is resolved into the longitudinal forces $F_z/4$ applied to each beam. This force is $RM/2DM_1$ times less than the force caused by the moment M_y, where D is the characteristic arm of force. The ratio of the nominal moments $M/M_1 \leq 1$, and the characteristic arm of force is generally greater than the half of the radius R of a sensor. Hence the effect of the force F_z is less than that of the moment M_y.

Linearity.

Under the loading with the longitudinal force F, the maximum relative mechanical nonlinearity κ of the beam deformation equals $2\delta F/(F_m l)$, where δ is a flexible deflection of the beam end due to the nominal torque, and $F_m = M/(4R)$ is the tangential force caused by this torque. The longitudinal force $F = M_1/(2R)$ is caused by the lateral moment. By substituting φ from (3.2.15) into the expression $\delta = \varphi R$, we obtain the maximum relative nonlinearity

$$\kappa = M_1 l^2/(Eah^3 R) \qquad (3.2.19)$$

3.2.4 Comparison to experiment

The parallelogram force module and the cruciform torque module have been manufactured and used as parts of multicomponent sensors designed in IPIT-IM. These modules are described in Section 2.3. Let us compare the design and experimentally determined parameters of two such modules.

Figure 2.1.10 shows the considered variation of a parallelogram elastic element sensor. In the sensor, the elastic beams are parts of the cylinder shell. The axis of the cylinder is parallel to the beams and passes through the center of the sensor. Denote by 2α the angular width of the beams as seen from the center of the sensor, by R, the inner radius of the cylinder, and by h, the thickness of the cylinder walls.

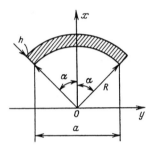

FIGURE 3.2.5
Beam cross-section for parallelogram element made of a hollow cylinder

Several parameters of the beam cross-section shown in Figure 3.2.5 need to be known in order to calculate characteristics of the sensor according to (3.2.7)–(3.2.12). We introduce a coordinate system with the origin at the center of the sensor. The axis Ox passes through the center of the cross-section, the axis Oy is normal to the axis Ox. Let us find the coordinate x_c of the center of mass of the cross-section:

$$x_c = \frac{1}{S} \int_{-\alpha}^{\alpha} \int_{R}^{R+h} r \cos \varphi r dr d\varphi = \frac{2 \sin \alpha (3R^2 + 3Rh + h^2)}{3(2R + h)\alpha} \quad (3.2.20)$$

The central moment of inertia with respect to the axis Oy is given by

$$I_0 = \int_{-\alpha}^{\alpha} \int_{R}^{R+h} r^2 \cos^2 \varphi r dr d\varphi - x_c^2 S$$

$$= \frac{h}{4}(2R + h)(2R^2 + 2Rh + h^2)(\alpha + \frac{1}{2}\sin 2\alpha)$$

$$- \frac{4}{9}h \frac{\sin 2\alpha}{\alpha} \frac{(3R^2 + 3Rh + h^2)}{2R + h}, \quad (3.2.21)$$

where S is the area of the cross-section.

We find the distance h_d from the neutral axis of the cross-section to the strain gauges location (the center line of the outer face) as $h_d = R + h - x_c$, where x_c is given by (3.2.20). We find the moment of inertia I_1 with respect to the axis Ox by assuming that the center of the cross-section has the coordinate $y_c = 0$

$$I_1 = \int_{-\alpha}^{\alpha} \int_R^{R+h} r^2 \sin^2 \varphi r\, dr\, d\varphi$$

$$= \frac{h}{4}(2R + h)(2R^2 + 2Rh + h^2)(\alpha - \frac{1}{2}\sin 2\alpha) \qquad (3.2.22)$$

We shall further substitute the linear width of the beam $a = 2R\sin\alpha$ for the angular width α.

One of the modules used in multicomponent sensors has the following dimensions: the beam length $l = 20$ mm, the beam width $a = 16$ mm, the beam thickness $h = 1.5$ mm, and the inner radius $R = 30$ mm. The module is manufactured from the aluminum alloy D16T ($E = 7 \cdot 10^{10}$ Pa). It is designed for the nominal loading $F = 50$ N, with the characteristic arm of force $D = 150$ mm (the characteristic moment $M = FD = 7.5$ N·m). We assume that the linear error of strain gauge placement is $\Delta = 0.25$ mm, and the maximum gauge factor variation $\eta = 5\%$ for the semiconductor strain gauges used. By substituting these values into the expressions (3.2.7)–(3.2.13) and (3.2.21), (3.2.22), we obtain the following design parameters. The average strain in the strain gauge bridge due to the nominal measured force is $e = 4 \cdot 10^{-4}$; the sensor stiffness $k = 2.3 \cdot 10^5$ N/m, the eigenfrequency with the mass $m = 5$ kg attached to the flange is $\nu = 47$ Hz. Lateral force coupling should not exceed 0.5% of the sensor output, and lateral moment coupling should not exceed 0.9%. The nonlinearity is $\kappa = 1\%$, the safety factor for lateral buckling is $n = 30$.

The experimentally measured parameters of the manufactured module are as follows: the average strain in the strain gauge bridge due to the nominal force equals $4 \cdot 10^{-4}$, the eigenfrequency with the mass $m = 5$ kg attached to the flange is 40 Hz, the coupling does not exceed 1%, the nonlinearity does not exceed 1%.

Consider now the torque sensing module shown in Figure 3.2.4. One of the manufactured modules has the following dimensions: the beam length $l = 26$ mm, the beam width $a = 10$ mm, the beam thickness $h = 1.2$ mm, the module radius $R = 35$ mm. The elastic element is made of the aluminum alloy D16T. The module is designed for a nominal torque $M = 1.5$ N·m. The characteristic force $F = 50$ N, hence the characteristic arm of force $D = 30$ mm. The module is designed for a characteristic lateral moment $M_1 = 7.5$ N·m. Assume the silicon strain gauges placement error $\Delta = 0.25$

mm, and their gauge factor variation $\eta = 5\%$. By substituting the above data into the expressions (3.2.14)–(3.2.19), we obtain the parameters of the module as follows. The average strain in the strain gauge bridge induced by the nominal torque is $e = 8 \cdot 10^{-4}$, the module torsional stiffness is $k_\varphi = 2.5 \cdot 10^2$ N·m/rad. The lateral force coupling does not exceed 1.2% of the sensor output, the lateral moment coupling does not exceed 1.5%, and the nonlinear coupling is no more than 2%.

In the experiments, we obtained the following parameters. The average strain in the strain gauge bridge due to the nominal torque was 10^{-3}, the torsional stiffness was $2 \cdot 10^2$ N·m/rad. Both linear and nonlinear couplings were less than 2% of the sensor output.

3.3   Six-Component Sensors with a Bending Elastic Element

Six-component sensors without mechanical selectivity show as much promise as modular force sensors. For such sensors, cross-coupling is electronically compensated by additional linear transformations of channel outputs.

This section considers strain gauge sensors with bending elastic elements. Sensors with elastic elements designed for compressive, tensile, and shear strain are studied in Section 3.4. Some of the material in this section was published in [80]. Related studies of design methods for six-component sensors with bending elements can be found in [179, 253, 265].

3.3.1   Design computations

We shall consider a multicomponent force sensor installed in the robot wrist, directly preceding the end-effector. This is the most common placement for a force sensor. The sensor has two rigid flanges, and the measured force is applied between them. One flange is attached to the robot arm, the other to the end-effector. The flanges are linked through an elastic element with strain gauges attached to it.

We attach a coordinate system to the arm of the robot, with the origin in the center of the undeformed sensor. Denote a force vector acting on the sensor by $\boldsymbol{F}$, and a torque vector by $\boldsymbol{M}$. Denote the nominal magnitude of the measured force by F_0, the nominal measured torque by M_0. These magnitudes are assumed to be the same for all components. The characteristic arm of the force $D = M_0/F_0$ coincides with a characteristic end-effector dimension – a distance from the end-effector to the center of the sensor.

The outputs of n sensor channels due to external loading can be described as an n-component column vector e of average strains in strain gauge bridges, $e = [e_1, \ldots, e_n]^T$. A displacement of the robot end-effector due to the sensor deformation is given by a small linear displacement vector δ and a small angular displacement vector φ. This section considers elastic elements comprising elastic beams subjected to bending and torsion and, possibly, rigid connecting elements. Such elastic elements are typically used for measuring moderate loads (less than 1000 N). Several physical dimensional parameters define the configuration of an elastic element. For the sensor type in question, let the characteristic beam length be L, beam cross-sectional dimension (thickness) be h, Young's modulus of the beam material be E. Denote by $\zeta = L/D$ the nondimensional ratio of the characteristic beam length to the end-effector size. We shall consider the sensitivity and stiffness (eigenfrequency with an attached load) of the sensor as its design goal parameters. The sensor channels do not need to be decoupled, because of the subsequent additional transformation of the primary outputs. However, it is important that the sensitivity matrix of the sensor should not be ill-conditioned. As to nonlinear deformation effects for the sensor under complex loading, it can be calculated by analogy with that of one-component sensors. The examples of such calculations are given in Section 3.2.

Sensitivity.

A vector of average strains in strain gauge bridges is linearly dependent on the components of the attached load. Provided that bending or torsion strain is measured, we can write

$$e = \frac{L}{Eh^3} S \begin{bmatrix} F \\ M/L \end{bmatrix}, \tag{3.3.1}$$

where S is the nondimensional sensitivity matrix of the sensor. The factor $L/(Eh^3)$ in the expression (3.3.1) describes the dependence of strain on the dimensional parameters of the elastic element. The nondimensional matrix $S = \{s_{ij}\}_{i,j=1}^{n,6}$ only depends on the design of the elastic element and a relative placement of strain gauges on the element. Note that the design sensitivity matrix can differ substantially from the experimentally defined one due to a strain gauge placement error and variation of parameters. However, these variations will not much change the "primary" elements of the matrix S that define the sensor sensitivity. They will mostly affect "secondary" elements defining the cross-coupling of sensor channels, which is unessential in this case.

Columns of the matrix S are the vectors of the sensor outputs induced by a given force or torque component. Usually, force sensors are designed so that these columns are orthogonal vectors. In particular, they are orthogonal in the two examples below. Sensor sensitivities to individual components of force and torque can be defined by computing a singular value decomposition of the matrix S. The squared singular values of S are eigenvalues of the matrix $S^T S$. The condition for the columns of S to be orthogonal has the form

$$\sum_{i=1}^{n} s_{i,j} s_{i,k} = 0, \quad \text{for } j \neq k$$

This means $S^T S$ is a diagonal matrix, and the squared singular values of S are the squared norms of its columns. Let us introduce coefficients that define the sensitivity to a given force or torque component through these singular values

$$\beta_{fi}^2 = \sum_{j=1}^{n} |s_{ij}|^2, \ \beta_{mi}^2 = \sum_{j=1}^{n} |s_{j,(i+3)}|^2 \quad (i = 1, 2, 3) \qquad (3.3.2)$$

Let us demand that the sensitivity of each channel be no less than the given strain e_0. In other words, the following conditions must be satisfied

$$e_f = \beta_f F_0 L / (E h^3) \geq e_0, \qquad (3.3.3)$$

$$e_m = \beta_m M_0 / (E h^3) \geq e_0, \qquad (3.3.4)$$

where

$$\beta_f = \min_{i=1,2,3} \beta_{fi}, \quad \beta_m = \min_{i=1,2,3} \beta_{mi} \qquad (3.3.5)$$

Since $M_0 = F_0 D$, then for $\zeta \leq \beta_m / \beta_f$, the inequality (3.3.4) is satisfied, if the inequality (3.3.3) is satisfied. For $\zeta > \beta_m / \beta_f$, it is sufficient that the inequality (3.3.4) is satisfied.

Stiffness.

The external force $\boldsymbol{F}$ and torque $\boldsymbol{M}$ cause linearly dependent on them small linear displacement vectors δ and angular displacement vectors φ in the elastic element of the sensor. Let $\xi = [\delta^T, \varphi^T L]^T$ be a six-dimensional vector of generalized coordinates, which defines a small end-effector displacement relative to the robot arm due to the sensor deformation. Then we can write

$$\xi = \frac{L^3}{Eh^4} K^{-1} \begin{bmatrix} F \\ M/L \end{bmatrix}, \tag{3.3.6}$$

where K is a nondimensional stiffness matrix. The inverse matrix K^{-1} is a nondimensional compliance matrix. The factor L^3/Eh^4 of the expression (3.3.6) describes the dependence of displacement on the dimensional parameters of the sensor design. The elements of the stiffness matrix K do not depend on these parameters, but rather depend on the configuration of the sensor elastic element and the relative sizes of its parts.

Eigenfrequency.

An elastic sensor and an inertial end-effector with a load attached to the sensor form together an oscillatory system with six degrees of freedom. In practice, it is most important to know the lowest of the six eigenfrequencies of such a system. Let us consider the dependence of this frequency on design parameters of the sensor.

Let Π be the potential energy of the deformed elastic element of the sensor. According to (3.3.6),

$$\Pi = Eh^4/(2L^3)\xi^T K\xi \tag{3.3.7}$$

If the sensor does not move, we can write the kinetic energy of the end-effector with an attached load in the form

$$T = \frac{1}{2}m\dot{\xi}^T A\dot{\xi}, \tag{3.3.8}$$

where m is the mass attached to the sensor, and A is a nondimensional matrix describing inertia properties of the end-effector with a load. We choose the end-effector dimension D as a characteristic length for moments of inertia in the matrix A. According to the definition of the vector ξ, the dimension L of the elastic element is used as a unit of length for representing the matrix A in the nondimensional form. Hence the elements of the matrix A depend on the nondimensional parameter $\zeta = L/D$. The submatrix of the matrix A, which describes the kinetic energy of translational motion, does not depend on the parameter ζ. The elements of the submatrix representing the energy of rotational motion are proportional to ζ^{-2}, because the moments of inertia of the end-effector with a load have the order of mD^2. By using expressions for the potential and kinetic energy (3.3.7) and (3.3.8), we can write the Lagrange equations of motion for the system in the form

$$mA(\zeta)\ddot{\xi} + [Eh^4/L^3]K\xi = 0 \qquad (3.3.9)$$

The oscillatory system (3.3.9) describes the motion of the inertial end-effector with a load, mounted on the elastic sensor. The squares of six eigenfrequencies of the system are the eigenvalues of the matrix

$$\Lambda = [Eh^4/(mL^3)]A^{-1}(\zeta)K \qquad (3.3.10)$$

Let us assume that the sensor configuration, beam shape, and the end-effector dimension are fixed and consider the dependence of eigenfrequencies on the dimension L of the sensor. In this case, the matrix K is constant, and the matrix A depends on L through the parameter ζ.

We choose the beam thickness h such that, with the rest of the parameters given, the required sensitivity of the sensor is provided in accordance with the conditions (3.3.3) and (3.3.4). By solving the equality (3.3.3) for h and using (3.3.10), we obtain

$$\Lambda = \Omega^2 \beta_f^{4/3} \zeta^{-5/3} A^{-1}(\zeta)K, \qquad (3.3.11)$$

where $\Omega = [F_0^4/(m^3 e_0^4 D^5 E)]^{1/6}$ is a characteristic frequency of the system.

If the sensitivity condition (3.3.4) is an equality, then

$$\Lambda = \Omega^2 \beta_m^{4/3} \zeta^{-3} A^{-1}(\zeta)K \qquad (3.3.12)$$

In accordance with the expressions (3.3.11) and (3.3.12), we can write the minimum eigenfrequency in the form

$$\nu = \Omega f/(2\pi) \qquad (3.3.13)$$

We shall call the value $f = f(\zeta)$ a nondimensional eigenfrequency.

The representation (3.3.13) allows us to compare eigenfrequencies for various sensor designs and different dimensions of elastic elements. In particular, it follows from (3.3.13) that eigenfrequency can be increased by using the material with a smaller Young's modulus E in the sensor elastic element.

The elements of the submatrix describing the kinetic energy of translational motions have the order of unity. The coefficients representing the kinetic energy of rotational motions have the order of ζ^{-2}. Thus, the minimum eigenvalue of the matrix A^{-1} has the order of ζ^2 for small values of the nondimensional parameter $\zeta = L/D$. Therefore, according to (3.3.11), the nondimensional frequency $f(\zeta)$ grows as $\zeta^{1/6}$ with the increase of ζ. The minimum eigenvalue of the matrix $A^{-1}(\zeta)$ has the order of unity for large values of ζ. In this case, according to the expression (3.3.12), $f(\zeta)$ decreases as $\zeta^{-3/2}$ with the increase of ζ. Hence the maximum value of the minimum eigenvalue (3.3.13) is achieved for a certain $\zeta = \zeta_0$ ($L = L_0$), where ζ_0 depends on the sensor design.

The optimal value ζ_0 for the specific matrices A and K can be defined, for example, numerically.

3.3.2 Examples of sensor designs

Suppose that the axes Ox, Oy, Oz of a coordinate system with the origin at the sensor center are parallel to the central axes of inertia of the end-effector with a load. The coordinates of the center of mass of the end-effector with a load are $(0, 0, D)$. Then the nondimensional matrix A describing the inertial properties of the system has the form

$$A(\zeta) = \begin{bmatrix} 1 & 0 & 0 & 0 & \zeta^{-1} & 0 \\ 0 & 1 & 0 & -\zeta^{-1} & 0 & 0 \\ 0 & 0 & 1 & 0 & 0 & 0 \\ 0 & -\zeta^{-1} & 0 & (1+i_x)\zeta^{-2} & 0 & 0 \\ \zeta^{-1} & 0 & 0 & 0 & (1+i_y)\zeta^{-2} & 0 \\ 0 & 0 & 0 & 0 & 0 & i_z\zeta^{-2} \end{bmatrix} \qquad (3.3.14)$$

Here, the nondimensional moments of inertia are: $i_x = I_x/(mD^2)$, $i_y = I_y/(mD^2)$, $i_z = I_z/(mD^2)$, where I_x, I_y, I_z are the principal central moments of inertia for the end-effector with a load. The numerical results below are obtained for the values $i_x = i_y = 0.03$; $i_z = 0.06$.

Let us consider the strain in the beams of the elastic element as pure bending and torsion. In the examples below, the beams have a square cross-section $h \times h$. For the calculations, we need to know the central axial moment of inertia for such a beam [260, 261]

$$I = h^4/12 \qquad (3.3.15)$$

and its torsional stiffness with a square cross-section [260, 261]

$$c = 0.1406 \cdot Gh^4 = \frac{0.1406 \cdot Eh^4}{2(1+\mu)}, \qquad (3.3.16)$$

where G is the shear modulus, E is Young's modulus, and μ is the Poisson's ratio for the material of the beam.

Let us consider two designs of the sensor elastic element in more detail.

1. Maltese cross elastic element

The Maltese cross elastic element is described in Section 2.3.1. It consists of four radial beams *1* of the length *l*, which connect a rigid hub *2* of the radius *r* with an outer flange *3* (see Figure 3.3.1). The outer edges of the beams rest upon the flange via membrane joints *4*, which restrain the beam ends from rotation and transversal displacement. The beams have a square cross-section of side *h*.

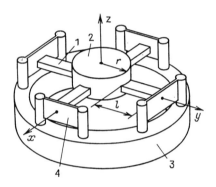

FIGURE 3.3.1
Maltese cross elastic element

Denote $L = r + l, \alpha = r/l$. The axis Oz is normal to the axes of all the beams and has its origin in the center of the elastic element. The axes Ox and Oy are directed along the beams. As in the previous example, we start with finding a stiffness matrix.

Stiffness.

By symmetry of the elastic element, its stiffness matrix is diagonal. We can assume that the outer end of each elastic beam is simply supported. Then the displacement of the end relative to the outer flange of the sensor under the action of the force F applied to the outer end of the beam (see Figure 3.3.2) is given by

$$\delta = \frac{F}{Eh^4} 4l^3 \qquad (3.3.17)$$

If the sensor hub moves in any direction parallel either to the axis Ox or Oy, then only the two beams, which are perpendicular to this direction,

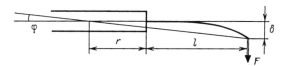

FIGURE 3.3.2
Beam deformation in the Maltese cross sensor

are deformed. If it is displaced along the axis Oz, all the four beams are deformed. Thus, in terms of (3.3.17), we obtain the relationship between the displacements and forces:

$$F_x = \delta_x Eh^4/(2l^3),$$

$$F_y = \delta_y Eh^4/(2l^3), \qquad (3.3.18)$$

$$F_z = \delta_z Eh^4/l^3$$

If the hub rotates about the axes Ox or Oy, the restoring torque is defined by bending of the beams normal to the respective axis, and by twist of the beams parallel to this axis. If the hub rotation angle is φ, the displacement of the beam ends relative to the hub is $\delta = L \cdot \varphi$. The restoring torque acting on the hub from the bent beams is $M = F \cdot L$, where the force F is given by (3.3.17). The restoring torque caused by the torsion of one beam is given by

$$M = \varphi \frac{c}{l} = \varphi \frac{0.1406}{l \cdot 2(1 + \mu)} Eh^4 \qquad (3.3.19)$$

If the hub rotates about the axis Oz, all the beams bend. Thus, we obtain

$$M_x = 2L\varphi_x L \frac{Eh^4}{4l^3} + 2\varphi_x \frac{0.1406}{l \cdot 2(1 + \mu)} Eh^4$$

$$= \varphi_x Eh^4 \left[\frac{L^2}{2l^3} + \frac{0.1406}{l(1 + \mu)} \right],$$

$$M_y = \varphi_y Eh^4 \left[\frac{L^2}{2l^3} + \frac{0.1406}{l(1 + \mu)} \right], \qquad (3.3.20)$$

$$M_z = \varphi_z E h^4 L^2 / l^3$$

Using (3.3.18), (3.3.19), and (3.3.6), we obtain a nondimensional stiffness matrix of the sensor

$$K = \operatorname{diag}\left\{ \frac{(1+\alpha)^3}{2}, \; \frac{(1+\alpha)^3}{2}, \; (1+\alpha)^3, \; \frac{(1+\alpha)^3}{2} + \frac{0.1406(1+\alpha)}{1+\mu}, \right.$$

$$\left. \frac{(1+\alpha)^3}{2} + \frac{0.1406(1+\alpha)}{1+\mu}, \; (1+\alpha)^3 \right\} \tag{3.3.21}$$

Sensitivity.

The sensor has eight output channels. Eight pairs of strain gauges a – h form eight two-active-arm strain gauge bridges. They are attached to the adjacent opposite surfaces of the beam at a distance $l/4$ from the hub. The placement of the strain gauges is shown in Figure 3.3.3. Let us find a nondimensional sensitivity matrix S appearing in (3.3.1).

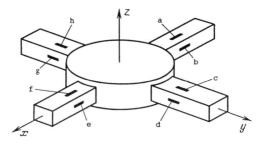

FIGURE 3.3.3
Strain gauge placement for the Maltese cross sensor

Under the action of the lateral force F_x, the two beams directed along the axis Oy bend, and the strain gauges e and h are strained. According to Figure 3.3.3, their strains are as follows:

$$\varepsilon_{e1} = -\varepsilon_{e2} = \frac{F_x}{2} \frac{3}{4} l \frac{h}{2} \frac{12}{Eh^4} = \frac{9F_x l}{4Eh^3}, \qquad \varepsilon_{h2} = -\varepsilon_{h1} = \frac{9F_x l}{4Eh^3}$$

Hence, given that $l = L/(1+\alpha)$ and using (3.1.7), we obtain the respective average strains

$$e_e = -e_h = \frac{F_x L}{Eh^3} \frac{9}{4(1+\alpha)} \tag{3.3.22}$$

The expressions for strains caused by the lateral force F_y can be derived similarly. The average strains in the bridges b and f are given by

$$e_b = -e_f = \frac{F_y L}{Eh^3} \frac{9}{4(1+\alpha)} \qquad (3.3.23)$$

The axial force F_z acts on the four beams instead of two. In this case, the strain gauges a, c, e, and g experience strain. The average strains in the respective half bridges are the same:

$$e_a = e_c = e_e = e_g = \frac{F_z L}{Eh^3} \frac{9}{8(1+\alpha)} \qquad (3.3.24)$$

The strain gauges c and g respond to the action of the lateral moment M_x. Due to this moment, the hub rotates through an angle φ_x relative to the outer flange, where φ_x can be found from (3.3.20). The hub rotation bends the beam and causes the displacement $\delta = \varphi_x L$ of the beam end with respect to the hub (see Figure 3.3.2). By relating the displacement to the force F acting on the end of the beam in accordance with (3.3.17), we obtain the strains in the strain gauges:

$$\varepsilon_{c1} = -\varepsilon_{c2} = \frac{12}{Eh^4} \frac{h}{2} \frac{3}{4} lF = \frac{M_x}{Eh^3} \frac{9}{4} \frac{4/l}{[L^2/l^2 + 0.2812/(1+\mu)]}$$

To calculate the average strains, we use (3.1.7) and take into account that $L = (1+\alpha)l$:

$$e_c = -e_g = \frac{M_x}{Eh^3} \frac{9(1+\alpha)}{4[(1+\alpha)^2 + 0.2812/(1+\mu)]} \qquad (3.3.25)$$

One can find the strains produced by the lateral moment M_y in a similar way:

$$e_a = -e_e = -\frac{M_y}{Eh^3} \frac{9(1+\alpha)}{4[(1+\alpha)^2 + 0.2812/(1+\mu)]} \qquad (3.3.26)$$

The action of the axial torque M_z resolves into four forces $F = M_z/(4L)$, applied to the end of each beam tangentially. By using this representation, the strains in the strain gauges b, d, f, and h can be easily found as

$$e_b = e_d = e_f = e_h = \frac{M_z}{Eh^3} \frac{9}{8(1+\alpha)} \qquad (3.3.27)$$

Thus, using (3.3.22)–(3.3.27), we obtain a nondimensional sensitivity matrix S from (3.3.1)

$$S = \begin{bmatrix} 0 & 0 & s_1 & 0 & s_2 & 0 \\ 0 & 2s_1 & 0 & 0 & 0 & s_1 \\ 0 & 0 & s_1 & s_2 & 0 & 0 \\ 2s_1 & 0 & 0 & 0 & 0 & s_1 \\ 0 & 0 & s_1 & 0 & -s_2 & 0 \\ 0 & -2s_1 & 0 & 0 & 0 & s_1 \\ 0 & 0 & s_1 & -s_2 & 0 & 0 \\ -2s_1 & 0 & 0 & 0 & 0 & s_1 \end{bmatrix}, \tag{3.3.28}$$

$$s_1 = 9/[8(1+\alpha)], \quad s_2 = \frac{9(1+\alpha)}{4[(1+\alpha)^2 + 0.2812/(1+\mu)]}$$

As one can see, the columns of S are orthogonal and we can now find nondimensional coefficients (3.3.5) of sensitivity:

$$\beta_f = 9/[4(1+\alpha)], \quad \beta_m = 9/[4(1+\alpha)] \tag{3.3.29}$$

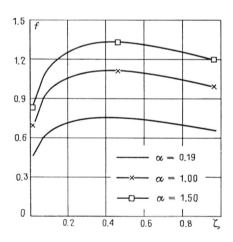

FIGURE 3.3.4
Plots of $f(\zeta)$ for the Maltese cross sensor

Figure 3.3.4 displays plots of dependencies $f = f(\zeta)$ computed in accordance with (3.3.13), (3.3.14), (3.3.21), and (3.3.29) for $\mu = 0.3$ and three values of the nondimensional parameter $\alpha = r/L$. For each α, the dependence $f(\zeta)$ reaches its maximum, when the ratio ζ of the sensor radius L to the end-effector size D is close to 0.4. The value of the parameter ζ at the maximum gives a guideline for the choice of the force sensor radius to achieve an optimal design with high eigenfrequency of vibrations. For each ζ, the value $f(\zeta)$ grows with the increase of α.

2. Elastic Cardan universal joint

The *elastic Cardan universal joint* is illustrated in Figure 3.3.5. It consists of a rigid hub *1* of the radius r and four radial beams *2* and *3* of the length l with a square cross-section $h \times h$. Two pairs of diametrically opposite beams *2* and *3* are fixed at both ends in the two opposite sensor flanges *4* and *5*. Denote by L the dimension $l+r$ of the elastic element. Let us introduce a nondimensional parameter of the design $\alpha = r/l$. Let the axis Oz be normal to the axes of the beams. The origin of the coordinate system is in the center of the elastic element. The axes Ox and Oy are directed along the elastic beams as shown in Figure 3.3.5. Design and some computations for such a sensor are described in [80, 265].

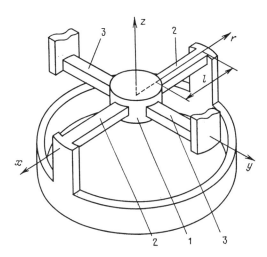

FIGURE 3.3.5
Elastic Cardan universal joint

For convenience, we shall first determine the stiffness matrix of this sensor, and then its sensitivity matrix.

Stiffness.

It is easier to derive a compliance matrix, which is inverse to the stiffness matrix. For convenience, we assume that the elastic element consists of two orthogonal beams of the length $2L$. Each beam is fixed at both ends in one of the flanges. The central hub is rigid and common for these two beams. By symmetry of the elastic element, each component of force or torque only causes displacement in the direction of its action, linear or angular. Therefore, the compliance matrix K^{-1} is diagonal.

By analogy with the derivation for the Maltese cross elastic element discussed above, the compliance matrix can be computed by using the Euler beam bending theory. One has to calculate displacements in the elastic element of the sensor produced by each force and torque component. We omit the calculations and only present the result:

$$K^{-1} = \mathrm{diag}\left\{ \frac{(1+\alpha)^{-3}}{2}, \; \frac{(1+\alpha)^{-3}}{2}, \; (1+\alpha)^{-3}, \right.$$

$$\frac{1}{1+\alpha}\left[\frac{3}{2(1+3\alpha+3\alpha^2)} + \frac{1+\mu}{0.1406} \right], \tag{3.3.30}$$

$$\left. \frac{1}{1+\alpha}\left[\frac{3}{2(1+3\alpha+3\alpha^2)} + \frac{1+\mu}{0.1406} \right], \; \frac{3}{(1+\alpha)(1+3\alpha+3\alpha^2)} \right\}$$

Sensitivity.

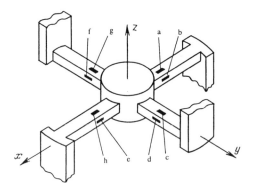

FIGURE 3.3.6
Strain gauge placement for the Cardan joint sensor

The sensor has eight output channels. Eight pairs of strain gauges $a - h$ are mounted on all pairs of opposite faces of four elastic beams at a distance $l/4$ from the hub 1 shown in Figure 3.3.6. The strain gauges $a1$, $c1$, $e1$, $g1$ are placed on the top faces, and the strain gauges $a2$, $c2$, $e2$, $g2$ − on the bottom faces. The strain gauges $b1$, $d1$, $f1$, $h1$ are attached to the clockwise lateral faces, the strain gauges $b2$, $d2$, $f2$, $h2$ − to the counterclockwise faces on the top view.

We can find strains in the strain gauges caused by each force or torque component by applying the Euler beam bending theory. Let us omit the calculations and write out the resulting expression for the nondimensional sensitivity matrix S (3.3.1):

$$S = \begin{bmatrix} 0 & 0 & s_1 & 0 & s_2 & 0 \\ 0 & s_1 & 0 & 0 & 0 & -s_2 \\ 0 & 0 & -s_1 & -s_2 & 0 & 0 \\ -s_1 & 0 & 0 & 0 & 0 & s_2 \\ 0 & 0 & s_1 & 0 & -s_2 & 0 \\ 0 & -s_1 & 0 & 0 & 0 & -s_2 \\ 0 & 0 & -s_1 & s_2 & 0 & 0 \\ s_1 & 0 & 0 & 0 & 0 & s_2 \end{bmatrix}, \qquad (3.3.31)$$

$$s_1 = 3/[4(1+\alpha)], \qquad s_2 = 3(5+6\alpha)/[8(1+3\alpha+3\alpha^2)]$$

This matrix has orthogonal columns. We obtain nondimensional sensitivity coefficients (3.3.3), (3.3.4).

$$\beta_f = \frac{3}{2(1+\alpha)}, \quad \beta_m = \frac{3(5+6\alpha)}{4(1+3\alpha+3\alpha^2)} \qquad (3.3.32)$$

Figure 3.3.7 displays plots of dependencies $f(\zeta)$ for the nondimensional frequency in (3.3.13). The plots are built in accordance with (3.3.14), (3.3.30), and (3.3.32) for $\mu = 0.3$. The ratio α of the hub radius r to elastic beam length l assumes three values in the plots. For each value, the frequency f grows monotonically with the increase of the elastic element dimension L. Note that $\zeta = L/D$.

The value of the nondimensional eigenfrequency $f(\zeta)$ at $\alpha \approx 1$ is approximately five times greater for the Maltese cross elastic element than for the Cardan universal joint elastic element. This can be attributed to large compliance of the latter with regard to lateral moments, because of low torsional stiffness of the beams. Unlike this, for the Maltese cross elastic element, stiffness with respect to lateral moments is defined solely by the bending of elastic beams.

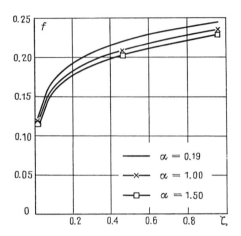

FIGURE 3.3.7
Plots of $f(\zeta)$ for the Cardan joint sensor

Comparison with experiment.

Force sensors with the above two types of elastic elements were developed and built by the IPIT–IM group. The Maltese cross elastic element is described in Section 2.3 (see Figure 2.3.10). Silicon strain gauges are used in both sensors. The sensors are not mechanically decoupled. Force and torque components acting on the sensor are computed by applying an appropriate linear transformation to the primary sensor outputs, as described in Section 1.2.

To identify the eigenfrequency experimentally, one flange of the sensor was fixed. A light thin-walled aluminum cylinder with a round flat load of mass $m = 1$ kg and radius $R = 45$ mm was attached to the other flange. The distance between the center of mass of the load and the center of the sensor was $D = 130$ mm. The sensor nondimensional moments of inertia in (3.3.14) were $i_x = i_y = (R/D)^2/4 = 0.03$, $i_z = (R/D)^2/2 \approx 0.06$. The oscillations in the described mechanical system were excited by an impact. The sensor output channels were sampled with frequencies ranging from 100 Hz to 1000 Hz in different cases. An example of the obtained signal u is displayed in Figure 3.3.8.

One can determine the lowest eigenfrequency by the amplitude range of the signal (see Figure 3.3.9), which was estimated using the discrete Fourier transform [15]. Let us compare theoretically computed and experimentally measured eigenfrequencies for the two sensors described above.

The sensor with a Maltese cross elastic element is manufactured from the aluminum alloy D16T designed for the average strain $e_0 = 10^{-4}$ in strain

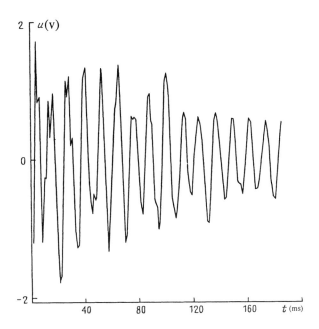

FIGURE 3.3.8
Sensor channel signal collected in the impact excitation experiment

gauge bridges at the nominal loading $F_0 = 50$ N. The beams have length $l = 20$ mm, the hub radius is $r = 30$ mm. Therefore, for $D = 130$ mm, the nondimensional parameters are $\alpha = 1.5$ and $\zeta = 0.44$. According to (3.3.3) and (3.3.29), the beams should be $h = 5.4$ mm thick in order to achieve the specified sensitivity. By using (3.3.13), we obtain the characteristic dimensional frequency $\Omega/(2\pi) = [F_0^4/(m^3 e_0^4 D^5 E)]^{1/6} \approx 85$ Hz. In accordance with Figure 3.3.4, we obtain $f = 1.3$, which yields $\nu = 110$ Hz. The experimentally measured frequency $\nu = 82$ Hz. The discrepancy can be attributed to the compliance of the base to which the sensor is attached, as well as to the fact that shear strains in the beams or finite stiffness of the membrane joints were not taken into account.

The sensor with the elastic Cardan universal joint is designed for the strain $e_0 = 10^{-4}$ at the loading $F_0 = 12.5$ N. The sensor is also manufactured from the aluminum alloy D16T. For $D = 130$ mm, the beam length $l = 48$ mm, and the hub radius $r = 9$ mm, we obtain the nondimensional parameters $\alpha = 0.19$ and $\zeta = 0.44$. The beam thickness $h = 4$ mm is chosen. For these parameters, we achieve the specified sensitivity of the sensor in agreement with (3.3.3) and (3.3.32).

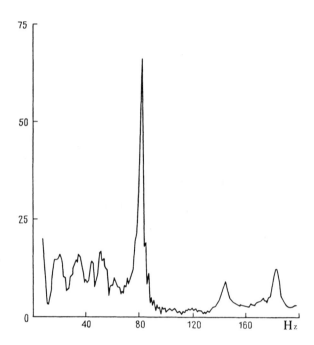

FIGURE 3.3.9
Spectrum of the signal in the impact excitation experiment

By substituting the above parameter values into the expression for Ω, we obtain $\Omega/(2\pi) = 34$ Hz for the characteristic dimensional frequency in (3.3.13). As seen in Figure 3.3.7, $f = 0.18$ for $\alpha = 1.5$ and $\zeta = 0.44$. Thus, the theoretically predicted frequency $\nu = f\Omega/(2\pi) = 6.1$ Hz closely agrees with the experimentally obtained $\nu = 6.0$ Hz.

3.4   Sensors with a Compressive-Tensile and Shear Elastic Element

In this section, we consider a sensor with an elastic element subjected to compressive-tensile and shear strains. Such elastic element comprises elastic struts connecting rigid flanges. We assume that strain in each strut subjected to measured force and torque components is uniform through its cross-section and length. Elastic elements of this type are generally employed for measuring comparatively large loads (exceeding 500 N).

3.4.1   Design computations

The following dimensional parameters define the configuration of an elastic element subjected to compressive-tensile and shear strains. Let L be the characteristic elastic element dimension, l and s, the characteristic length and cross-sectional area of the elastic struts, and E, the Young's modulus of the strut material. By analogy with Section 3.3, we denote by ζ the nondimensional ratio L/D of the characteristic sensor dimension to that of the end-effector with a load. In the same manner as for sensors with bending elastic elements, we shall consider the sensitivity and stiffness (eigenfrequency with an attached load) of the sensor to be design parameters.

Sensitivity.

The vector e of average strains in $(n \times 1)$ strain gauge bridges linearly depends on components of the force vector F acting on the force sensor, and the vector of moment M. Assuming that the strains in the elastic struts are uniform, we can write

$$e = \frac{1}{Es} S \begin{bmatrix} F \\ M/L \end{bmatrix}, \tag{3.4.1}$$

where, similarly to (3.3.1), S is a $n \times 6$ nondimensional sensitivity matrix of the sensor. The matrix S only depends on the configuration of the elastic element and strain gauge placement, and does not depend on dimensions or material of the struts.

The coefficient $1/(Es)$ is factored out in (3.4.1), because the strains in the struts are supposed to be uniform (shear or compressive-tensile). The expression (3.4.1) differs from (3.3.1), since it is assumed in (3.3.1) that the beams in the elastic element experience bending strain.

Let us introduce nondimensional coefficients β_f and β_m, defining the sensor sensitivity to force and moment components, respectively, in accordance with (3.3.2) and (3.3.5). Then, in order to ensure that the sensor sensitivity to measured forces is no less than specified, the following condition must be satisfied

$$e_f = \beta_f F_0/(Es) \geq e_0 \tag{3.4.2}$$

Since the sensitivity to moments should be no less than specified, we demand that

$$e_m = \beta_m M_0/(LEs) \geq e_0 \tag{3.4.3}$$

Since $M_0 = F_0 D$, then, for $\zeta \le \beta_m/\beta_f$, the inequality (3.4.3) is satisfied, if the inequality (3.4.2) is satisfied. For $\zeta > \beta_m/\beta_f$, it is sufficient that the inequality (3.4.3) is satisfied. If the sensor only measures either force or moment components, then for any ζ, only (3.4.2) or (3.4.3) must be satisfied, respectively.

Stiffness.

Small linear displacements δ and angular displacements φ in the elastic element depend on components of the external force F and moment M. Hence the six-dimensional vector $\xi = [\delta^T, \varphi^T L]^T$ of generalized coordinates that defines small end-effector displacements due to strain in the elastic element can be written in the form

$$\xi = \frac{1}{Es} K^{-1} \begin{bmatrix} F \\ M/L \end{bmatrix}, \tag{3.4.4}$$

In the above formula, as in (3.3.6), K is a nondimensional stiffness matrix, K^{-1} is a nondimensional compliance matrix. The factor $l/(Es)$ describes the dependence of displacement on dimensional parameters.

Eigenfrequency.

Let us consider the dependence of the lowest eigenfrequency of the system on the sensor parameters. By (3.4.4), the potential energy of the sensor under strain is given by

$$\Pi = Es/(2l)\xi^T K\xi \tag{3.4.5}$$

We can write the kinetic energy of the end-effector with an attached load, mounted to the sensor, in the form (3.3.8). The formula for small oscillations of the system, written in accordance with the expressions for kinetic and potential energy (3.4.5) and (3.3.8), is analogous to (3.3.9). Then the squares of six eigenfrequencies are the eigenvalues of the following matrix:

$$\Lambda = \frac{Es}{ml} A^{-1}(\zeta)K \tag{3.4.6}$$

Assuming the sensor configuration, beam shape, and the end-effector dimension to be fixed, consider the dependence of eigenfrequencies on the dimension L of the sensor. The matrix K does not depend on L, while the matrix A depends on L through ζ. We choose the cross-sectional area s of the elastic struts such that the given sensor sensitivity requirements (3.4.2) and (3.4.3) are satisfied.

On solving (3.4.2) for s, the matrix Λ in (3.4.6) can be presented as

$$\Lambda = \Omega^2 \beta_f A^{-1}(\zeta)K, \qquad (3.4.7)$$

where $\Omega = [F_0/mle_0]^{1/2}$ is a characteristic dimensional frequency of the system.

If the expression (3.4.3) is an equality, then

$$\Lambda = \Omega^2 \beta_m \zeta^{-1} A^{-1}(\zeta)K \qquad (3.4.8)$$

In accordance with (3.4.7) and (3.4.8), the minimum eigenfrequency has the form

$$\nu = \Omega f/(2\pi), \qquad (3.4.9)$$

where, in the same manner as in (3.3.13), $f = f(\zeta)$ is a nondimensional eigenfrequency. Note that unlike (3.3.13), the dimensional frequency Ω in (3.4.9) does not depend on the Young's modulus (of the elastic element material).

For "small" magnitudes of the nondimensional parameter ζ, the minimum eigenvalue of the matrix A^{-1} has the order of ζ, for "large" ζ, it has the order of unity. If a sensor only measures forces, then, according to (3.4.7), the eigenfrequency grows as $\zeta^{1/2}$ with an increase of the elastic element dimension for small ζ, and does not depend on ζ for large ζ. If a sensor only measures moments, the lowest eigenfrequency, according to (3.4.8), does not depend on ζ for small ζ, and decreases as $\zeta^{-1/2}$ for large ζ.

If a sensor measures both forces and moments, then the lowest eigenfrequency grows as $\zeta^{1/2}$ for small ζ and decreases as $\zeta^{-1/2}$ for large ζ. Hence, for such a sensor, a dimension $L = \zeta D$ of the elastic element can be found such that its lowest eigenfrequency is maximum.

3.4.2 Example

Consider a sensor with the above described elastic element. Let us assume that the axes Ox, Oy, Oz of the coordinate system with the origin O in the sensor center are collinear to the principal axes of inertia of the end-effector with a load. The coordinates of the center of mass of the end-effector with a load are $(0, 0, D)$. Then the nondimensional matrix A describing the kinetic energy has the form (3.3.14).

The design of the sensor is shown in Figure 3.4.1. Two round flanges *1* are linked by four elastic struts *2* uniformly positioned along the circumference. The length of each strut is l, the cross-sectional area is s, and the distance

$2L$ between the diametrically opposite struts is much greater than the strut
dimensions.

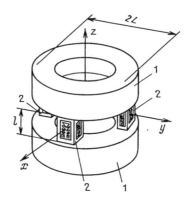

FIGURE 3.4.1
Sensor with compressive-tensile and shear elastic element

Sensitivity.

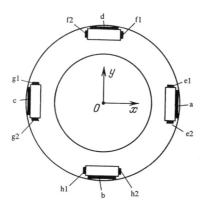

FIGURE 3.4.2
Strain gauge placement for the tensile-compressive and shear sensor

The sensor has eight output channels. A strain gauge rosette is attached
to each strut for measuring shear strain in the tangential direction. The
placement of the rosettes a, b, c, d is shown in Figure 3.4.2. Two strain
gauges in each rosette are connected into adjacent active arms of the strain

gauge bridge, with passive resistors in the other two arms (see Figure 3.1.3). The strain gauges are connected so that each bridge output is positive under the action of a positive axial torque.

Two more strain gauges are mounted to each strut for measuring its longitudinal strain. The strain gauges $e1$ and $e2$, $f1$ and $f2$, $g1$ and $g2$, $h1$ and $h2$ are connected as opposite active arms of the four strain gauge bridges, with dummy gauges in the other two arms of each bridge (see Figure 3.1.4). The strain gauges are connected so that the outputs of all the four bridges are positive under the action of a positive axial component of the force F_z.

Let us find the average strains in the bridges for each force and torque component and write out a nondimensional sensitivity matrix of the sensor. Under the action of the lateral force F_x, all the four struts experience shear strains. Since all the cross-sections have equal areas s, each of them is subjected to the shear force $F_x/4$. This force produces shear stress $\tau = F_x/(4s)$. Hence the strut shear strain $\gamma = \tau/G = 2(1 + \mu)\tau/E$, where G is the shear modulus of the sensor material, E is its Young's modulus, and μ is the Poisson's ratio. The average strains e_b and e_d in the bridges b and d are nonzero, since the rosettes in these bridges are mounted so that the force F_x is parallel to their plane. According to (3.1.8), we obtain

$$e_b = -e_d = -(1 + \mu)F_x/(4Es) \qquad (3.4.10)$$

Under the action of the force F_y, the average strains in the bridges a and c will be nonzero. By analogy with (3.4.10), we obtain for them

$$e_a = -e_c = (1 + \mu)F_y/(4Es) \qquad (3.4.11)$$

Under the action of the axial force F_z, all the four struts experience tensile strain $\varepsilon = F_z/(4Es)$. The average strains in the bridges a, b, c, d are zero. The average strains in the bridges e, f, g, h, according to (3.1.9), are given by

$$e_e = e_f = e_g = e_h = F_z/(4Es) \qquad (3.4.12)$$

Under the action of the lateral moment M_x, applied to the center of the inner flange of the sensor, two diametrically opposite struts are subjected to compression and tension, while the strains in the other two struts are negligible. The average strains in the bridges f and h are nonzero. The moment M_x produces a couple of forces of magnitude $M_x/(2L)$, which act on the struts in the opposite directions parallel to the axis Oz. The strain in the respective struts along the axis Oz is $M_x/(2LEs)$. Hence, according to (3.1.9), we obtain the average strains in the bridges

$$e_f = -e_h = M_x/(2LEs) \tag{3.4.13}$$

By analogy, under the action of the moment M_y, the average strains in the bridges e and g will be nonzero

$$e_e = -e_g = -M_y/(2LEs) \tag{3.4.14}$$

The action of the torque M_z resolves into four forces $M_z/(4L)$, which act on each of the struts tangentially. Each force creates a shear strain

$$\gamma = M_z/(4LGs),$$

where $G = E/[2(1+\mu)]$. By using (3.1.8), we find the average strains

$$e_a = e_b = e_c = e_d = M_z(1+\mu)/(4LsE) \tag{3.4.15}$$

The expressions (3.4.10)–(3.4.15) define columns of the sensitivity matrix. By factoring out the dimensional parameters according to (3.4.1), we obtain the nondimensional sensitivity matrix

$$S = \begin{bmatrix}
0 & (1+\mu)/4 & 0 & 0 & 0 & (1+\mu)/4 \\
-(1+\mu)/4 & 0 & 0 & 0 & 0 & (1+\mu)/4 \\
0 & -(1+\mu)/4 & 0 & 0 & 0 & (1+\mu)/4 \\
(1+\mu)/4 & 0 & 0 & 0 & 0 & (1+\mu)/4 \\
0 & 0 & 1/4 & 0 & -1/2 & 0 \\
0 & 0 & 1/4 & 1/2 & 0 & 0 \\
0 & 0 & 1/4 & 0 & 1/2 & 0 \\
0 & 0 & 1/4 & -1/2 & 0 & 0
\end{bmatrix} \tag{3.4.16}$$

From (3.4.16), we obtain the nondimensional coefficients (3.3.5), defining the sensitivity of the sensor

$$\beta_f = 1/2, \quad \beta_m = (1+\mu)/2 \tag{3.4.17}$$

Stiffness.

By symmetry of the sensor configuration, its stiffness matrix is diagonal. One can easily calculate its elements by referencing the above results for the sensor sensitivity. Indeed, linear displacement is given by εl or γl in each elastic strut subjected to a force or torque component, where ε and γ

are the above computed longitudinal and shear strain, and l is the length of the strut. Therefore, we only present here the final result of the stiffness matrix calculation

$$K^{-1} = \text{diag}\left\{ \frac{2}{(1+\mu)}, \ \frac{2}{(1+\mu)}, \ 4, 2, 2, \ \frac{2}{(1+\mu)} \right\} \qquad (3.4.18)$$

Figure 3.4.3 displays a plot of dependence of the nondimensional frequency $f(\zeta)$ from (3.4.9) on ζ. The plot is computed in accordance with (3.4.7)–(3.4.9), (3.3.14), (3.4.16), and (3.4.17). As illustrated in Figure 3.4.3, for the realizable sizes of the sensor elastic element, $\zeta \leq 1$ ($L \leq D$), the frequency f grows nearly linearly with the increase of ζ (increase of the sensor dimension L).

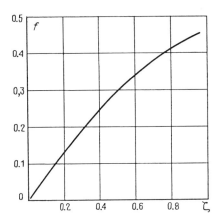

FIGURE 3.4.3
Plot of $f(\zeta)$ for the tensile-compressive and shear sensor

We have assumed that one of the flanges is fixed when calculating the eigenfrequency. In the experiments for measuring eigenfrequency, stiffness of the fixture used to secure the sensor flange must be much greater than that of the elastic element. Sensors with a bending beam elastic element, considered in Section 3.3, can be fixed rigidly enough. Therefore, the theoretical computation agrees well with the experiment in Section 3.3. Compressive-tensile and shear elastic elements, considered in the present section, possess a higher stiffness in comparison with bending beam elastic elements. It is very difficult to provide a sufficiently rigid fixture for such a sensor. Because of this, the lowest eigenfrequency values calculated in keeping with the method described in this section tend to be much overestimated. Thus, the experimentally measured frequency is usually two to three times lower than the calculated one for the type of sensor considered in the present section (see Figure 2.3.11). In this case, the nonrigid mounting is due to the fact that the cross-sectional area of screws securing the sensor is three to four times less than that of its elastic struts.

Part II.

Mathematical Models and Control

To analyze performance and design a controller for any system, it is beneficial to develop its mathematical model. This part of the book develops mathematical models for robotic systems with force sensors. We use these models to design control laws and analyze different operation modes and stability issues for such systems. We start from a simple model and then add various real-life features, such as feedback loop delay, structure compliance, etc., to our analysis. The part begins with a brief description of a research manipulator system in Chapter 4. Many mathematical models and control laws discussed in this monograph have been developed for this system. The system was also used in the experiments described in the book.

Chapter 4

Research Robotic System

This chapter describes an experimental robotic system that was used in the experiments presented in the book. The research system is described in detail in [347].

Robotic system capabilities depend on kinematics and dynamics of the manipulator in use, computer system, and available sensor systems. In the considered robotic system, a general-purpose computer controls a two-armed manipulator and processes the information from sensors. Therefore, unlike industrial robots using proprietary controllers and software, the research robotic system allows implementation of almost any control scheme.

The robotic system was designed in view of experimentally testing various control algorithms and evaluating the performance of specific sensing devices. The majority of experiments presented in this book were carried out with this system. It was used to study some assembly operations, workpiece surface machining, part retrieving and grasping tasks. To carry out the mentioned operations, several interchangeable specialized end-effectors, or hands, attachable to the manipulator arm were designed. The end-effectors have rotational degrees of freedom, a gripper or some other tool, and are instrumented with force sensors and infrared proximity sensors.

4.1 Manipulator Design and Kinematics

The manipulator UM-1.25 used in the experimental robotic system was designed in the Moscow Research Institute for Metal-Working Machines. The manipulator has a gantry type configuration with two arms.

Gantry manipulators are preferable to articulated arms in many tasks. The use of a gantry configuration helps to ensure a high stiffness of the structure and to balance the arm. Kinematic and dynamic computations required for motion control of such a manipulator tend to be significantly simpler than for an articulated arm. Though the importance of computational simplicity decreases with the current growth of computational capacity, it is still very useful for tutorial purposes of this book. Gantry industrial robots have been used in different manufacturing tasks, in particular, for assembly. The assembly robots produced by companies such as Olivetti and IBM had such configuration.

The manipulator shown in Figure 4.1.1 consists of a rigid box frame mounted on a base, and two arms, identical in kinematics and design. The frame height is 0.95 m, its horizontal dimensions are 1.4 × 0.76 m.

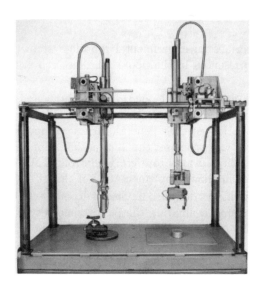

FIGURE 4.1.1
General view of the research manipulator system

The manipulator arm is moved along the horizontal axes Ox and Oy by a carriage riding on a bridge. The first degree of freedom corresponds to the linear motion of the bridge riding on runway rails mounted on the manipulator frame along the axis Ox. Another set of runways is installed on the bridge normal to the axis Ox. The second degree of freedom corresponds to the linear motion along the axis Oy of the carriage carrying the manipulator arm (mast) of length 0.7 m in the runways on the bridge. The vertical mast can slide up and down on the carriage. The vertical motion of the mast along the axis Oz corresponds to the third degree of freedom.

Drive reduction gears of the three translational degrees of freedom are connected through cylindrical pinions with racks mounted on the runways. The racks are attached to the mast, bridge, and frame, respectively. Maximum linear displacements of the arm (mast) along each degree of freedom are about 0.4 m. The translational motions are limited by emergency switches.

Consider in more detail the transmission kinematics for the third degree of freedom moving the mast along the axis Oz (the drives of the first and second degrees of freedom are analogous to that of the third). Figure 4.1.2 shows the motor *1* with the rotor axis to which the pinion *2* is attached. The pinion *2* is in gear with the pinion *3*. The next gear pair consists of the pinions *4* and *5*. The output power gear *6* is connected to the rack *7* attached to the mast *8*.

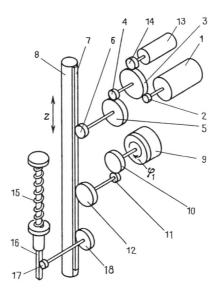

FIGURE 4.1.2
Kinematic scheme of the transmission for the manipulator translational degree of freedom

The reduction ratio of the first gear pair is $i_1 = 7.6$, of the second, $i_1 = 5.6$; the overall reduction ratio is $i = i_1, i_2 = 42.6$. To increase the positioning accuracy, a backlash in the pinion gears can be compensated by adjusting eccentric bushings that support gear axes.

The displacement z of the mast *8* relates to the rotation angle φ of the motor shaft *1* as follows:

$$z = \frac{r}{i}\varphi = \rho\varphi \quad \left(\rho = \frac{r}{i}\right),$$

where $r = 12.5$ mm is the radius of the output power gear *6*.

The position of the mast is measured by the transducer *9*, which is connected to the rack *7* via pinions *10, 11, 12*. The reduction ratio of the pinion gear *10* and *11* is 0.2. The linear displacement z of the mast is related to the rotation angle φ_1 of the transducer axis by the formula

$$z = \frac{r_1}{i_g}\varphi_1, \tag{4.1.1}$$

where $r_1 = 4.1$ mm is the radius of the pinion *12*, and $i_g = 0.2$.

The angular velocity sensor (tachometer) *13* is connected through the pinion *14* of the radius 4.3 mm to the pinion *3* of the radius 23 mm, and through it, to the pinion *2* mounted to the rotor axis of the motor *1*.

To balance the weight of the mast with the attached end-effector, a spring *15* pulls the mast up through the rack-and-pinions *16, 17*, and *18*. Compression of the spring may vary to suit the weight of the mast with the end-effector. In the drive, a DPM-35-N1-01 DC motor is used. Characteristics of this motor are given in [169]. The output nominal torque of the gear train can be determined as $M = i\eta M_n$, where M_n is the nominal torque at the motor shaft, η is an efficiency factor for the gear train. For the DPM-35-N1-01 motor, $M_n = 0.0147$ N·m, $\eta \approx 0.85$. Given the radius r of the output pinion gear, we can find the nominal force exerted by the motor on the beam: $F = M/r = 43$ N.

All translational degrees of freedom of the described research manipulator have standard gear trains with the same parameters, and the same displacement and velocity transducers are used. Potentiometric transducers are used to measure position. DC motors DPM-20-N1-11 are used as velocity sensors (tachometers). Each arm ends with a flange to which various end-effectors may be attached.

4.2   Control System

The manipulator control system developed in house in IPIT–IM has two feedback loops. The outer loop is controlled by a general-purpose microcomputer. The inner loop consists of an analog servo-system hardware. Figure 4.2.1 shows a block diagram of the control system. An output signal from each servo-system is fed through a Power Width Modulator (PWM) to a power amplifier, which outputs the control voltage to the DC motor.

The outputs of the control system are signals of potentiometric position sensors in the manipulator degrees of freedom and tachometers measuring rotation velocities of the DC motors. Position and velocity signals can be fed to the servo-system inputs for the respective degrees of freedom, or to the Analog-Digital Converter (ADC) of the digital control computer. The computer control signals are output by the Digital-Analog Converter (DAC) and fed to the respective servo-system inputs.

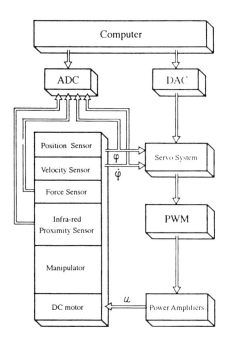

FIGURE 4.2.1
Block diagram of the research manipulator control system

Output signals of the external sensors (force or infrared proximity sensors) are sampled by the computer ADC.

The servo-systems have switches controlling their mode of operation. The switches allow open or closed position and velocity feedback loops. The servo-system gains were set by manual tuning. The Power Width Modulator (PWM) outputs the signals supplied to the input of the power amplifier that powers the DC motors.

Figure 4.2.2 displays the feedback loop schematics for a single translational degree of freedom. Denote by z the respective coordinate of this degree of freedom. The angle of rotation of the position transducer *1* is proportional to the coordinate z (see (4.1.1)). The angular velocity measured by the tachometer *2* is proportional to the velocity $\dot{z}$.

Signals of the position transducer and tachometer are added with gains by the amplifier *3* and then fed to the amplifier *4*. If the switch $S2$ is open, then only the velocity signal $\dot{z}$ with a certain gain g' gets to the amplifier *3* and then, to *4*. The control voltage calculated by the computer *6* is applied via the DAC *5* to the other output of the summing amplifier *4*. The amplifier *4* computes the difference between this voltage and the output voltage of the amplifier *3*. The difference of the signals is fed to the PWM *7* and then to the power amplifier *8* which outputs the voltage u applied to the DC motor *9*. Thus, if the switch $S1$ is in the position II and the switch $S2$ is open, then the system becomes a velocity servo system, and the voltage applied to the DC motor can be represented in the form

$$u = g(v_d - \dot{z}), \tag{4.2.1}$$

where v_d is the commanded velocity calculated by the computer, g is the gain in the velocity feedback loop. This is the most commonly used mode of operation for the servo-system. If the switch $S1$ is in the position I, then the PWM receives the voltage from the DAC rather than the output voltage from the servo-system.

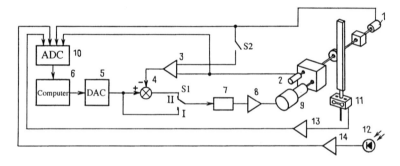

FIGURE 4.2.2
Schematics of the feedback loops for a single degree of freedom

If the switch $S2$ is closed, then the output signal at the summing amplifier *3* depends both on the velocity $\dot{z}$ and the coordinate z. This signal, $z + g'\dot{z}$, is subtracted from the command obtained from the DAC. The voltage u applied to the DC motor depends on the difference in signals (the switch $S1$ is in the position II). In this case,

$$u = g(z_d - z - g'\dot{z}), \tag{4.2.2}$$

where z_d is the desired value of the coordinate z proportional to the computer-calculated voltage, g is the gain in the position feedback loop.

In all modes of control system operation, the computer receives signals from the position and velocity sensors through the ADC. Signals from force sensors *11* and infrared proximity sensors *12* pass through the strain-gauge amplifiers *13* and infrared proximity sensor amplifiers *14*, respectively, to the ADC to the computer. For the translational degrees of freedom, given the nominal DC motor power voltage 27 V and the overall gain of $g \approx 270$ V·s/mm in the velocity feedback loop, the servo-system is able to track arm velocities in the range from 0.1 to 200 mm/s.

4.3 End-effectors and Sensors

The manipulator design described in Section 4.1 allows movement of the arm translationally along three orthogonal axes. This number of degrees of freedom is sufficient to perform many operations. However, such tasks as handling randomly oriented parts, assembly, or machining often require more degrees of freedom. To research these operations, several hand-like devices having rotational degrees of freedom and attaching to the manipulator arm, have been developed and experimentally tested. These devices are generally referred to as end-effectors. Several types of grippers for the end-effectors have also been developed.

An end-effector shown in Figure 4.3.1 has three rotational degrees of freedom and is equipped with a separate sensory probe and a gripper. The end-effector structure consists of two rigid brackets – the outer *1* and the inner *2*. The outer bracket is linked to the flange of the manipulator mast *4* via the drive *3*. This drive rotates the bracket *1* and thereby, the entire end-effector, about the vertical axis. The inner bracket is powered by the drive *6* and connected to the outer bracket through a joint with the horizontal axis *5*. The third drive (not visible in the photo) rotates the gripper *7* about the axis coplanar to the gripper jaws and normal to the horizontal axis *5*. All three rotation axes have a common intersection point. The reduction gears in the drives are not back-drivable. The reduction ratios are 270 for the pitch drive *6* and about 230 for the two other drives. Rotation angles in the joints vary from $-165°$ to $165°$, maximum rotation velocities are about 3.5 rad/s. DPR-42 motor is used in the pitch drive, DPM-20-N1-11 electromotors, in the other two drives. A rotation angle sensor is installed in each joint. The sensor *8* measuring the end-effector pitch angle is shown in Figure 4.3.1.

The jaws of the gripper *7* are actuated by a single drive and perform a parallel motion. The maximum opening of the jaws is 100 mm, the clamping force is about 50 N. The infrared proximity sensors *9*, which enable the manipulator to monitor objects in its workspace, are installed in the jaw tips.

FIGURE 4.3.1
The end-effector with the parallel jaw gripper

The sensory probe *10*, shown in Figure 4.3.1 as well as in Figure 2.2.7, is mounted on the end-effector. The probe, described in detail in Section 2.2, measures three components of the force. It is attached to the flange of the drive *1* through a spring to provide for angular compliance. The probe tip contacting external objects is connected to the rest of the probe through another spring. The sliding connection allows the tip to slide 15 mm up and down along the probe. While searching for a workpiece or tracking a part contour, the probe is protracted by the drive *11* so that its tip emerges below the tips of the jaws. When the jaws grasp a workpiece, the probe is retracted and allows the jaws to close. The probe can move up and down with the velocity of 15 mm/s.

We have used the above-described end-effector with the probe in the tasks of rotating a steering wheel (see Section 6.6), opening a hatch lid (see Section 7.6), and searching for a workpiece and determining its position for grasping.

One more way to install force sensors is to integrate them into gripper fingers, i.e. to design finger sensors. Figure 4.3.2 displays an end-effector with force sensors integrated into the gripper fingers. The kinematic scheme

FIGURE 4.3.2
The end-effector with the force sensing fingers

of this end-effector differs from the scheme shown in Figure 4.3.1. The end-effector in Figure 4.3.2 has two brackets connected through a joint. The bracket *1* is rigidly attached to the manipulator arm. The drive enabling the bracket *2* to pitch in the range of ±90° around the horizontal axis *3* is built into the side wall of the bracket *1*. The motor *4* rotates the link *5* about the axis *6*, which is perpendicular to the axis *3*. Its gear train is placed inside the side wall of the bracket *2*. The gear train of the motor *7* is built into the link *5*, and this motor rotates the gripper *8* about the axis *9* perpendicular to the axis *6*. In the nominal configuration of the end-effector, the fingers are in the vertical plane. In this configuration the three rotation axes for the described end-effector are orthogonal, which ensures that a one-to-one mapping exists between the rotation angles of the wrist degrees of freedom and the orientation of the gripper.

Spur pinions are used to gear down the drives for the rotational degrees of freedom of the wrist of the end-effector, the reduction ratio for each of the three drives is 270. DC motors DPM-30-N1-11 are used in the drives.

Maximum rotation velocities are about 3.5 s^{-1} for all the degrees of freedom. A position sensor and a velocity sensor (a tachometer) is installed
in each degree of freedom. Potentiometric transducers PTP-11 are used
to measure rotation angles, electromotors DPM-20-N1-11 are converted to
tachometers and used as velocity sensors.

The gripper has two separately actuated fingers *10*. Each finger is actuated by a separate DC motor through a worm gear. The maximum opening
of the fingers is 120 mm, their velocities are in the range from 1 to 20 mm/s,
the maximum force exerted by the gripper is about 40 N. Each finger is at
the same time a three-component force sensor. Such sensing fingers are
described in Section 2.2, and their close-up view is shown in Figure 2.2.2.

The end-effector with force sensing fingers was used to perform a peg-in-
hole insertion (see Section 10.2) and to grasp randomly placed workpieces
(see Section 11.3).

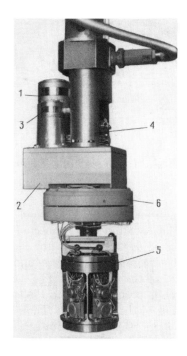

FIGURE 4.3.3
The end-effector with the screwdriver

Figure 4.3.3 illustrates an end-effector designed for studying threaded
connection assembly – a screwdriver. In the figure, *1* is a DPM-30-N1-11
electromotor, *2* is a reduction gear, *3* is a velocity sensor, *4* is a rotation
angle sensor, *5* is a four-component force sensor, *6* denotes sliding contacts

that transmit electrical signals through a joint with an infinite rotation. The screwdriver has a single degree of freedom and rotates the workpiece attached to the force sensor about the vertical axis. The rotation of the lower part of the screwdriver is unbounded. A position transducer is available to measure the rotation angle. The maximum rotation velocity of the device is 7 rad/s, the nominal torque is 3 N·m.

The force sensor *5* mounted to the output shaft of the drive is analogous to the sensor shown in Figure 2.3.9. The sensor measures three force components and a vertical torque component. Voltage signals to and from the sensor are transmitted to the control system through the sliding contacts *6*. To make the device more reliable, parallel contact clusters are employed. Each cluster consists of a brass groove and a spring-loaded silvered rod. The contact pressure of the rod against the groove can be adjusted. In Section 10.4, we study a threaded connection assembly using the described screwdriver.

FIGURE 4.3.4
The end-effector with the grinding tool

Figure 4.3.4 displays an end-effector for studying surface machining operations, such as grinding, fettling, or polishing. The end-effector has an electric drive *1*, to the axis of which a grinding tool *2* or a milling cutter can be mounted. The drive is connected to the manipulator arm through a two-component force sensor *3*. A DC motor with the nominal torque 0.04 N·m and the nominal rotation velocity 480 rad/s is used in the drive. The tool rotates about the vertical axis. The sensor measures two horizontal force components. It consists of two identical single-component modules, similar in design to the module shown in Figure 2.1.9.

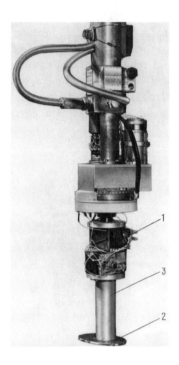

FIGURE 4.3.5
The end-effector with a workpiece rotating about the vertical axis

Our experiments in force control of grinding are described in Chapter 9. In the experiments, in addition to the above-described device, the end-effector shown in Figure 4.3.5 is used. It allows the manipulator to move and rotate a workpiece about a stationary grinding tool. The end-effector uses the same drive as the screwdriver. A two-component force sensor *1* (see also Figure 4.3.4) is mounted to the output shaft of the drive. A workpiece *2* is attached to the rod *3*. This device enables the workpiece to rotate about the vertical axis.

Chapter 5

Control of Manipulator Contact With an Object

The analysis of manipulation tasks performed by humans suggests that their motions may be decomposed into primitive ones. In general, results of physiological and biomechanical research [70] demonstrate that a large portion of human and animal motor activity is composed from relatively standard motor procedures. It seems advantageous to design control of complex motions for robotic systems as a composition of primitive motions, as well. These primitive motions can be called "basic" motions.

By studying such tasks as probing, grasping an object, inserting a peg into a hole, assembling a threaded joint, etc., one may notice that a human keeps contact with manipulated objects during the manipulation. Clearly, the task of keeping contact can be recognized as one of the basic tasks. This task is an important component of many complex manipulation motions and can be considered as a key task.

This chapter designs control for keeping a manipulator in contact with a stationary or moving object and for contact transition at initial impact with an object. It considers mathematical models of one-dimensional motion.

This and the three subsequent chapters derive and study mathematical models for robotic systems with force sensors, and describe approaches to control design for such systems.

5.1 Mathematical Model for One-degree-of-freedom Manipulator Motion

We shall study a one-dimensional translational motion of a manipulator. The manipulator with translational degrees of freedom has been described in the preceding, fourth chapter of the book (see Figure 4.1.1). The mathematical models and methods of this chapter are also applicable, though in a somewhat modified form, for manipulators with a different kinematic configuration, such as articulated arms.

Let us consider a linear horizontal motion of the manipulator arm. We assume that the manipulator (with a single translational degree of freedom) is driven along the axis Ox by a single motor. A single-component force sensor is mounted to the lower end of the manipulator arm and measures the horizontal force component acting on it. Force sensing fingers or a wrist force sensor can serve the purpose. Figure 5.1.1 illustrates the schematics of a one-dimensional motion of the arm establishing contact with an object. The sensor is shown as a thin elastic plate (rod).

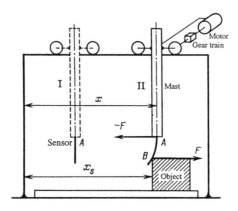

FIGURE 5.1.1
Schematics for one-dimensional motion of the manipulator

In the figure, the manipulator position prior to the contact with an object is labeled as I, and the position in contact, as II. Denote by x the coordinate of the point A projection onto the axis Ox, where the sensor is attached to the manipulator. Denote by x_s the coordinate of the point B of the object edge closest to the sensor. In what follows, we assume that the coordinate of the point A is x_s when the sensor establishes or loses contact with the object.

Let us now derive the equations of motion for the system. The equation for the torque balance with respect to the rotor shaft of the manipulator driving motor has the form

$$J\ddot{\varphi} = M \tag{5.1.1}$$

In this equation, φ is a revolution angle of the motor rotor with respect to the reference orientation; J is an equivalent moment of inertia of the moving parts (the motor rotor, gears, and manipulator arm) with respect to the motor shaft. The torque M is the sum of the motor torque M_G and the torque M_a produced by all other forces:

$$M = M_G + M_a \tag{5.1.2}$$

The linear displacement x of the manipulator arm with respect to the initial position (see Figure 5.1.1) is related to the revolution angle φ of the motor as (see Section 4.1):

$$x = \frac{r}{i}\varphi = \rho\varphi \quad (\rho = \frac{r}{i}), \tag{5.1.3}$$

where i is a gear train reduction ratio, and r is a radius of the outer pinion in the gear train.

If we neglect the masses of the pinion gears, then the overall moment of inertia can be presented as

$$J = J_G + \frac{mr^2}{i^2} = J_G + m\rho^2, \tag{5.1.4}$$

where J_G is the moment of inertia of the motor rotor, and m is the mass of the manipulator arm including the mass of the motor. For "large" gear ratios, we obtain from (5.1.4)

$$J \approx J_G, \tag{5.1.5}$$

that is, the equivalent moment of inertia with respect to the motor shaft is largely defined by the moment of inertia of the motor rotor. The approximate equality (5.1.5) is valid for the analysis of many robotic systems.

For the translational degrees of freedom of the experimental robotic system described in Chapter 4, the gear ratio is $i = 42.6$, and the radius r of the outer pinion is 12.5 mm ($\rho = 0.294$ mm). For the rotor of the DPM-35-01 motor, the moment of inertia $J_G = 6.1 \cdot 10^{-6}$ kg·m^2 [169]. For the mass $m = 10$ kg of the manipulator arm with a load, $J = J_G + m\rho^2 = 6.1 \cdot 10^{-6}$ kg·m^2 + $0.86 \cdot 10^{-6}$ kg·m^2 = $6.96 \cdot 10^{-6}$ kg·m^2, and J exceeds J_G by less than 15%.

The motor torque M_G is directly proportional to the electrical current in the rotor armature. By neglecting the armature inductance (electromagnetic time constant in the rotor circuit), we can write this torque in the form [65]

$$M_G = c_1 u - c_2 \dot{\varphi}, \tag{5.1.6}$$

where u is the voltage supplied to the motor, and $c_2\dot{\varphi}$ is the moment of the back e.m.f.

The positive constants c_1 and c_2 for a specific motor can be calculated by using the values for the starting torque M_1, nominal voltage u, nominal torque M_2, and nominal angular velocity $\dot{\varphi}$ given as performance specifications:

$$c_1 = \frac{M_1}{u}, \quad c_2 = \frac{M_1 - M_2}{\dot{\varphi}}$$

For example, for the DPM-35-01 motor used in the robotic system described in Chapter 4, the handbook [169] gives $M_1 = 0.0687$ N·m, $M_2 = 0.0147$ N·m, $\dot{\varphi} = 942$ s^{-1}, $u = 27$ V. Using these data, we obtain $c_1 = 2.64$ ·10^{-3} N·m/V, $c_2 = 6\cdot10^{-5}$ N·m·s. The electromechanical time constant $T = J_G/C_2 = 0.105$ s. The values of the electromagnetic time constant $9 \cdot 10^{-4}$ to $15 \cdot 10^{-4}$ given in [169] are significantly less than T. Hence the assumptions of (5.1.6) are justified. Chapter 8 considers dynamical models of the manipulator, taking into account discrete character of control caused by the use of a digital computer in the control loop. The sampling time considered there is of the order of 10^{-2} s. The electromagnetic time constant is an order of magnitude less. Therefore, we can also neglect armature inductance when developing discrete dynamical models in Chapter 8.

By substituting (5.1.2), (5.1.6) into (5.1.1), we obtain

$$J\ddot{\varphi} + c_2\dot{\varphi} = c_1 u + M_a \qquad (5.1.7)$$

Denote by F_a the sum of external forces applied to the manipulator arm. Then the torque M_a produced by these forces with respect to the axis of the motor rotor is given by

$$M_a = F_a \frac{r}{i} = F_a \rho$$

The equivalent mass $\mathcal{M}$ of all the moving parts of the manipulator as involved in the arm motion is given by

$$\mathcal{M} = \frac{J}{\rho^2} = \frac{J_G}{\rho^2} + m$$

By introducing the notations

$$d_1 = \frac{c_1}{\rho}, \quad d_2 = \frac{c_2}{\rho^2},$$

we can rewrite (5.1.7) in the form

$$M\ddot{x} + d_2\dot{x} = d_1 u + F_a \qquad (5.1.8)$$

We shall model a single-component force sensor by a massless lumped elastic element (see Section 1.2). Figure 5.1.1 depicts this element as a thin elastic plate, which is attached to the manipulator arm at point A and deflects in the plane of the figure. The plate is assumed to be massless. If the sensor is installed in the wrist preceding the gripper, then we also neglect the mass of the gripper. In the absence of deformation, the plate is vertical (see position I in Figure 5.1.1). Thus, when the sensor establishes or loses contact with the object, the coordinate of the point A coincides with the coordinate x_s of the point B. If the contact is maintained at the point B, i.e., if $x > x_s$, the plate bends and applies a certain force to the object. Denote by F the horizontal component of this force. Since the plate is assumed to be massless, the force F of the same magnitude, but opposite in direction, acts on the manipulator arm from the plate at the point A. If the stiffness of the sensor is large and its deformations are small, then, in accordance with (1.2.4), by neglecting energy dissipation in the sensor, we consider the force F to be directly proportional to the difference $x - x_s$. When the sensor is not in contact with the object, i.e., if $x < x_s$, the force is $F = 0$. Thus, we come to the formula

$$F = \begin{cases} 0 & \text{if } x < x_s, \\ k(x - x_s) & \text{if } x \geq x_s, \end{cases} \qquad (5.1.9)$$

where the positive coefficient k defines the lumped sensor stiffness. Note that the stiffness coefficient generally depends on the distance between the point where the force sensor is attached to the manipulator arm, and the point of contact with the object. For a parallelogram elastic element (see Figure 2.1.1), we can consider k to be independent of this distance. In view of the aforesaid, let us present another schematics of a manipulator with a force sensor (see Figure 5.1.2). In the figure, a linear spring that models the sensor is attached to a perfectly rigid holder $ACDE$. The spring can deform horizontally.

The transducers mounted on the elastic element measure its deformations. According to (1.2.5)–(1.2.7), we assume that the deformation $x - x_s$ is measured, which means that the force F acting on the object is also measured.

One may consider a force sensor as an element that can very accurately detect the respective positions of the manipulator and the manipulated object in contact. For instance, if the sensor stiffness $k = 3 \cdot 10^4$ N/m (which is the case for the force sensing fingers described in Section 2.2

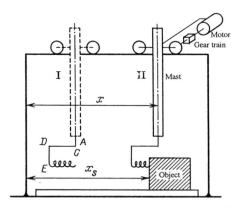

FIGURE 5.1.2
Schematics of the manipulator with a force sensor

and shown in Figure 2.2.6), and the force F is measured with the (coarse) resolution of 0.3 N, then the deformation $x - x_s$ can be determined with the resolution of 0.01 mm.

Displacement sensors measuring relative displacement of the manipulator links also allow one to determine the coordinates of the gripper and its position relative to the object, provided that the object coordinates are known. Yet these sensors are considerably less accurate than force sensors, because of lower resolution, compliance and backlash in the manipulator structure. Besides, force sensors do not require the knowledge of the manipulated object coordinates. However, one must keep in mind that force sensors cannot be used to determine an object position relative to the manipulator in the absence of contact with the object.

Let the external force acting on the manipulator be $F_a = -F + F_e$, where F_e may, for instance, be a Coulomb friction force in the motor or gear train. Then we can rewrite (5.1.8), (5.1.9) in the form

$$M\ddot{x} + d_2\dot{x} = d_1 u + F_e \qquad\qquad \text{if } x \leq x_s \qquad (5.1.10)$$

$$M\ddot{x} + d_2\dot{x} + k(x - x_s) = d_1 u + F_e \qquad \text{if } x \geq x_s \qquad (5.1.11)$$

The control voltage u applied to the motor is bounded by a certain constant u_0

$$|u| \leq u_0 \qquad\qquad\qquad (5.1.12)$$

According to (1.2.3), we may also introduce into (5.1.9) a term for a force of viscous friction caused by the force sensor deformation. In this case, we have to add a viscous friction factor to the coefficient d_2 in (5.1.10), (5.1.11). However, this will not change the structure of the equations.

The equations (5.1.10), (5.1.11) can be used as the simplest model of the one-dimensional motion of a manipulator with a force sensor. This model does not take into account a number of factors, such as, for example, a manipulator structure elasticity or backlash in the gear train. Even so, it offers an explanation for many features of manipulator motion with force control and allows analysis of various control methods. For a wrist sensor, the gripper mass is neglected, as it is assumed in (5.1.10). The equation (5.1.11) holds without such assumption.

For convenience, we shall further use nondimensional variables for the analysis of (5.1.10), (5.1.11). By computing the nondimensional variables from the formulae

$$\tau = \sqrt{\frac{k}{\mathcal{M}}}t, \quad \chi = \frac{k}{d_1 u_0}x, \quad \chi_s = \frac{k}{d_1 u_0}x_s, \quad w = \frac{u}{u_0}, \quad f_e = \frac{F_e}{d_1 u_0}, \quad (5.1.13)$$

we can write (5.1.10), (5.1.11) in the form

$$\ddot{\chi} + \alpha\dot{\chi} = w + f_e, \qquad\qquad \text{if } \chi \leq \chi_s \qquad\qquad (5.1.14)$$

$$\ddot{\chi} + \alpha\dot{\chi} + \chi - \chi_s = w + f_e, \quad \text{if } \chi \geq \chi_s \qquad\qquad (5.1.15)$$

$$|w| \leq 1 \qquad\qquad\qquad\qquad (5.1.16)$$

Here, dots denote derivatives with respect to the nondimensional time τ, and the nondimensional parameter α is given by

$$\alpha = \frac{d_2}{\sqrt{\mathcal{M}k}} = \frac{c_2}{\rho\sqrt{Jk}} = \frac{c_2}{\rho\sqrt{k(J_G + m\rho^2)}} \qquad (5.1.17)$$

We shall further use the equations of the form (5.1.14)– (5.1.16) to analyze some problems.

5.2   Problem Statement for Keeping Contact with an Object

The manipulator motion is controlled by applying a voltage u to the motor. Control information is provided by a sensor, measuring a force F, and a tachometer, measuring a velocity $\dot{x}$ of the manipulator motion. Let us formulate the problem of keeping in contact with a stationary object ($x_s = $ const) for the single-dimensional manipulator motion described above as a control problem.

Problem.   Find a control u dependent on the force F and the velocity $\dot{x}$ such that the system achieves a steady state given by

$$F = F_d, \quad \dot{x} = 0, \tag{5.2.1}$$

where a positive constant F_d is a desired (commanded) force with which the manipulator contacts the object.

In other words, we should design a control such that the manipulator, upon completion of the transient process, contacts the object with a given force, and the transient oscillations die out sufficiently fast. The commanded force F_d can be defined, for example, by a threshold frictional force F_θ, which restrains the object placed on the manipulator table from moving. If the object cannot be moved, then F_d must be less than F_θ. On the other hand, F_d certainly has to exceed the force sensor sensitivity threshold F_0, i.e.,

$$F_0 < F_d < F_\theta \tag{5.2.2}$$

The force F_d corresponds to the manipulator coordinate $x_d = x_s + F_d/k$ given by (5.1.9). To solve our problem, we should demand that

$$F_d < d_1 u_0 \quad (x_d < x_s + d_1 u_0/k) \tag{5.2.3}$$

The coordinate x measured by a position sensor is useless unless the coordinate x_s of the object is known a priori, before the manipulator contacts the object. When in contact, the manipulator position relative to the object is determined by a force sensor, so a position sensor is not needed, either. However, a position sensor can be used to determine the coordinate x_s of the object through the coordinate x of the manipulator at the instant of contact. A force sensor signal greater than its sensitivity threshold F_0 shows that the contact is established.

If the coordinate x_s is known, then, in the absence of contact, measuring x can be useful for control. We use this data to solve the problem of keeping in contact with a moving object (see Section 5.9).

The simplest method of establishing and keeping contact with an object is to apply a constant voltage u_1 to the motor. This approach does not require any information on forces, velocity, or manipulator arm position. Let us consider a manipulator motion for $u(t) = u_1$, where $0 < u_1 \leq u_0$.

Note that this approach is an implicit force control method, as discussed in Section 1.3.

We assume that there the manipulator is out of contact with the object at the initial instant of time, i.e., $x(0) < x_s$ (see Position I in Figures 5.1.1, 5.1.2). Then the motion of the manipulator arm prior to the contact is described by (5.1.10) or (5.1.14). If $F_e \equiv 0$, then for $u(t) \equiv u_1$, we obtain

$$\dot{x}(t) = \left(\dot{x}(0) - \frac{d_1 u_1}{d_2} \right) e^{t d_2 / \mathcal{M}} + \frac{d_1 u_1}{d_2}$$

For $t \to \infty$, the velocity $\dot{x}(t)$ tends to the steady-state value

$$\dot{x}_c = \frac{d_1}{d_2} u_1 = \frac{\rho c_1}{c_2} u_1 \tag{5.2.4}$$

At certain time ϑ, the force sensor contacts the object, which means $x(\vartheta) = x_s$. For $t \geq \vartheta$, the motion before the instant of contact loss is described by (5.1.11) or (5.1.15). If $F_e \equiv 0$, then the solution of (5.1.11) for $u(t) \equiv u_1$, $x(\vartheta) = x_s$, $\dot{x}(\vartheta) = \dot{x}_c$ has the form

$$x(t) = x_s + \frac{d_1 u_1}{k} + \frac{d_1 u_1}{k} e^{-\frac{\alpha}{2}(t-\vartheta)}$$

$$\times \left\{ \frac{1}{\omega \alpha} (1 - \frac{\alpha^2}{2}) \sin[\omega(t - \vartheta)] - \cos[\omega(t - \vartheta)] \right\}, \tag{5.2.5}$$

$$\dot{x}(t) = \frac{d_1 u_1}{\sqrt{\mathcal{M}k}} e^{-\frac{\alpha}{2}(t-\vartheta)} \left\{ \frac{1}{2\omega} \sin[\omega(t - \vartheta)] + \frac{1}{\alpha} \cos[\omega(t - \vartheta)] \right\}$$

$$(\omega^2 = 1 - \frac{\alpha^2}{4})$$

We only consider here the case of $\alpha < 2$, where all eigenvalues of (5.1.15) are complex.

The times at which $\dot{x}(t)$ becomes zero are

$$t_n = \vartheta - \frac{1}{\omega} \arctan \frac{2\omega}{\alpha} + \pi n \quad (n = 1, 2, ...)$$

By substituting t_1 into (5.2.5), (5.1.9), we obtain

$$F(t_1) = d_1 u_1 \left\{ 1 + \frac{1}{\alpha} \exp\left[-\frac{\alpha}{\sqrt{4 - \alpha^2}} \left(\pi - \arctan \frac{\sqrt{4 - \alpha^2}}{\alpha} \right) \right] \right\} \quad (5.2.6)$$

It is easy to verify that (5.2.6) defines the maximum contact force between the manipulator and the stationary object for $u(t) \equiv u_1$, $x(\vartheta) = x_s$, $\dot{x}(\vartheta) = \dot{x}_c$.

By substituting t_2 into (5.2.5), (5.1.9), we obtain

$$F(t_2) = d_1 u_1 \left\{ 1 - \frac{1}{\alpha} \exp\left[-\frac{\alpha}{\sqrt{4 - \alpha^2}} \left(2\pi - \arctan \frac{\sqrt{4 - \alpha^2}}{\alpha} \right) \right] \right\}$$

$$= d_1 u_1 \phi(\alpha) \tag{5.2.7}$$

The function $\phi(\alpha)$ has a sole zero $\alpha_0 \approx 0.38$. The inequality $\phi(\alpha) > 0$ holds for $\alpha > \alpha_0$. Therefore, if $\alpha > \alpha_0$, then the manipulator does not lose contact with the object over the whole interval $[\vartheta, \infty]$. For $\alpha < \alpha_0$, the contact is lost at a certain instant of time $t > \vartheta$ (if $\dot{x}(\vartheta) = \dot{x}_c$).

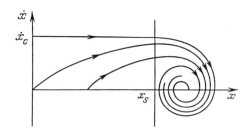

FIGURE 5.2.1
Transient processes in the phase plane x, $\dot{x}$

If $\dot{x}(0) < \dot{x}_c$, then $\dot{x}(\vartheta) < \dot{x}_c$. Then, for $u(t) \equiv u_1$, the maximum force of contact is less than (5.2.6). If

$$\frac{c_2}{\rho\sqrt{k(J_G + m\rho^2)}} > 0.38,$$

then the manipulator, upon establishing contact with the object, does not lose it any longer. The transient processes in the phase plane $(x, \dot{x})$ are shown in Figure 5.2.1 for $\alpha > \alpha_0$.

For the above-mentioned parameters of the research manipulator, $c_2 = 6 \cdot 10^{-5}$ N·m·s, $\rho = 2.94 \cdot 10^{-4}$ m, $k = 3 \cdot 10^4$ N/m, $J = 6.96 \cdot 10^{-6}$ kg·m^2, $\alpha = 0.46$, which gives us $\phi(\alpha) > 0$.

If $u(t) \equiv u_1$, then, for $t \to \infty$, the variables $x(t) - x_s$ and $F(t)$ converge to the values

$$x - x_s = \frac{d_1}{k}u_1 = \frac{c_1}{\rho k}u_1, \quad F = d_1 u_1 = \frac{c_1}{\rho}u_1 \qquad (5.2.8)$$

For $u(t) = u_1$, the maximum force (5.2.6) of the manipulator contact with an object can be too large and may even damage the sensor. To reduce this force, we can diminish u_1, but then the manipulator velocity at establishing contact with the object will decrease. Given Coulomb friction, the voltage u_1 should be sufficiently large to move the manipulator from its initial position. The use of information on the force of contact between the manipulator and the object for control allows reduction of the maximum value of the contact force without reducing the approach velocity. In the subsequent chapters, we solve the above problem by using force feedback control.

The problem of keeping a desired force of contact with an object, formulated in this section, does not only arise in control design for an autonomous manipulator. It is also relevant for teleoperator design. In a teleoperator system, a human operator controls the motion of an object or a "slave" manipulator using a master manipulator. The master manipulator should reflect the force of the object resistance to the motion to the operator's hand. In other words, the operator should sense the force of environmental resistance to the object motion. This raises the question of building an "inner" control loop for the master manipulator to ensure the required force of contact with the operator's hand. If the master manipulator is instrumented with force sensors, then the control methods considered below in the present chapter can be used to design a control system for it.

5.3   Linear Control

The simplest force feedback control for solving the above problem is linear measurement of feedback with constant gains

$$u = -g_1(F - F_d) - g_2\dot{x} \quad (F_d > 0, \ g_1 > 0, \ g_2 > 0) \quad (5.3.1)$$

Here and further, the signs in control laws are chosen under the assumption that the object is to the right from the manipulator arm, as shown in Figures 5.1.1 and 5.1.2. Since $g_1 > 0$, the first term in (5.3.1) is positive for $F < F_d$, and negative for $F > F_d$. Therefore, for $F < F_d$, the force $d_1 g_1 (F_d - F)$ exerted by the electric motor "pushes" the manipulator to the right, i.e., towards the object (presses it against the object), and for $F > F_d$, to the left, i.e., from the object. The control law (5.3.1) cannot be implemented for all F and $\dot{x}$ because of the constraint (5.1.12).

For the control law (5.3.1), taking into account (5.1.9), we can rewrite (5.1.10), (5.1.11) in the form

$$\mathcal{M}\ddot{x} + (d_2 + d_1 g_2)\dot{x} + d_1 g_1 k(x_s - x_d) = F_e, \qquad \text{if } x \leq x_s \quad (5.3.2)$$

$$\mathcal{M}\ddot{x} + (d_2 + d_1 g_2)\dot{x} + k(x - x_s) + d_1 g_1 k(x - x_d) = F_e, \quad \text{if } x \geq x_s \quad (5.3.3)$$

Let the force of friction be constant ($F_e = \text{const}$) and not too large in magnitude, so that

$$d_1 g_1 F_d + F_e > 0 \quad (d_1 g_1 k(x_d - x_s) + F_e > 0) \quad (5.3.4)$$

Then, since $d_2 + d_1 g_2 > 0$, the velocity $\dot{x}$ defined by (5.3.2) becomes positive from some time instant on, regardless of the initial conditions. Hence, even if the manipulator arm does not contact the object in the beginning of the motion ($x(0) < x_s$), under the control (5.3.1), it will contact the object at some time instant. The motion of the manipulator in contact with the object is described by (5.3.3). The third term in the left part of the equation describes the direct influence of the sensor on the manipulator motion. The sensor in contact with the object acts on the arm as a mechanical spring. Close examination of the third term in (5.3.3) shows that when in contact, the force feedback is analogous to position feedback and similarly affects the manipulator motion as an ordinary mechanical spring of the length x_d. Therefore, introducing a force feedback to the system is equivalent to introducing an additional stiffness. This stiffness is artificial, as distinct from the natural one, which is caused by the elasticity of the force sensor and manipulator links (assuming they are elastic). Unlike natural stiffness, one can "electrically" control such artificial stiffness by changing the gain g_1. For $F_e = $ const, (5.3.3) has a single equilibrium point

$$x = x_s + \frac{d_1 g_1}{1 + d_1 g_1}(x_d - x_s) + \frac{1}{1 + d_1 g_1}\frac{F_e}{k}, \quad \dot{x} = 0, \qquad (5.3.5)$$

which is located to the right of the point x_s provided that (5.3.4) holds.

In a steady-state (equilibrium) condition,

$$F = \frac{d_1 g_1}{1 + d_1 g_1}F_d + \frac{1}{1 + d_1 g_1}F_e \qquad (5.3.6)$$

A steady-state error in tracking the commanded contact force F_d is

$$\Delta F = F - F_d = \frac{1}{1 + d_1 g_1}(-F_d + F_e) \qquad (5.3.7)$$

This error approaches zero if the force feedback gain $g_1 \to \infty$. Since the coefficients d_1, d_2, g_1, g_2 are positive, the conditions

$$d_2 + d_1 g_2 > 0, \quad 1 + d_1 g_1 > 0 \qquad (5.3.8)$$

are satisfied, and the stationary regime (5.3.5) is asymptotically stable.

For $F_e = $ const, we can write the system (5.3.2), (5.3.3) in the form

$$\mathcal{M}\ddot{x} + (d_2 + d_1 g_2)\dot{x} + \varphi(x) = 0, \qquad (5.3.9)$$

where the nonlinear (piecewise linear) function $\varphi(x)$ has the form

$$\varphi(x) = \begin{cases} d_1 g_1 k(x_s - x_d) - F_e & \text{if } x \leq x_s, \\ k(x - x_s) + d_1 g_1 k(x - x_d) - F_e & \text{if } x \geq x_s \end{cases} \qquad (5.3.10)$$

Figure 5.3.1 displays the plot of the function (5.3.10), where x_c denotes the equilibrium point (5.3.5).

Under the conditions (5.3.4), (5.3.8)

$$\Phi(x) = \int_{x_c}^{x} \varphi(x)dx$$

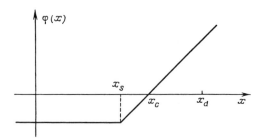

FIGURE 5.3.1
Force function

is a positive definite function unbounded for $|x| \to \infty$ and zero only for $x = x_c$.

The Lyapunov function candidate

$$V(x, \dot{x}) = \frac{1}{2}\mathcal{M}\dot{x}^2 + \Phi(x)$$

is also positive definite and unbounded. If the first inequality in (5.3.8) holds, then for the total derivative of this function along the trajectories of the system (5.3.9), (5.3.10), it can be shown that $\dot{V} < 0$ for $\dot{x} \neq 0$, and $\dot{V} = 0$ for $\dot{x} = 0$. There are no complete trajectories of the system along the axis $\dot{x} = 0$, except for the equilibrium state (5.3.5). It follows from the Barbashin's total stability theorems [21] that the equilibrium state (5.3.5) of the system (5.3.9), (5.3.10) is globally asymptotically stable. It means that for any initial conditions, the solution of the system (5.3.9), (5.3.10) converges to the stationary regime (5.3.5) for $t \to \infty$.

Alternatively, we can prove that the system (5.3.9), (5.3.10) is globally asymptotically stable by examining its phase portrait. In so doing, we shall first prove that the system solution is bounded in the phase plane for any initial conditions. Then we can prove that the system has no periodic solution.

Thus, the change of the system structure caused by establishing and losing contact with the object does not make the system unstable, i.e., it is not a destabilizing factor.

In view of the constraint (5.1.12), we should rewrite (5.3.1) in the form

$$u = -\text{sat}\,[g_1(F - F_d) + g_2\dot{x}], \tag{5.3.11}$$

where sat $y = y$ for $|y| \leq u_0$, sat $y = u_0$ sgn y for $|y| \geq u_0$.

If the maximum control force $d_1 u_0$ is sufficiently large in comparison with the friction force F_e, then (5.3.5) is an equilibrium state for the system (5.1.10), (5.1.11) with the control (5.3.11). This state is asymptotically stable.

One can prove that the system is asymptotically stable with the control (5.3.11) for any initial conditions, i.e., the system (5.1.10), (5.1.11) is globally stable.

The linear with saturation control (5.3.11) solves the formulated problem of keeping contact with a stationary object. When the gain g_1 is small, the system response is slow, and the static tracking error ΔF of the contact force is large. Coulomb friction further increases this error. The static error ΔF decreases with the increase of g_1. If g_2 remains constant, then the system becomes oscillatory when g_1 exceeds a certain value. The oscillations can be damped by simultaneously increasing g_2.

The control (5.3.11) was experimentally tested with the robotic system presented in Chapter 4. The experiments prove the efficiency of the linear (with saturation) control and confirm the results of the above theoretical analysis.

One possible method of implementing the control (5.3.1) is to use a digital computer and analog velocity servo-system hardware (see Chapter 4). To illustrate this strategy, let us write (5.3.1) in the form

$$u = g_2(v_d - \dot{x}), \tag{5.3.12}$$

where the commanded velocity v_d is defined by the expression

$$v_d = g(F_d - F), \quad g = \frac{g_1}{g_2} \tag{5.3.13}$$

Using this method, the computer receives a measurement F from the force sensor and calculates v_d by the formula (5.3.13). Then the signal v_d is sent to one of the inputs of the velocity servo system. Another input receives a signal for the velocity $\dot{x}$ from the tachometer. The feedback amplifier of the servo system determines the control voltage u.

Note that the force $F' = -F$ acts on the manipulator arm from the force sensor. In view of that, we can rewrite (5.3.13) in the form $v_d = g(F' - F'_d)$. In this form, it conforms to the expressions for commanded velocity in other chapters of the book.

Let us rewrite the system (5.1.10),(5.1.11) with the linear control (5.3.1) for $F_d = 0$ ($F_e = 0$) in the form

$$\mathcal{M}\ddot{x} + (d_2 + d_1 g_2)\dot{x} = -(1 + d_1 g_1)F \tag{5.3.14}$$

We assume that the force $F' = -F$ is applied to the force sensor. This force is constant for all movements of the manipulator (the strain in the sensor does not change during the motion). Then, after the transient process, the manipulator achieves a constant velocity of motion

$$\dot{x} = -\frac{1 + d_1 g_1}{d_2 + d_1 g_2} F = \frac{1 + d_1 g_1}{d_2 + d_1 g_2} F' \qquad (5.3.15)$$

Thus, the applied force causes the manipulator to move at a velocity that grows as g_1 increases and g_3 decreases. This behaviour of the manipulator, analogous to that of a body in a viscous medium [155, 163], is called *active accommodation* in [291]. Such behaviour is subjectively perceived as "responsive". If one touches the sensor with a finger, the manipulator starts moving in the direction of the finger pressure. Note that with a large gain g_1 and small g_2, the manipulator moves fast even for a small force F'.

Linear control, which includes force and velocity feedback as well as position feedback, has the form

$$u = -g_1(F - F_d) - g_2 \dot{x} - g_3(x - x_g), \qquad (5.3.16)$$

where g_3 is a constant positive gain, x_g is a given position. The system (5.1.10), (5.1.11) with the control (5.3.16) for $F_d = F_e = 0$ has the form

$$M\ddot{x} + (d_2 + d_1 g_2)\dot{x} + d_1 g_3(x - x_g) = -(1 + d_1 g_1)F \qquad (5.3.17)$$

If the force F' applied to the force sensor is constant, then the manipulator position after the transient process can be written as

$$x - x_g = -\frac{1 + d_1 g_1}{d_1 g_3} F = \frac{1 + d_1 g_1}{d_1 g_3} F' \qquad (5.3.18)$$

Thus, with the control (5.3.16), the applied force causes the manipulator to move by a distance that grows with the increase of g_1 and decrease of g_3. In other words, with the control (5.3.16), the manipulator behaves as a spring with a stiffness factor $\frac{1 + d_1 g_1}{d_1 g_3}$. This property is called *artificial* or *active compliance* [133, 222].

Unlike natural compliance that is caused by structure elasticity and backlash in the joints, the active compliance is caused by a combined force and position feedback and can be controlled.

Accommodation and compliance of a robotic system may be useful for a number of industrial tasks, such as automated assembly. These properties can relax the requirements to the accuracy of positioning when mating (contacting) a manipulated object with other objects.

5.4 Switching Control

For large gains g_1, g_2, linear with saturation control (5.3.11) closely approximates switching control

$$u = -u_0 \operatorname{sgn} [g_1(F - F_d) + g_2 \dot{x}] \qquad (5.4.1)$$

The control (5.4.1) may essentially be implemented using relays as well as linear elements with large gains.

Since $g_2 > 0$, we can rewrite (5.4.1) as

$$u = -u_0 \operatorname{sgn} [g_1(F - F_d) + \dot{x}] \qquad (g = g_1/g_2) \qquad (5.4.2)$$

or

$$u = u_0 \operatorname{sgn} (v_d - \dot{x}), \qquad (5.4.3)$$

where v_d is given by (5.3.13).

By taking (5.1.9) into account, we can write (5.4.2) in the form

$$u = -u_0 \operatorname{sgn} \begin{cases} gk(x_s - x_d) + \dot{x} & \text{if } x \leq x_s, \\ gk(x - x_d) + \dot{x} & \text{if } x \geq x_s \end{cases} \qquad (5.4.4)$$

The switching curve in the control law (5.4.4) is defined by the following broken line:

$$S = gk(x_s - x_d) + \dot{x} = 0 \quad \text{if } x \leq x_s,$$

$$S = gk(x - x_d) + \dot{x} = 0 \quad \text{if } x \geq x_s \qquad (5.4.5)$$

In view of (5.4.5), we can rewrite (5.4.4) in the form

$$u = -u_0 \operatorname{sgn} S$$

With the control (5.4.3), the equation (5.1.10) has the form

$$\mathcal{M}\ddot{x} + d_2\dot{x} = d_1 u_0 \operatorname{sgn} (v_d - \dot{x}) + F_e \qquad (5.4.6)$$

We assume that $v_d = $ const and $|F_e - d_2 v_d| < d_1 u_0$. Then, for a certain ε, $0 < \varepsilon < d_1 u_0$, we have

$$|F_e - d_2 v_d| < d_1 u_0 - \varepsilon \qquad (5.4.7)$$

Hence for $\dot{x} > v_d$,

$$\ddot{x} < -\frac{\varepsilon}{M}$$

and for $\dot{x} < v_d$,

$$\ddot{x} > \frac{\varepsilon}{M}$$

Therefore, the system achieves the exact tracking of the commanded velocity

$$\dot{x} = v_d \qquad (5.4.8)$$

in a finite time, regardless of the initial velocity $\dot{x}(0)$.

The solution (5.4.8) of the equation (5.4.6) defines a *sliding mode* [244, 267, 62, 63, 67]. The equivalent control u [267, 62, 63, 67] for this solution is given by the formula

$$d_1 u = d_2 v_d - F_e \qquad (5.4.9)$$

Now let us assume that $v_d \neq$ const, but the condition (5.4.7) holds. We also assume that there exists a constant $0 < \delta < \frac{\varepsilon}{M}$ such that at any time instant

$$|\dot{v}_d| < \frac{\varepsilon}{M} - \delta$$

Then (5.4.8) will still be an asymptotically stable solution of (5.4.6). The above consideration is true for $F_e = $ const as well as for $F_e = F_e(t)$ or $F_e = F_e(\dot{x})$.

If the threshold F_e' of the Coulomb friction force F_e is less than $d_1 u_0$ ($F_e = -F_e'$ sgn $\dot{x}$), and the commanded velocity v_d is small, then the condition (5.4.7) is satisfied. It follows that the switching control law allows tracking of small values of commanded velocity without errors. For a large gain g_2, the linear with saturation control (5.3.11), (5.3.12)

$$u = \text{ sat } [g_2(v_d - \dot{x})] \qquad (5.4.10)$$

is equivalent to the switching control (5.4.3). In the experiments with the robotic system described in Chapter 4, the control (5.4.10), for $g_2 \approx 7 \cdot 10^3$ V·s/mm, allowed to track velocities in the range of 0.1 – 200 mm/s (i.e., the control dynamical range is 2000).

Thus, switching control allows tracking of the commanded velocity v_d without errors for a certain range of system parameters and external influences. Therefore, switching control makes the system invariant with respect to external influences and system parameter variations.

By substituting the switching control (5.4.4) into (5.1.10) (upper line), we obtain an expression for the manipulator motion out of contact with an object, for $x \leq x_s$:

$$M\ddot{x} + d_2\dot{x} = d_1 u_0 \operatorname{sgn}\left[gk(x_d - x_s) - \dot{x}\right] + F_e \qquad (5.4.11)$$

In this case, the inequality (5.4.7) takes the form

$$|F_e + d_2 gk(x_s - x_d)| < d_1 u_0 - \varepsilon \qquad (5.4.12)$$

With the constraint (5.4.12), the equation (5.4.11) has an asymptotically stable solution

$$\dot{x} = gk(x_d - x_s) \quad (\dot{x} = gF_d) \qquad (5.4.13)$$

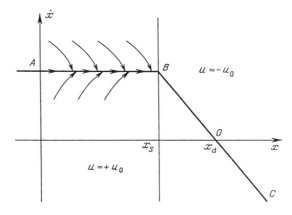

FIGURE 5.4.1
Switching control phase trajectories out of contact with the object

The equation (5.4.13) defines the segment AB (see Figure 5.4.1) of the switching line (5.4.5). It means that this segment is a trajectory of the system (5.1.10), (5.1.11), (5.4.4).

By substituting the switching control (5.4.4) into (5.1.11) (lower line), we obtain an expression for the manipulator motion in contact with the object, for $x \geq x_s$:

$$\mathcal{M}\ddot{x} + d_2\dot{x} + k(x - x_s) = d_1u_0 \operatorname{sgn}\left[gk(x_d - x) - \dot{x}\right] + F_e \quad (5.4.14)$$

Among the motions defined by (5.4.14), there is a *sliding* motion [267, 62, 63, 67] along the switching line

$$\dot{x} + gk(x - x_d) = 0 \quad (5.4.15)$$

To find control in the sliding mode (equivalent control [267, 62, 63, 67]), let us differentiate the function S from (5.4.5) with respect to time along the trajectories of (5.4.11)

$$\frac{dS}{dt} = \ddot{x} + gk\dot{x} = \frac{1}{\mathcal{M}}[F_e + k(x_s - x) - d_2\dot{x} + d_1u] + gk\dot{x}$$

$$= \frac{1}{\mathcal{M}}[F_e + k(x_s - x) + d_1u] + \left(gk - \frac{d_2}{\mathcal{M}}\right)\dot{x} \quad (5.4.16)$$

On the switching line, i.e., for $S = 0$

$$\frac{dS}{dt} = \frac{1}{\mathcal{M}}[F_e + k(x_s - x) + d_1u] + \left(gk - \frac{d_2}{\mathcal{M}}\right)gk(x_d - x) \quad (5.4.17)$$

For the sliding motion, $S = 0$, and, consequently, $\frac{dS}{dt} \equiv 0$. Therefore, we obtain the control formula

$$u = \frac{1}{d_1}\left[(\mathcal{M}gk - d_2)gk(x - x_d) + k(x - x_s) - F_e\right]$$

$$= \frac{1}{d_1}\left[(\mathcal{M}(gk)^2 - d_2gk + k)x + gk(d_2 - \mathcal{M}gk)x_d - kx_s - F_e\right] \quad (5.4.18)$$

Clearly, one does not need to compute (5.4.18) in order to implement switching control in practice. It will suffice to switch when crossing the line (5.4.15). We need (5.4.18) for the analytical study, while searching for a sliding mode segment on the switching line (5.4.15).

To simplify the expressions below, let us introduce nondimensional parameters as in (5.1.13), (5.1.17)

$$\chi = -\frac{k}{d_1 u_0}x, \quad \chi_s = \frac{k}{d_1 u_0}x_s, \quad \chi_d = \frac{k}{d_1 u_0}x_d, \quad w = \frac{u}{u_0},$$

$$f_e = -\frac{F_e}{d_1 u_0}, \quad \alpha = \frac{d_2}{\sqrt{u_0}}, \quad \lambda = g\sqrt{Mk} \qquad (5.4.19)$$

Using the nondimensional parameters (5.4.19), the expression (5.4.18) can be written in the form

$$w = (\lambda^2 - \alpha\lambda + 1)\chi + \lambda(\alpha - \lambda)\chi_d - \chi_s - f_e \qquad (5.4.20)$$

The segment of the straight line (5.4.15), where the equality (5.4.20) holds, is defined by the constraint (5.1.16). From (5.4.20) and (5.1.16), we obtain

$$|(\lambda^2 - \alpha\lambda + 1)\chi + \lambda(\alpha - \lambda)\chi_d - \chi_s - f_e| \leq 1 \qquad (5.4.21)$$

The sliding mode exists precisely in the segment (5.4.15), (5.4.21). For the nondimensional force feedback coefficient λ such that

$$\lambda^2 - \alpha\lambda + 1 > 0, \qquad (5.4.22)$$

the sliding mode segment is defined by the inequalities

$$\frac{\lambda(\lambda - \alpha)\chi_d + \chi_s + f_e - 1}{\lambda^2 - \alpha\lambda + 1} \leq \chi \leq \frac{\lambda(\lambda - \alpha)\chi_d + \chi_s + f_e + 1}{\lambda^2 - \alpha\lambda + 1} \qquad (5.4.23)$$

If

$$\lambda^2 - \alpha\lambda + 1 < 0, \qquad (5.4.24)$$

then the signs in the inequalities should be changed for the opposite:

$$\frac{\lambda(\lambda - \alpha)\chi_d + \chi_s + f_e - 1}{\lambda^2 - \alpha\lambda + 1} \geq \chi \geq \frac{\lambda(\lambda - \alpha)\chi_d + \chi_s + f_e + 1}{\lambda^2 - \alpha\lambda + 1} \qquad (5.4.25)$$

If

$$\lambda^2 - \alpha\lambda + 1 = 0, \qquad (5.4.26)$$

and

$$|\lambda(\lambda - \alpha)\chi_d + \chi_s + f_e| \leq 1, \tag{5.4.27}$$

then the sliding mode exists along the whole switching line (5.4.15). Note that by using (5.4.26), we can rewrite (5.4.27) in the form

$$|\chi_s - \chi_d + f_e| \leq 1 \quad (|F_e - F_d| \leq d_1 u_0) \tag{5.4.28}$$

· If damping in the motor is not too large, and $\alpha < 2$, then the inequality (5.4.22) holds for all λ. For $\alpha \geq 2$, (5.4.26) has two real positive roots λ_1, λ_2 ($\lambda_1 \leq \lambda_2$). Then (5.4.22) holds for $\lambda < \lambda_1$ or $\lambda > \lambda_2$, while (5.4.24) holds for $\lambda_1 < \lambda < \lambda_2$. In case the feedback coefficient $\lambda = \lambda_1$ or $\lambda = \lambda_2$ and the condition (5.4.28) is satisfied, then the sliding mode occurs along the whole switching line.

For $F_e = \text{const}$, the sliding mode segment does not change with time. For $F_e = F_e(t)$, the sliding mode segment changes with time. If F_e is the Coulomb friction force, the sliding mode segment does not change with time.

The point $O(x_d, 0)$ belongs to the sliding mode segment (5.4.23) or (5.4.25) if and only if the condition (5.4.28) is satisfied. Given the strict inequality

$$|F_e - F_d| < d_1 u_0 \quad (|\chi_s - \chi_d + f_e| < 1) \tag{5.4.29}$$

the point $O(x_d, 0)$ lies inside the sliding mode segment.

If the point $O(x_d, 0)$ belongs to the sliding mode segment or line, then it is an equilibrium point, i.e., the system (5.1.11), (5.4.4), (5.4.18) has the solution

$$x = x_d \quad (F - F_d), \quad \dot{x} = 0 \tag{5.4.30}$$

The solution (5.4.30) is asymptotically stable if and only if the following two requirements are satisfied. The first requirement is that the sliding mode segment be an *attractor set*. It means that the inequality $\frac{dS}{dt} < 0$ holds in the vicinity of the sliding mode segment for $S > 0$, while $\frac{dS}{dt} > 0$ holds for $S < 0$. This requirement is satisfied, if, for the points $(x, \dot{x})$ inside the sliding mode segment, $\frac{dS}{dt} < 0$ for $u = -u_0$, and $\frac{dS}{dt} > 0$ for $u = u_0$. It is easy to verify that these inequalities hold for (5.4.14), and the phase trajectories "hit" the sliding mode segment DE (see Figure 5.4.2).

On reaching the segment DE, the phase point moves along it for a finite or infinite period of time. The second requirement is that the phase point in the sliding mode should asymptotically approach the point O from

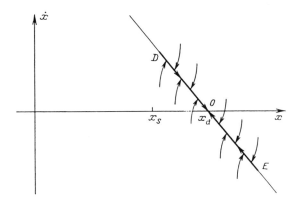

FIGURE 5.4.2
Phase trajectories in the vicinity of the sliding mode segment

(5.4.30), and, therefore, the phase point should belong to the segment DE indefinitely. Since the sliding mode is given by (5.4.15), the second require-ment is equivalent to the inequality $g > 0$ $(\lambda > 0)$. The solution of (5.4.15) has the form

$$x(t) = x_d + [x(0) - x_d]e^{-gt}$$

It follows from the inequalities (5.4.23), (5.4.25), and (5.4.28) that for $\lambda = 0$ $(g = 0), \lambda = \lambda_1$, or $\lambda = \lambda_2$, as well as for any λ approaching zero, λ_1, or λ_2, the abscissa of the left end of the sliding mode segment is less than x_s. It follows from the inequality (5.4.23) that for $\lambda \to \infty$, the sliding mode segment DE contracts into the point O. Hence for a sufficiently large λ, the entire segment DE lies to the left of the straight line $x = x_s$. If the abscissa of the left end of the sliding mode segment is less than x_s, then the entire part ABO of the switching line is the phase trajectory of the system (5.1.10), (5.1.11), (5.4.4). Along this trajectory, the phase point in the sliding mode comes to the equilibrium for $t \to \infty$. The trajectory ABO is an attractor of the system. Figure 5.4.3 presents a phase portrait of the system (5.1.10), (5.1.11), (5.4.4) for this case.

As can be seen from Figure 5.4.3, the phase point gets on the broken line ABE (shown in bold) from any initial state in a finite period of time. Then, along this line, the phase point asymptotically approaches the equilibrium O in the sliding mode. Figure 5.4.4 illustrates the case, where the sliding mode segment DE lies to the right of the straight line $x = x_s$. In this case, the phase point asymptotically comes to the equilibrium O from any initial

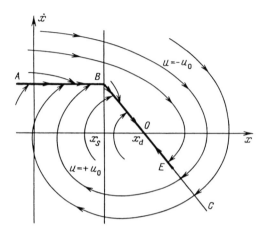

FIGURE 5.4.3
Phase portrait of the system for small values of λ

state, as well. Thus, the equilibrium state (5.4.30) is globally stable in both cases.

Transient processes settle fast in the sliding mode with large values of g. However, the sliding mode segment is small in this case, and the system makes many oscillations before it enters the sliding mode. For small g, system oscillations are small, but transient processes in the sliding mode last longer. If $\alpha > 2$, then it seems best to choose g close to $\lambda_2/\sqrt{\mathcal{M}k}$, where λ_2 is the largest root of (5.4.26).

Under the switching control (5.4.1), the equilibrium state (5.4.30), with zero deviation of the force F from its desired value F_d, is independent of perturbations, such as the system deviation from nominal parameters or external forces. The trajectories ABO, EO do not depend on perturbations, either. Therefore, switching control makes the system invariant in a certain sense, as discussed above.

The analysis performed above shows that both linear control with large gains and purely switching control resolve the problem of keeping contact with a stationary object. The contact is kept without a tracking error for the commanded contact force, even in the presence of perturbations of motion. The latter is an advantage of the switching control as compared to linear one. Experiments prove the efficiency of the switching control. Some experimental results are presented in the following section.

For $g_2 = 0$, (5.4.2) has the form

$$u = u_0 \ \mathrm{sgn} \ (F_d - F) \qquad (5.4.31)$$

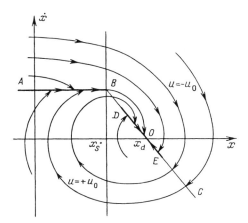

FIGURE 5.4.4
Phase portrait of the system for large values of λ

In the experiments, sustained oscillations appear when the manipulator contacts the stationary object with the control (5.4.31). It follows from the theoretical results of Sections 5.7 and 5.8, obtained taking into account the elasticity in the manipulator structure, that the system is unstable under the control (5.4.31) (see (5.7.16), (5.7.30)). However, with the control (5.4.31), the manipulator is most "responsive". It can be felt especially well for $F_d = 0$. If a finger touches the sensor (for $F_d = 0$), the manipulator arm immediately responds by moving in the direction of the finger pressure. It is not so easy to start and keep moving the manipulator arm in the absence of force feedback. A significant force should be applied to the arm in order to move it, because of the friction in the drive. Force control produces the same result as a reduction of friction in the drive. Thus, force feedback can effectively "remove" friction in the gear train of the drive and make it "back-drivable".

5.5 Contact Transition Control

Linear and switching control laws considered in the preceding sections solve the problem of keeping the manipulator in contact with a stationary object. These laws also solve the problem of contact transition where the manipulator does not contact the object at the beginning of the motion, but the direction of motion to the object is given. This is a consequence of the global stability of the system. However, the velocity of the manipulator approaching the object with the control (5.4.3) or (5.4.10), where v_d is given by (5.3.13), may be too small. According to (5.3.13), if $F = 0$ (for $x \leq x_s$), then $v_d = gF_d$. To increase the operation speed, we have to set $v_d = v_0 = \text{const}$, where $v_0 > gF_d$, in the beginning of the motion, and then calculate the velocity v_d by (5.3.13) after the manipulator first touches the object. In so doing, we obtain a significant gain in motion time, if the manipulator is far from the object in the beginning of the motion.

The relationship

$$v_d = \begin{cases} v_1 & \text{if } F \leq F_d - \Delta F_1, \\ g(F_d - F) & \text{if } F_d - \Delta F_1 \leq F \leq F_d + \Delta F_2, \\ -v_2 & \text{if } F_d + \Delta F_2 \leq F \end{cases} \qquad (5.5.1)$$

where $\Delta F_1, \Delta F_2, v_1, v_2$ are constants, is more general than (5.3.13). For $\Delta F_1 > F_d$ and $\Delta F_2 = \infty$, the velocities v_d, calculated by the formulae (5.3.13) and (5.5.1), coincide. In the experiments, we used the relationship (5.5.1), while the values $\Delta F_1, \Delta F_2, v_1, v_2$ were experimentally chosen to improve the transient process. The values $\Delta F_1, \Delta F_2$ define the range, where the commanded velocity v_d linearly depends on force tracking error. The increase of v_1, v_2 in a certain range contributes to the operation speed for the system.

Figure 5.5.1 displays a plot $F(t)$ logged for the control laws (5.4.10), (5.5.1) applied in the experiment with the robotic system of Section 4. In the experiment, the force sensor is built into the gripper finger that touches the object. The gain g_2 is chosen to be large, at $g_2 \approx 7 \cdot 10^3$ V·s/mm. The other parameter values in the control law are as follows: $F_d = 1$ N, $\Delta F_1 = \Delta F_2 = 0.5$ N, $v_1 = v_2 = 3$ mm/s, $g = 1$ mm/(N·s), $v_0 = 50$ mm/s.

In the beginning of the motion, the contact force F is close to zero (see Figure 5.5.1), because the finger does not contact the object at that time ($F(t) \not\equiv 0$ due to a noise in the force measurement channel). On establishing contact, the force F sharply grows and approaches its maximum of about 3.5 N, then abruptly falls and, after some oscillations, settles close to F_d. The force F does not drop to zero in the transient process, therefore, the manipulator arm does not lose the contact with the object, once it has been established. The considerable oscillations observed in the transient process may be attributed to the elastic structure of the manipulator, which is not taken into account in the above mathematical model. The elasticity of the manipulator structure is considered in the models of Sections 5.7, 5.8, and Chapter 8. Though the voltage applied to the motor is reversed at the moment the arm contacts the object, the force F abruptly increases (a force splash occurs) due to the insufficient operation speed of the drive. If the maximum contact force F exceeds the threshold force F_θ of the object

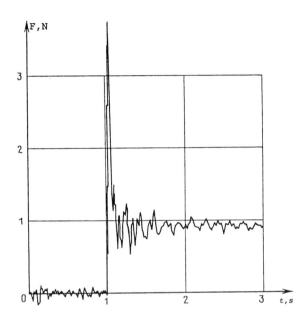

FIGURE 5.5.1
The time-history of force in the contact transition experiment

friction against the surface where it is placed, then the object is moved out of its position (see the inequality (5.2.2)). In this case, the velocity v_0 has to be reduced, leading to a decrease in operation speed.

5.6 Influence of Delay in Feedback Loop on the Stability of Contact

The experiments prove that both linear (5.3.1) and switching (5.4.1) control laws solve the problem of keeping the manipulator in contact with a stationary object. However, the experiments also show that sustained oscillations arise in the system for a large force feedback gain. Such oscillations cannot be explained by the second order mathematical model developed in Sections 5.1 and 5.3. These oscillations may be related to a delay in the feedback loop, compliance and free play in the manipulator structure, and other causes, which this mathematical model does not take into account.

In this section, the influence of delay in the feedback loop on the stability of keeping the manipulator in contact with a stationary object is studied. We consider the influence of compliance and backlash in the manipulator structure in the two following Sections 5.7 and 5.8.

One possible way of implementing the control (5.3.1) is to use a digital computer, which receives measurements of the force F and velocity $\dot{x}$ from the force sensor and tachometer. The time required for the computer to compute the control signal u, causes this signal to get to the manipulator drive with a delay. Besides, the control signal produced by the computer is a piecewise-constant (zero-order hold) function of time. Below, we present some relationships, which only allow evaluation the influence of the first cause, i.e., pure delay. The motion of the manipulator with a digital computer in the feedback loop, taking into account the discrete-time character of control, is considered further in Chapter 8. Note that the issue of stability for general form linear systems with a computer in the feedback loop is studied, for example, in [144].

By taking a pure delay of the control signal computation into account, we can write (5.3.1) in the form

$$u(t) = -g_1[F(t - \theta) - F_d] - g_2\dot{x}(t - \theta), \qquad (5.6.1)$$

where $\theta > 0$ is the constant delay time.

By substituting (5.6.1) for (5.3.1) into (5.1.11), we obtain, instead of (5.3.3), the expression

$$\mathcal{M}\ddot{x}(t) + d_2\dot{x}(t) + k[x(t) - x_s] + d_1g_2\dot{x}(t - \theta)$$

$$+ d_1g_1k[x(t - \theta) - x_d] = F_e \qquad (5.6.2)$$

The characteristic equation corresponding to the differential equation (5.6.2) has the form

$$\mathcal{M}p^2 + d_2p + k + d_1(g_2p + g_1k)e^{-p\theta} = 0 \qquad (5.6.3)$$

or

$$1 + W(p)e^{-p\theta} = 0,$$

where $W(p)$ is the transfer function

$$W(p) = \frac{d_1(g_2p + g_1k)}{\mathcal{M}p^2 + d_2p + k}$$

For the asymptotic stability conditions (5.3.8), the hodograph of the vector $W(i\omega)$ does not encircle the point $(-1, 0)$. However, if there exists

ω such that $|W(i\omega)| > 1$, then for some θ (5.6.3) has roots in the right half-plane [181, 217]. For example, under the condition

$$d_1 g_1 > 1 \quad (g_1 c_1 i > r)$$

a delay θ exists such that the system is unstable. By plotting the hodograph of the vector $W(i\omega)$, one can find the values of θ for which the system becomes unstable. To evaluate these values approximately, assume that θ is small and substitute $\frac{1}{\theta p + 1}$ for $e^{-p\theta}$. Then we can replace the characteristic equation (5.6.2) by the approximation

$$\mathcal{M}\theta p^3 + (\mathcal{M} + d_2\theta)p^2 + (d_2 + d_1 g_2 + k\theta)p + k(1 + d_1 g_1) = 0 \quad (5.6.4)$$

The problem of replacing a characteristic quasipolynomial by an ordinary polynomial for feedback systems is considered, for instance, in [145].

If the conditions (5.3.8) hold, then the only stability condition for (5.6.4) is the Hurwitz inequality

$$(\mathcal{M} + d_2\theta)(d_2 + d_1 g_2 + k\theta) - \mathcal{M}\theta k(1 + d_1 g_1) > 0, \qquad (5.6.5)$$

which we can write in the form

$$d_2 + d_1(g_2 - g_1 k\theta) + \frac{d_2\theta}{\mathcal{M}}(d_2 + d_1 g_2 + k\theta) > 0, \qquad (5.6.6)$$

If either the force feedback gain g_1, or sensor stiffness k increases, the inequality (5.6.5) and with it, the system stability, may be violated. In the experiments, it is large g and k that lead to sustained oscillations. If θ is small and g_1, g_2 are large, then the inequality (5.6.6) may be replaced by the approximation

$$g_1 k\theta < g_2 \qquad (5.6.7)$$

In any case, if the condition (5.6.7) is satisfied, then (5.6.6) holds as well.

Above, we have considered one possible way to implement the control (5.3.1) by using only a digital computer. The other approach, which employs both a digital computer and velocity servo-system hardware, is presented in Section 5.3 (see (5.3.12), (5.3.13)). With this approach, the signal from the tachometer "bypasses" the computer. The delay in the analog servosystem takes microseconds, while the calculation time θ takes milliseconds for the digital computer. By neglecting the delay in the servosystem, we obtain the following control formula (instead of (5.6.1)):

$$u(t) = -g_1[F(t-\theta) - F_d] - g_2\dot{x}(t) \tag{5.6.8}$$

By substituting (5.6.8) into (5.1.11), we obtain

$$\mathcal{M}\ddot{x}(t) + (d_2 + d_1g_2)\dot{x}(t) + k[x(t) - x_s]$$

$$+ d_1g_1k[x(t-\theta) - x_d] = F_e \tag{5.6.9}$$

The characteristic equation corresponding to (5.6.9) has the form

$$\mathcal{M}p^2 + (d_2 + d_1g_2)p + k + d_1g_1ke^{-p\theta} = 0 \tag{5.6.10}$$

We assume that θ is small and substitute $\frac{1}{\theta p + 1}$ for $e^{-p\theta}$. Then, instead of the exact characteristic equation (5.6.10), we obtain an approximate equation

$$\mathcal{M}\theta p^3 + [\mathcal{M} + (d_2 + d_1g_2)\theta]p^2 + (d_2 + d_1g_2 + k\theta)p$$

$$+ k(1 + d_1g_1) = 0 \tag{5.6.11}$$

If the conditions (5.3.8) hold, then the only stability condition for (5.6.11) is the inequality

$$[\mathcal{M} + (d_2 + d_1g_2)\theta](d_2 + d_1g_2 + k\theta) - \mathcal{M}\theta k(1 + d_1g_1) > 0 \tag{5.6.12}$$

By comparing the conditions (5.6.5) and (5.6.12), we conclude that if the inequality (5.6.5) holds, then (5.6.12) also holds. Therefore, the condition (5.6.12) is less strict than (5.6.5). It follows that if the signal $\dot{x}$ received from the tachometer "bypasses" the computer, then the system is "more" stable.

The inequality (5.6.12) can be written in the form

$$d_2 + d_1(g_2 - g_1k\theta) + \frac{\theta}{\mathcal{M}}(d_2 + d_1g_2)(d_2 + d_1g_2 + k\theta) > 0 \tag{5.6.13}$$

The inequality (5.6.13) holds under the condition (5.6.7).

5.7 Influence of Transmission Compliance on the Stability of Contact

In the experiments, sustained oscillations arise when the force feedback gain is large. In this and the following sections, we shall demonstrate that these oscillations may be attributed to compliance in the manipulator structure, as well as to the feedback loop delay. The present section studies the influence of compliance in the drive transmission (gear train) on the stability of contact keeping. We defer the consideration of a joint influence of delay and structure compliance to Chapter 8.

Here and further in Section 5.8, we adopt an approximate model of *lumped elastic compliance with damping* for the manipulator. As in Section 5.1, we study a one-dimensional motion of the manipulator (see Figure 5.1.1). Unlike Section 5.1, we assume that the motor rotor, gear train, and manipulator arm form a system with *two degrees of freedom* instead of one. The second, unactuated degree of freedom corresponds to a lumped elastic compliance of the structure "between" the motor and the arm, such as a compliance in the gear train. We shall model the structure compliance as a spring of stiffness k_c and a dashpot of damping β_c. Figure 5.7.1 illustrates the schematics of the two-degree-of-freedom system in hand. This model only differs from the system shown in Figure 5.1.1 in that it has a lumped visco-elastic element representing compliance in the structure.

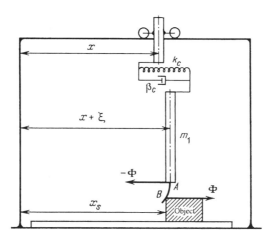

FIGURE 5.7.1
Two-degree-of-freedom model for a manipulator with lumped structure compliance

The model includes the manipulator moving parts "before the compliance". These parts have the mass $\mathcal{M}$, which includes the equivalent mass of the rotating parts, namely, the motor rotor and gear train pinions (see Section 5.1). Similar to Section 5.1, we denote the displacement of these moving parts by x. We denote the manipulator arm mass "after the compliance" by m_1, and its horizontal displacement relative to the mass $\mathcal{M}$, by ξ. Hence, the absolute displacement of the mass m_1 is $x + \xi$. As mentioned above, the force F_a acting on the mass $\mathcal{M}$ from m_1 is given by the formula

$$F_a = k_c \xi + \beta_c \dot{\xi} \tag{5.7.1}$$

The equation of motion for $\mathcal{M}$, derived in view of (5.1.8), (5.7.1), has the form

$$\mathcal{M}\ddot{x} + d_2\dot{x} - \beta_c\dot{\xi} - k_c\xi = d_1 u \tag{5.7.2}$$

Same as in Section 5.1, the force sensor is assumed to be an infinitely thin massless elastic plate. The force Φ acting on the object from the sensor is given by the formula (see 1.2.3)

$$\Phi = \begin{cases} 0 & \text{if } x + \xi \leq x_s, \\ k(x + \xi - x_s) + \beta(\dot{x} + \dot{\xi}) & \text{if } x + \xi \geq x_s \end{cases} \tag{5.7.3}$$

In (5.7.3), same as in (5.1.9), x_s is the coordinate of the front edge of the object, and k is the sensor stiffness. The positive damping constant β determines the sensor energy dissipation. The term for the sensor energy dissipation is missing in (5.1.9). As mentioned in Section 5.1, by adding this term, we only affect the value of d_2 in (5.1.10), (5.1.11) and leave the form of these equations unchanged. The form of the equations considered in this section changes for $\beta \neq 0$.

The equation of motion with respect to m_1 has the form

$$m_1(\ddot{x} + \ddot{\xi}) + \beta_c\dot{\xi} + k_c\xi + \Phi = 0 \tag{5.7.4}$$

For the linear (5.3.1) or switching (5.4.1) control law, F is determined by the force sensor output, which is proportional to its strain. We assume, in the same manner as in Section 5.1, that the sensor measures the force

$$F = \begin{cases} 0 & \text{if } x + \xi \leq x_s, \\ k(x + \xi - x_s) & \text{if } x + \xi \geq x_s \end{cases} \tag{5.7.5}$$

If the manipulator is in contact with the object, then the equations of motion (5.7.2)–(5.7.4) with the linear control (5.3.1) can be written as

$$\mathcal{M}\ddot{x} + d_2\dot{x} - \beta_c\dot{\xi} - k_c\xi + d_1g_1k(x + \xi) + d_1g_2\dot{x}$$

$$= d_1g_1(kx_s + F_d), \qquad (5.7.6)$$

$$m_1(\ddot{x} + \ddot{\xi}) + \beta_c\dot{\xi} + k_c\xi + k(x + \xi) + \beta(\dot{x} + \dot{\xi}) = kx_s$$

The steady state of the system (5.7.6) is given by the expressions

$$x = x_s + \frac{d_1g_1}{1 + d_1g_1}F_d\left(\frac{1}{k} + \frac{1}{k_c}\right), \qquad (5.7.7)$$

$$\xi = -\frac{1}{k_c}\frac{d_1g_1}{1 + d_1g_1}F_d \qquad (5.7.8)$$

In the stationary regime, the contact force F applied by the arm to the object and the force F_a developed in the damping elastic element are equal in magnitude:

$$F = \frac{d_1g_1}{1 + d_1g_1}F_d, \qquad F_a = k_c\xi = -F \qquad (5.7.9)$$

Let us rewrite the equations (5.7.6) in the variations from the steady-state values (5.7.7), (5.7.8), denoting the variations by x and ξ:

$$\mathcal{M}\ddot{x} + (d_2 + d_1g_2)\dot{x} + d_1g_1kx - \beta_c\dot{\xi} + (d_1g_1k - k_c)\xi = 0,$$

$$m_1\ddot{x} + \beta\dot{x} + kx + m_1\ddot{\xi} + (\beta_c + \beta)\dot{\xi} + (k_c + k)\xi = 0 \qquad (5.7.10)$$

The characteristic equation of the system (5.7.10) has the form

$$a_0p^4 + a_1p^3 + a_2p^2 + a_3p + a_4 = 0, \qquad (5.7.11)$$

where

$$a_0 = m_1\mathcal{M}, \qquad a_1 = m_1(d_2 + d_1g_2 + \beta_c) + \mathcal{M}(\beta_c + \beta),$$

$$a_2 = m_1k_c + \mathcal{M}(k_c + k) + (\beta_c + \beta)(d_2 + d_1g_2), \qquad (5.7.12)$$

$$a_3 = \beta k_c + \beta_c k(1 + d_1 g_1) + (k_c + k)(d_2 + d_1 g_2),$$

$$a_4 = kk_c(1 + d_1 g_1)$$

For $g_1 = g_2 = 0$, the system (5.7.10) is asymptotically stable, as a natural system with a full dissipation (the full mechanical energy of the system is its Lyapunov function). For $g_1 = 0$, $g_2 > 0$, the system (5.7.10) also has a full dissipation, and, consequently, it is asymptotically stable. Let us study the system stability for $g_1 > 0$.

In the asymptotic stability domain, the leading Hurwitz determinant is $a_1 a_2 a_3 - a_1^2 a_4 - a_0 a_3^2 > 0$ and the free term $a_4 > 0$ [135]. The stability boundary is determined by the equalities

$$a_1 a_2 a_3 - a_1^2 a_4 - a_0 a_3^2 = 0 \quad \text{and} \quad a_4 = 0 \qquad (5.7.13)$$

Since always $a_i > 0$ $(i = 0, \ldots, 4)$, the condition for the positive leading Hurwitz determinant can be written as

$$a_2 > \frac{a_1 a_4}{a_3} + \frac{a_0 a_3}{a_1} \qquad (5.7.14)$$

By substituting (5.7.12) into the inequality (5.7.14), we obtain

$$m_1 k_c + \mathcal{M}(k_c + k) + (\beta_c + \beta)(d_2 + d_1 g_2)$$

$$> \frac{kk_c(1 + d_1 g_1)[m_1(d_2 + d_1 g_2 + \beta_c) + \mathcal{M}(\beta_c + \beta)]}{\beta k_c + \beta_c k(1 + d_1 g_1) + (k_c + k)(d_2 + d_1 g_2)}$$

$$+ \frac{m_1 \mathcal{M}[\beta k_c + \beta_c k(1 + d_1 g_1) + (k_c + k)(d_2 + d_1 g_2)]}{m_1(d_2 + d_1 g_2 + \beta_c) + \mathcal{M}(\beta_c + \beta)} \qquad (5.7.15)$$

The left-hand part of (5.7.15) does not depend on the force feedback gain g_1. Its right-hand part is a strictly monotonically increasing unbounded function g_1. Since the system is asymptotically stable ((5.7.15) is satisfied) for $g_1 = 0$, the first equality in (5.7.13) holds for a certain $g_1^0 > 0$. Therefore, the steady state (5.7.7), (5.7.8) is asymptotically stable for

$$0 \leq g_1 < g_1^0, \qquad (5.7.16)$$

and unstable for $g_1 > g_1^0$. The first equation in (5.7.13) that defines the *critical* value g_1^0 for g_1 is quadratic. The expression for g_1^0 is rather cumbersome, so we do not present it here. In particular, g_1^0 depends on g_2. If the gains g_1, g_2 are large and have the same order of magnitude, then one can derive (5.7.27) and (5.7.30) for the critical gain g_1 from the inequality (5.7.15). Therefore, if the gains in the linear control law (5.3.1) are large, then we can use (5.7.30) to obtain an approximate evaluation of the stability range.

Thus, it follows from (5.7.16) that the process of keeping the manipulator in contact with the object with linear control (5.3.1) is asymptotically stable only for a range of force feedback gains that is *bounded from above*. This conclusion agrees with the experimental results.

Let us now investigate the issue of asymptotic stability for *switching control* (5.4.1) or (5.4.2), which is the same. For $x + \xi \geq x_s$, the equations of motion (5.7.2), (5.7.4) with the switching control (5.4.1) have the form

$$\mathcal{M}\ddot{x} + d_2\dot{x} - \beta_c\dot{\xi} - k_c\xi = -d_1 u_0 \, \text{sgn} \, \{g_1[k(x + \xi - x_s) - F_d] + g_2\dot{x}\},$$

$$m_1(\ddot{x} + \ddot{\xi}) + \beta_c\dot{\xi} + k_c\xi + k(x + \xi) + \beta(\dot{x} + \dot{\xi}) = kx_s \qquad (5.7.17)$$

In the stationary regime

$$x = x_s + \frac{F_d}{k} + \frac{F_d}{k_c}, \quad \xi = -\frac{F_d}{k_c} \quad (F = F_d) \qquad (5.7.18)$$

For the stationary regime (5.7.18) to exist and be inside the sliding mode region, it is necessary and sufficient that

$$F_d < d_1 u_0 \qquad (5.7.19)$$

Let us rewrite the equations (5.7.17) in the variations from the steady-state values (5.7.18), still denoting the variations by x and ξ:

$$\mathcal{M}\ddot{x} + d_2\dot{x} - \beta_c\dot{\xi} - k_c\xi = -d_1 u_0 \, \text{sgn} \, [g_1 k(x + \xi) + g_2\dot{x}],$$

$$m_1(\ddot{x} + \ddot{\xi}) + \beta_c\dot{\xi} + k_c\xi + k(x + \xi) + \beta(\dot{x} + \dot{\xi}) = 0 \qquad (5.7.20)$$

The switching of the control takes place on the hyperplane in the four-dimensional phase space

$$S = g_1 k(x + \xi) + g_2\dot{x} = 0 \qquad (5.7.21)$$

In the hyperplane (5.7.21), trajectories of the sliding mode motions of the system (5.7.20) fill a region, which is easy to define [67, 71]. This region, under the condition (5.7.19), contains the coordinate origin that represents a steady state (5.7.18) in the new coordinates. One of the asymptotic stability conditions for this steady state is that the phase trajectories should *converge* to the switching hyperplane (5.7.21) in its vicinity [67]. This condition is satisfied, if the full derivative $\frac{dS}{dt}$ computed along the trajectories of the system (5.7.20) is positive in the coordinate origin for $S = -0$, and negative for $S = +0$. In this case, the whole region of sliding motions is an *attractor set*. It is easy to demonstrate [67] that the above-mentioned first condition for the asymptotic stability of (5.7.20) has the form

$$g_2 > 0 \tag{5.7.22}$$

The second asymptotic stability condition for the steady state is that it is asymptotically stable in the *sliding motion* [67]. To obtain the equations for the sliding motion [67, 71, 267], it suffices to substitute (5.7.21) for the first equation in (5.7.20)

$$g_2 \dot{x} + g_1 k x + g_1 k \xi = 0,$$

$$m_1 \ddot{x} + \beta \dot{x} + k x + m_1 \ddot{\xi} + (\beta_c + \beta)\dot{\xi} + (k_c + k)\xi = 0 \tag{5.7.23}$$

The characteristic equation of the system (5.7.23) has the form

$$b_0 p^3 + b_1 p^2 + b_2 p + b_3 = 0, \tag{5.7.24}$$

where

$$b_0 = m_1 g_2, \quad b_1 = g_2(\beta_c + \beta),$$

$$b_2 = g_1 k \beta_c + g_2(k_c + k), \quad b_3 = g_1 k k_c \tag{5.7.25}$$

Since $b_i > 0 \ (i = 0, \dots, 3)$, the only condition for the asymptotic stability of (5.7.24) is the inequality

$$b_1 b_2 - b_0 b_3 > 0 \tag{5.7.26}$$

By substituting (5.7.25) into the inequality (5.7.26) and using (5.7.22), we obtain

$$g_1 k[\beta_c(\beta_c + \beta) - m_1 k_c] + g_2(\beta_c + \beta)(k_c + k) > 0 \tag{5.7.27}$$

It follows from (5.7.27) that if

$$\beta_c(\beta_c + \beta) > m_1 k_c, \tag{5.7.28}$$

then the system (5.7.23) is asymptotically stable for all $g_1 \geq 0$. For

$$\beta_c(\beta_c + \beta) < m_1 k_c, \tag{5.7.29}$$

the system (5.7.23) is asymptotically stable if the condition (5.7.16) holds, where

$$g_1^0 = g^0 g_2 = \frac{(\beta_c + \beta)(k_c + k)}{k[m_1 k_c - \beta_c(\beta_c + \beta)]} g_2 \tag{5.7.30}$$

The steady state (5.7.18) of the system (5.7.17) is asymptotically stable under the condition (5.7.16), (5.7.22). It means that $g_1^0 = \infty$ in the case of (5.7.28),and g_1^0 is defined by (5.7.30) in the case of (5.7.29).

Thus, given (5.7.29), the range of the force feedback gains for which the process of keeping the manipulator in contact with the object is asymptotically stable is bounded from above for the switching control (5.4.1), in the same manner as for the linear control (5.3.1). The critical value of the force feedback gain g_1^0 decreases with the growth of sensor stiffness k. In other words, the system is more "stable" with a "soft" sensor than with a rigid one. A similar assertion is made in [213, 291].

Note that one can obtain (5.7.27), and, consequently, (5.7.28)–(5.7.30), from (5.7.15) for $g_2 \to \infty$. When passing to the limit, one has to keep in mind that for the switching control (5.4.1), the stability is determined by the ratio g_1/g_2.

The results derived in the present section may be used if the manipulator structure compliance has more generic rheological roperties. The compliance can be ascribed to various elastic elements with dissipation such as a backlash with Coulomb friction, etc. If ξ is the variation in such an element, then, given sustained oscillations, the characteristic $F_a(\xi, \dot{\xi})$ of the element can be linearized by replacing it with the function of the form (5.7.1), for instance, using the harmonic balance method [213]. Then the above-obtained results can be applied to the critical value of g_1^0 for which sustained oscillations are possible.

In [78], the influence of compliance on the manipulator motion out of contact with an object is also considered. The paper studies a wrist force sensor and takes into account the mass of the gripper. The stationary regime in question is a uniform motion. The characteristic equation corresponding to this stationary regime has the fifth order with linear feedback, and the fourth order with switching feedback. These characteristic equations are analytically studied in [78] using approximate methods.

5.8   Influence of Manipulator Base Compliance on the Stability of Contact

In this section, we continue our study of the influence of the manipulator structure compliance on the stability of contact keeping.

Unlike the preceding section, let us assume that the lumped compliance of the manipulator structure is concentrated in its base structure, or frame. We consider the motor rotor, gear train, and manipulator arm to be rigidly connected, as in Section 5.1. We denote the mass of the motor, gear train with pinions, and arm by m.

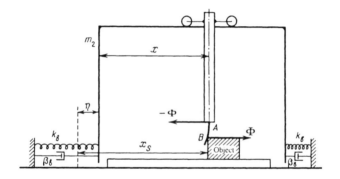

FIGURE 5.8.1
Manipulator with base (frame) compliance

We shall model the frame compliance by a spring with stiffness k_b and a dashpot with damping β_c. Figure 5.8.1 displays the schematics of the two-degree-of-freedom system in hand. This model only differs from the system shown in Figure 5.1.1 in that it has an elastic element with damping, which connects the frame of the mass m_2 with the fixed base. We denote the displacement of the base structure relative to base by η. As before, we denote the displacement of the manipulator arm relative to the frame by x.

As mentioned above, the force F_b acting from the fixed base on the base structure is given by the formula

$$F_b = -k_b\eta - \beta_b\dot{\eta} \qquad (5.8.1)$$

The force sensor exerts force, in one direction, on the object, and in the other (reverse) direction, on the manipulator arm. The force Φ acting on the object from the sensor is given by analogy with (5.7.3) as

$$\Phi = \begin{cases} 0 & \text{if } \eta + x \le x_s, \\ k(\eta + x - x_s) + \beta(\dot{\eta} + \dot{x}) & \text{if } \eta + x \ge x_s \end{cases} \qquad (5.8.2)$$

In this formula, k and β are, respectively, the sensor stiffness and damping, and x_s is the coordinate of the front edge of the object relative to the base (see Figure 5.8.1).

We shall use a Lagrange's formulation for the equations of motion [76, 196]. The expression for the kinetic energy of the system has the form

$$T = \frac{1}{2}[m_2\dot{\eta}^2 + m(\dot{\eta} + \dot{x})^2 + \frac{J_G}{\rho^2}\dot{x}^2], \qquad (5.8.3)$$

where J_G is the equivalent moment of inertia of the rotating parts, i.e., the motor rotor and gears, with respect to the motor shaft.

The expression for the potential energy of the system in view of (5.8.1) has the form

$$\Pi = \frac{1}{2}k_b\eta^2 \qquad (5.8.4)$$

The dissipative function, according to (5.1.6), (5.1.8), (5.8.2), can be written as

$$R = \frac{1}{2}(\beta_b\dot{\eta}^2 + d_2\dot{x}^2) \qquad (5.8.5)$$

The generalized force Φ performs work on possible displacements $\delta\eta$ and δx. The value $d_1 u$, proportional to the voltage u supplied to the electro-motor, is the generalized force, which does the work through the possible displacement δx.

We derive the equations of motion for the system under consideration by using (5.8.3)–(5.8.5):

$$\left(m + \frac{J_G}{\rho^2}\right)\ddot{x} + m\ddot{\eta} + d_2\dot{x} + \Phi = d_1 u,$$

$$m\ddot{x} + (m + m_2)\ddot{\eta} + \beta_b\dot{\eta} + k_b\eta + \Phi = 0 \qquad (5.8.6)$$

Let us introduce the notations

$$M = m + \frac{J_G}{\rho^2}, \quad M_1 = m + m_2, \tag{5.8.7}$$

the first of which coincides with the notation introduced in Section 5.1. By taking into account (5.8.2), the equations of motion for the manipulator in contact with the object (see (5.8.6)), in view of (5.8.7), have the form

$$M\ddot{x} + m\ddot{\eta} + d_2\dot{x} + \beta(\dot{x} + \dot{\eta}) + k(x + \eta - x_s) = d_1 u,$$

$$m\ddot{x} + M_1\ddot{\eta} + \beta_b\dot{\eta} + \beta(\dot{x} + \dot{\eta}) + k_b\eta + k(x + \eta - x_s) = 0 \tag{5.8.8}$$

Given $\eta \equiv 0$, and $\beta = 0$, the first equation in (5.8.8) coincides with (5.1.11) for $F_e \equiv 0$.

It is worth noting that though (5.8.6) or (5.8.8) is a *two*-degree-of-freedom system, it has *three* equivalent masses m, M, M_1. This is caused by the fact that this mechanical system has translational and rotational (drive) motions. At the same time, the *one*-degree-of-freedom system considered in Section 5.1, where the manipulator arm moves translationally, while the motor rotor rotates, has *one* equivalent mass. The *two*-degree-of-freedom system with translationally moving arm and rotating motor rotor, also considered in Section 5.1, has only *one* equivalent mass. The system with *two* degrees of freedom considered in Section 5.7 has *two* equivalent masses. For $m_2 = 0$ ($M_1 = m$), the system (5.8.6) or (5.8.8) coincides, including notation, with the system (5.7.2), (5.7.4).

We shall assume that the force sensor measures the force

$$F = \begin{cases} 0 & \text{if } \eta + x \le x_s, \\ k(\eta + x - x_s) & \text{if } \eta + x \ge x_s \end{cases} \tag{5.8.9}$$

The expression (5.8.9) is analogous to (5.7.5).

With the linear control (5.3.1), the steady state for the system (5.8.8) is given by the expressions analogous to (5.7.7)–(5.7.9). Asymptotic stability conditions for this steady state may be written as Hurwitz inequalities, though they prove to have a cumbersome form. Let us now consider the switching control (5.4.1). With this control, the stability conditions form is simpler. They coincide with the stability conditions obtained for the linear control (5.3.1), if the feedback gains g_1 and g_2 are assumed to be "large".

With the switching control (5.4.1), the equations of motion (5.8.8) have the form

$$M\ddot{x} + m\ddot{\eta} + d_2\dot{x} + \beta(\dot{x} + \dot{\eta}) + k(x + \eta - x_s)$$

$$= -d_1 u_0 \text{ sgn } \{g_1[k(x + \eta - x_s) - F_d] + g_2 \dot{x}\}, \qquad (5.8.10)$$

$$m\ddot{x} + \mathcal{M}_1 \ddot{\eta} + \beta_b \dot{\eta} + \beta(\dot{x} + \dot{\eta}) + k_b \eta + k(x + \eta) = kx_s$$

In the stationary regime

$$x = x_s + \frac{F_d}{k} + \frac{F_d}{k_b}, \quad \eta = -\frac{F_d}{k_b} \quad (F = F_d) \qquad (5.8.11)$$

For the stationary regime (5.8.11) to exist and belong to the sliding mode region, it is necessary and sufficient that the equality (5.7.19) holds. The relationship (5.8.11) is analogous to (5.7.18).

Let us rewrite the equations (5.8.10) in the variations from the stationary values (5.8.11), still denoting the variations by x and η:

$$\mathcal{M}\ddot{x} + m\ddot{\eta} + d_2\dot{x} + \beta(\dot{x} + \dot{\eta}) + k(x + \eta)$$

$$= -d_1 u_0 \text{ sgn } [g_1 k(x + \eta) + g_2 \dot{x}], \qquad (5.8.12)$$

$$m\ddot{x} + \mathcal{M}_1 \ddot{\eta} + \beta_b \dot{\eta} + \beta(\dot{x} + \dot{\eta}) + k_b \eta + k(x + \eta) = 0$$

The switching of the control takes place on the hyperplane

$$g_1 k(x + \eta) + g_2 \dot{x} = 0 \qquad (5.8.13)$$

The coordinate origin in the plane (5.8.13) lies inside the sliding mode region. It is easy to demonstrate [67] that this region is an attractor set if and only if the inequality (5.7.22) holds. To write an equation for the sliding motion, it is sufficient to replace the first equation in (5.8.12) by (5.8.13)

$$g_2 \dot{x} + g_1 kx + g_1 k\eta = 0$$

$$m\ddot{x} + \beta\dot{x} + kx + \mathcal{M}_1\ddot{\eta} + (\beta_b + \beta)\dot{\eta} + (k_b + k)\eta = 0 \qquad (5.8.14)$$

We can write the characteristic equation for the system (5.8.14) in the form (5.7.24), where

$$b_0 = \mathcal{M}_1 g_2, \quad b_1 = g_2(\beta_b + \beta) + g_1 m_2 k,$$

$$b_2 = g_1 k\beta_b + g_2(k_b + k), \quad b_3 = g_1 k k_b \qquad (5.8.15)$$

Since $b_i > 0$ $(i = 0, \ldots, 3)$, the only condition for the asymptotic stability is the inequality (5.7.26), which in the terms of (5.8.7), (5.8.15) has the form

$$m_2 k^2 \beta_b g_1^2 + [m_2 k - mk_b + \beta_b(\beta_b + \beta)]k g_1 g_2$$

$$+(k_b + k)(\beta_b + \beta)g_2^2 > 0 \qquad (5.8.16)$$

If

$$m_2 k - mk_b + \beta_b(\beta_b + \beta) \geq 0,$$

then the inequality (5.8.16) holds for any $g_1, g_2 > 0$.
   If

$$m_2 k - mk_b + \beta_b(\beta_b + \beta) < 0, \qquad (5.8.17)$$

but

$$[m_2 k - mk_b + \beta_b(\beta_b + \beta)]^2 - 4m_2(k_b + k)(\beta_b + \beta) < 0,$$

then the inequality (5.8.16) also holds for any $g_1, g_2 > 0$.
   If the inequality (5.8.17) is satisfied, and

$$[m_2 k - mk_b + \beta_b(\beta_b + \beta)]^2 - 4m_2(k_b + k)(\beta_b + \beta) > 0, \quad (5.8.18)$$

then the gains $g' > 0$ and $g'' > g'$ exist such that the inequality (5.8.16) holds for

$$0 \leq g_1 < g' g_2, \quad g'' g_2 < g_1 < \infty \qquad (5.8.19)$$

and does not hold for

$$g' g_2 \leq g_1 \leq g'' g_2 \qquad (5.8.20)$$

The gains g' and g'' satisfy the quadratic equation

$$m_2 k^2 \beta_b g^2 + [m_2 k - mk_b + \beta_b(\beta_b + \beta)]k g + (k_b + k)(\beta_b + \beta) = 0$$

The conditions (5.8.17), (5.8.18) can hold for large values of frame stiffness k_b and relatively small damping β_b.

Thus, with compliance in the manipulator frame, the stability of contact keeping may only break in a bounded range of force feedback gains g_1 given by (5.8.20). Unlike that, with compliance in the gear train, the stability is lost for all g_1 exceeding a certain value (see (5.7.16), (5.7.30)).

5.9 Keeping Contact with a Moving Object

In the previous sections, we addressed the issue of keeping contact with a *stationary* object ($x_s = $ const). In the present section, we shall consider the case where the contacted object is moving, i.e., $x_s = x_s(t) \neq$ const. The point of contact between the manipulator and the object may also move in the direction of the axis OX, if the manipulator arm moves orthogonally to this axis. The coordinate of the contact point x_s may change due to such motion. We begin our study of control laws for a manipulator with a moving contact point in this section, and continue it further in Sections 8.6–8.8.

Let us recall the linear (5.3.1) and switching (5.4.1) control laws designed in the preceding sections for the task of keeping contact with a stationary object ($x_s = $ const).

Consider linear control (5.3.1). The motion of the manipulator arm out of contact with the object ($x < x_s, F_e = 0$, see (5.3.2)), is given by the equation

$$\mathcal{M}\ddot{x} + (d_2 + d_1 g_2)\dot{x} - d_1 g_1 F_d = 0, \qquad (5.9.1)$$

and in contact with the object ($x \geq x_s$, see (5.3.3)), by

$$\mathcal{M}\ddot{x} + (d_2 + d_1 g_2)\dot{x} + k(1 + d_1 g_1)[x - x_s(t)] - d_1 g_1 F_d = 0 \qquad (5.9.2)$$

The steady-state velocity of the arm motion out of contact with the object, according to (5.9.1) is given by

$$\dot{x} = \frac{d_1 g_1 F_d}{d_2 + d_1 g_2} \qquad (5.9.3)$$

We assume that the object moves at a constant velocity c, i.e.,

$$x_s = x_s(t) = ct + e \quad (e = \text{const}) \qquad (5.9.4)$$

If $c > 0$, then the object moves away from the manipulator, if $c < 0$, towards the manipulator (see Figure 5.1.1).

To ensure that the manipulator can establish contact with the object moving at the velocity c (catch up with the object, if $c > 0$), the velocity (5.9.3) must exceed c, i.e.,

$$d_1 g_1 F_d > c(d_2 + d_1 g_2) \qquad (5.9.5)$$

We now assume that the manipulator arm moves in contact with the object controlled by (5.9.4). Then the equation (5.9.2) that describes such motion has the form

$$M\ddot{x} + (d_2 + d_1 g_2)\dot{x} + k(1 + d_1 g_1)(x - ct - e) = d_1 g_1 F_d \qquad (5.9.6)$$

The steady-state solution for (5.9.6) has the form

$$x = at + b \qquad (5.9.7)$$

By substituting (5.9.7) into (5.9.6), we obtain the expressions for $b - e$ and a. The contact force in the stationary regime is given by

$$F = k(x - x_s) = k(b - e) = \frac{d_1 g_1 F_d - c(d_2 + d_1 g_2)}{1 + d_1 g_1} \qquad (5.9.8)$$

Under the condition (5.9.5), the force (5.9.8) is positive, hence the control (5.3.1) ensures the manipulator motion in contact with the object. The tracking error for the commanded contact force F_d is

$$\Delta F = F - F_d = -\frac{F_d + c(d_2 + d_1 g_2)}{1 + d_1 g_1} \qquad (5.9.9)$$

The error (5.9.9) grows with the increase of the object velocity c. If c is so large that the inequality (5.9.5) is violated, then it is impossible to keep contact with the object.

Let us now consider the switching control (5.4.1) or (5.4.2), which is the same. The manipulator motion out of contact with the object ($F_e = 0$, see (5.4.11)), is given by

$$M\ddot{x} + d_2 \dot{x} = d_1 u_0 \ \text{sgn} \ (g F_d - \dot{x}), \qquad (5.9.10)$$

and in contact with the object (see (5.4.14)), by

$$M\ddot{x} + d_2 \dot{x} + k[x - x_s(t)] = d_1 u_0 \ \text{sgn} \ [g F_d + g k(x_s(t) - x) - \dot{x}] \qquad (5.9.11)$$

According to (5.9.10), the steady-state velocity of the manipulator motion out of contact with the object is given by

$$\dot{x} = gF_d \qquad (5.9.12)$$

To ensure that the manipulator can establish contact with the object moving with the control (5.9.4), the velocity (5.9.12) must exceed c:

$$gF_d > c \qquad (5.9.13)$$

Note that (5.9.13) follows from (5.9.5). Therefore, the range of object velocities c at which the manipulator can establish contact with the object is larger for the switching control (5.4.1), than for the linear control (5.3.1).

Let us now assume that the manipulator arm moves in contact with the object controlled by (5.9.4). Then steady-state solution for (5.9.11) has the form (5.9.7). In the stationary regime, the expression in square brackets in the right part of the equation is zero. The contact force in the stationary regime is given by

$$F = F_d - \frac{c}{g} \qquad (5.9.14)$$

Under the condition (5.9.13), the force (5.9.14) is positive, hence the manipulator motion in contact with the object is possible. The error in tracking the commanded contact force F_d is

$$\Delta F = F - F_d = -\frac{c}{g} \qquad (5.9.15)$$

The expressions (5.9.12)–(5.9.15) can also be derived from (5.9.3), (5.9.5), (5.9.8), and (5.9.9), by assuming that $g_1 = gg_2$ and g_2 tends to infinity.

Thus, both linear control (5.3.1) and switching control (5.4.1) solve the problem of keeping the manipulator arm in contact with an object moving at a constant (a priori unknown) velocity c, provided this velocity satisfies (5.9.5) or (5.9.13). The experimental results demonstrate that linear and switching control laws can ensure contact keeping even if the object velocity slowly changes. Yet, if the velocity of the object motion is negative $(c < 0)$, then the contact force applied to the object by the sensor may be too large. For example, if the object is moved by a conveyer, and the contact force exceeds the frictional force between the object and the conveyer surface, then the manipulator will move the object out of its position, which is undesirable. As the gain g increases, the contact force (5.9.14) approaches its commanded value, while the steady-state error (5.9.15) diminishes. As mentioned above, the system may develop sustained oscillations for large gains g, which means this gain cannot be too large. Since the commanded values of the contact force F_d, as well as gains, are bounded from above (see (5.2.2)), the velocities that satisfy (5.9.5), (5.9.13), are also bounded from above. Therefore, the linear and switching control laws can ensure contact keeping in the given range (5.2.2) provided that velocities of the object motion are limited. In the experiments, the values $F_d = 1$ N, $g = 3$ mm/(N·s) were used. For these parameters, the velocity was $gF_d = 3$ mm/s.

Now we shall modify the linear and switching control laws designed in Sections 5.3 and 5.4 by introducing into them the information on the velocity of the object motion into them. We present the linear (with saturation) and switching control in the form (see (5.4.10), (5.4.3))

$$u = \text{ sat } [g_2(v_d - \dot{x})], \tag{5.9.16}$$

$$u = u_0 \text{ sgn } (v_d - \dot{x}) \tag{5.9.17}$$

As for v_d, we shall substitute the following expression for (5.3.13)

$$v_d = g(F_d - F) + \hat{v}, \tag{5.9.18}$$

where $\hat{v}$ is an estimate of the object motion velocity.

To begin with, let us assume that the object motion velocity is a priori known and constant, such as, for example, the velocity of a conveyer where the object is placed. Let $\hat{v} = c$ in (5.9.18). Then, with the linear control (5.9.16), (5.9.18), we obtain, instead of (5.9.8), (5.9.9), the following expressions for the force and the steady-state error in force tracking

$$F = \frac{d_1 g_1 F_d - cd_2}{1 + d_1 g_1}, \quad \Delta F = -\frac{F_d + cd_2}{1 + d_1 g_1}$$

With the switching control (5.9.17), (5.9.18), we obtain $F = F_d, \Delta F = 0$ instead of (5.9.14), (5.9.15).

We shall further consider the object motion velocity to be a priori unknown. Let us write the velocity estimate $\hat{v}$ as follows

$$\hat{v}(t) = \begin{cases} \hat{v}(t_{i-1}) & \text{if } t_{i-1} \leq t < t_i, \quad i = 1, 2, \ldots, \\ \frac{x(t_i) - x(t_{i-1})}{t_i - t_{i-1}} & \text{if } t = t_i, \end{cases} \tag{5.9.19}$$

where t_0 is the time when the process begins. We consider the object to be stationary at the time t_0, hence $\hat{v}(t_0) = 0$. We compute the time

t_i $(i = 1, 2, \ldots)$ as the least root of the equation

$$|x(t) - x(t_{i-1})| = d \quad (d = \text{const}) \tag{5.9.20}$$

It means that by the time t_i, the coordinate x of the manipulator arm changes by the commanded value d as compared to the coordinate $x(t_{i-1})$. According to (5.9.19), (5.9.20), the velocity $\hat{v}(t)$ takes the values $\pm d/(t_i - t_{i-1})$.

In order to use (5.9.19), (5.9.20) for control, the manipulator should be instrumented with a position sensor for measuring the coordinate x. With a digital computer in the control loop, we take the time t_i to be the first instant of time when the following inequality is satisfied:

$$|x(t) - x(t_{i-1})| \geq d \tag{5.9.21}$$

The distance d can be determined experimentally by using the following guidelines. On the one hand, it is desirable to choose d to be small. Then the "new" velocity estimate $\hat{v}$ may be sampled more frequently during the motions of the object and the manipulator arm following it, and the manipulator will respond to the object motions faster. On the other hand, the distance d should exceed the position sensor error, as well as be greater than the manipulator motions required for keeping contact with the stationary object during the transient process. Otherwise, the controller can perceive these errors and motions as object motions.

The "new" velocity estimate $\hat{v}$ is computed at the time t_i $(i = 1, 2, \ldots)$ and does not change on the time interval $[t_i, t_{i+1})$.

It is easy to demonstrate that with the object moving at an a priori unknown uniform velocity, the system (5.1.10), (5.1.11) moves at the same velocity in the stationary regime, with both the control (5.9.16) and control (5.9.17). The contact force F and the steady-state error ΔF are the same whether or not the velocity of the object is known a priori. The experiments prove that the control laws (5.9.16) and (5.9.17), where the commanded velocity is given by (5.9.18) and $\hat{v}$ is computed by (5.9.19), work well if the object velocity changes slowly enough.

We shall study the manipulator motion with the control (5.9.17)–(5.9.20), assuming that the object is initially at rest. Then, at time $t = t_0$, the object instantly acquires a velocity greater than gF_d and breaks away from the manipulator arm that follows it. By neglecting the transient process with respect to velocity, we can assume that the arm immediately starts moving with the velocity gF_d. According to (5.9.18)–(5.9.20), the estimate $\hat{v}$, which initially is zero, becomes gF_d in time $d/(gF_d)$, and the commanded velocity v_d becomes $2gF_d$. By neglecting the transient process, we can assume that the manipulator velocity at the time $t_0+d/(gF_d)$ increases to $2gF_d$ in a step-wise manner. In time $d/(2gF_d)$, provided that the arm does not contact the object, $\hat{v}$ will equal $2gF_d$, and the commanded velocity v_d will be $3gF_d$. At the time $t_0+(1+1/2)d/(gF_d)$, the manipulator velocity out of contact with the object will reach $3gF_d$; at the time $t_0+(1+1/2+1/3)d/(gF_d)$, it will reach $4gF_d$; at the time $t_0+(1+1/2+\ldots+1/n)d/(gF_d)$, $(n+1)gF_d$. The plot of dependencies of the commanded velocity v_d and the manipulator velocity on time is shown in Figure 5.9.1.

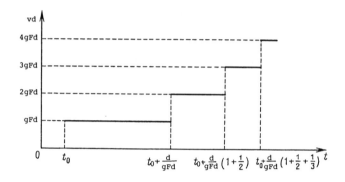

FIGURE 5.9.1
Time dependency of the commanded manipulator velocity

The velocity $(n+1)gF_d$ is attainable as long as it does not exceed the maximum possible velocity d_1u_0/d_2. Consequently, if the object moves at a velocity less than d_1u_0/d_2, then the manipulator arm will catch up with it in a finite time. Let the object motion velocity c remain constant and nonzero after the contact has been established ($c = \mathrm{const} \neq 0$). Then, by neglecting the transient processes, one can demonstrate that, in time d/c, the manipulator moving at the same velocity c will contact the object with the commanded force F_d. Now let the object motion velocity start to decrease at a certain time instant. Assume that this velocity is a piece-wise constant time function, and the object moves with a uniform velocity through distances greater than d. The velocities $\hat{v}, v_d$ and the arm motion velocity will decrease. By neglecting the transient processes, we can consider the commanded velocity v_d and the arm motion velocity to be piece-wise time functions. If the object velocity reduces to zero, i.e., if the object comes to a halt at a certain time instant, then the manipulator halts as well. However, $\hat{v}$ will not go to zero, and the contact force F will not equal the commanded force F_d. It follows from the analysis of (5.9.19), (5.9.20) that if $\hat{v}$ is nonzero at a certain time, then it will never reduce to zero.

The velocity $\hat{v}$ may go to zero, if the time t_i $(i = 1, 2, \ldots)$ is computed from the equation

$$t = t_{i-1} + T \quad (t_i = t_{i-1} + T), \tag{5.9.22}$$

rather than (5.9.20).

Using computer control, we consider the time t_i to be the first time instant when the following inequality holds:

$$t - t_{i-1} \geq T \tag{5.9.23}$$

The time T should be determined experimentally. On the one hand, it is desirable to choose T to be rather small, so that the manipulator arm responds to the object motions faster. On the other hand, the time T should be greater than the arm eigenfrequency period in keeping contact with the stationary object. Otherwise, the controller will perceive the arm motions in the transient process as object motions.

One can compute the time t_i as the smallest root of the equation (5.9.20), (5.9.22). In this case, the time t_i is computed as the first time instant when at least one of the inequalities (5.9.21), (5.9.23) is satisfied.

Let us consider a different computation method for $\hat{v}$ in (5.9.18) than by (5.9.19):

$$\hat{v} = g_3 \int_0^t [F_d - F(\tau)] d\tau \tag{5.9.24}$$

With the integral control (5.9.24) used out of contact with the object, for both the control (5.9.16) and (5.9.17), the steady-state velocity of the manipulator motion is a linearly-increasing time function. When the voltage u equals u_0, this velocity takes its maximum constant value $d_1 u_0 / d_2$.

The linear manipulator motion with the control (5.9.16), (5.9.18), and (5.9.24) in contact with the object is defined by the following integro-differential equation:

$$\mathcal{M}\ddot{x} + (d_2 + d_1 g_2)\dot{x} + k(1 + d_1 g_1)[x - x_s(t)]$$

$$-d_1 g_1 F_d + g_3 \int_0^t [k(x - x_s(\tau)) - F_d] d\tau = 0 \tag{5.9.25}$$

It is easy to show that if the object moves at a constant velocity c with the control (5.9.4), then the manipulator in the stationary regime moves with the same velocity c, and the contact force equals the commanded force (there is no steady-state error). The same is true for the switching control (5.9.17), (5.9.18), and (5.9.24).

The homogeneous differential equation corresponding to (5.9.25) has the form

$$\mathcal{M}\ddddot{x} + (d_2 + d_1 g_2)\ddot{x} + k(1 + d_1 g_1)\dot{x} + g_3 k x = 0 \qquad (5.9.26)$$

If the coefficients of this equations are positive, then the only stability condition for the stationary regime, i.e., the Hurwitz condition, has the form

$$(d_2 + d_1 g_2)(1 + d_1 g_1) > \mathcal{M} g_3 \qquad (5.9.27)$$

It follows from (5.9.27) that the integral feedback gain g_3 should not be too large. It is commonly known in control theory that the introduction of integral feedback can make the process unstable.

We further continue to examine the control laws (5.9.16) and (5.9.17). In this case, we do not compute the commanded value of v_d using either (5.3.13) or (5.9.18), (5.9.19), or (5.9.18), (5.9.24), but in a different way. Consider the function

$$x_1(t) = \begin{cases} x(t) & \text{if } F \geq F_0, \\ x_1(t - \Delta) & \text{if } F < F_0, \end{cases} \qquad (5.9.28)$$

where F_0 is the threshold of the sensor sensitivity (see (5.2.2)). We assume that the manipulator is in contact with the object if $F \geq F_0$, and out of contact, if $F < F_0$. The delay Δ is the interval (cycle time) for the run-time control software. It follows from (5.9.28) that if the manipulator arm is in contact with the object, the coordinate x_1 coincides with the moving arm coordinate x; if the arm is out of contact, x_1 coincides with the arm coordinate x at the last contact instant. Thus, in contact, the coordinate x_1 changes together with the coordinate x, and out of contact, remains constant. Let us define the commanded velocity v_d by the formula

$$v_d = \begin{cases} g(F_d - F) & \text{if } x < x_1 + d, \\ v_1 & \text{if } x \geq x_1 + d \end{cases} \qquad (5.9.29)$$

In the above expression, the distance d is an experimentally determined value in the neighborhood of the point x_1 along the coordinate x, where the commanded velocity v_d is given by (5.3.13); $v_1 = \text{const}$ is the commanded value of the manipulator velocity at the loss of contact, where v_1 is greater than $g F_d$.

We assume that at the initial time instant $t = t_0$, the manipulator arm is at a certain distance from the object. Let $x_1(t_0)$ be much less than $x(t_0)$. According to (5.9.29), the commanded velocity in the beginning of the motion is v_1. When the arm contacts the object ($F \geq F_0$), x_1 coincides with x, and the commanded velocity v_d is equal to $g(F_d - F)$. The velocity v_d still equals $g(F_d - F)$, in spite of errors in the measurement of the manipulator position, provided the relationship $x < x_1 + d$ holds. Unless the manipulator moves through the distance d, the equality also holds even if the condition $F \geq F_0$ is temporarily violated.

If the object which has initially been in contact with the manipulator, moves from it at a distance greater than d, then, at first, the commanded velocity is gF_d. After the manipulator has moved through the distance d, the commanded velocity takes the value v_1.

In addition to $x_1(t)$, let us consider the function

$$x_2(t) = \begin{cases} x(t) & \text{if } F \leq F_1, \\ x_2(t - \Delta) & \text{if } F > F_1, \end{cases} \qquad (5.9.30)$$

where F_1 is a certain constant value of the contact force ($F_1 > F_d > F_0$).

It follows from (5.9.30) that if the contact force exceeds F_1, then x_2 remains constant and coincides with the manipulator coordinate x at an instant preceding such excess.

Let us write the commanded velocity v_d in the form

$$v_d = \begin{cases} g(F_d - F) & \text{if } x > x_2 - d, \\ -v_2 & \text{if } x \leq x_2 - d, \end{cases} \qquad (5.9.31)$$

where $v_2 = \text{const}$ is the commanded velocity of the manipulator exceeding $g(F_1 - F_d)$.

If the contact force $F \leq F_1$, then $x_2 = x$ and $v_d = g(F_d - F)$. The latter equality holds in spite of errors in the measurement of the coordinate x, provided they do not exceed d. This equality also holds with the condition $F \leq F_1$ temporarily violated, unless the manipulator moves through the distance d.

Suppose that the object "advances" towards the manipulator with a velocity greater than $g(F_1 - F_d)$. Then the contact force F starts increasing, becomes equal to F_1 at a certain instant of time, and subsequently exceeds it. From this time on, x_2 stops changing together with the arm coordinate x and remains constant (see (5.9.30)). After the manipulator, "pushed" by the object, backs off at the distance d from the point where the inequality $F \leq F_1$ has first been violated, the commanded velocity v_d equals $-v_2$ (see (5.9.31)). At the instant the decreasing contact force F becomes equal to F_1, the variable x_2 coincides with the arm coordinate x, and the commanded velocity v_d equals $g(F_1 - F_d)$.

By combining (5.9.29) and (5.9.31), we obtain the expression for the commanded velocity

$$v_d = \begin{cases} v_1 & \text{if } x \geq x_1 + d, \\ g(F_d - F) & \text{if } x_2 - d < x < x_1 + d, \\ -v_2 & \text{if } x \leq x_2 - d, \end{cases} \tag{5.9.32}$$

which allows the manipulator "to adapt" both to an "advancing" and "retreating" object.

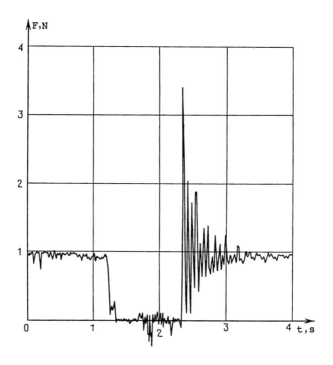

FIGURE 5.9.2
The time-history of contact force as an object starts moving

Figure 5.9.2 displays a time-history of the contact force between the manipulator arm and the object, obtained in one of the experiments with the control (5.9.16), (5.9.32). We used a force sensing finger (Section 2.2) as a force sensor for these experiments. The control parameters in the experiment were as follows: $u_0 = 27$ V, $g_2 = 7 \cdot 10^3$ V·s/mm, $g = 3$ mm/(N·s), $F_0 = 0.5$ N, $F_d = 1$ N, $F_1 = 1.5$ N, $v_1 = v_2 = 50$ mm/s, $d = 2$

mm. The experiment begins with the arm in contact with a stationary object. The object starts moving away from the arm with a high velocity at a certain time instant, then stops and remains stationary. Figure 5.9.2 shows that, at first, the contact force is constant, which corresponds to the contact with a stationary object; then it reduces below its threshold value ($F(t) \not\equiv 0$ because of the noise in the force measurement channel), which corresponds to the loss of contact. After losing contact, the arm initially moves with the velocity $gF_d = 3$ mm/s (the transient process with respect to velocity settles fast) and beyond the distance $d = 2$ mm, with the velocity $v_1 = 50$ mm/s. Having caught up with the object, the arm establishes contact with it. After the transient process, the contact force is close to the commanded force.

It is advisable to compute the velocity using (5.9.32), if the object is stationary most of the time, i.e., if it moves comparatively rarely. If the object moves frequently or continuously, computing the commanded velocity by (5.9.32) does not yield a high performance. For example, if the object moves away from the manipulator with a constant velocity greater than gF_d, but less than v_1, then the manipulator motion velocity switches abruptly from low gF_d to high v_1, then to low again, etc. In this case, it is advisable to compute the commanded velocity using (5.9.18), (5.9.19) or (5.9.18), (5.9.24).

Note that the manipulator response to the object motions is the fastest with the control (5.4.1) for $g_2 = 0$, i.e., with the control (5.4.31). However, contacting a stationary object with the control (5.4.31), the manipulator arm experiences sustained oscillations.

Chapter 6

Two-degree-of-freedom Motion in Contact With an Object

In this chapter, we design control for a manipulator following a planar contour of an object. In other words, we consider the problem of keeping a manipulator in contact with an object as the manipulator moves along the contour of this object. Such motion allows one to determine an object contour that is a priori unknown. The designed control is useful for tracking part joints in welding, grinding, or polishing. By following an object contour, it is possible to avoid obstacles. Such motion may be considered as a motion along a mechanical constraint.

The contour following motion of a manipulator is studied in papers [89, 139, 248, 292, 295], among many others. In this book, a control for such motion is built on the basis of a *systematic* approach to the problem of control synthesis for robotic systems with force sensing. This approach designs robotic system motions along constraints by representing complex motions as a superposition of primitive (basic) ones. The control-keeping motion is a basic motion and it can be controlled as described in the preceding, fifth chapter.

6.1 Mathematical Model for Two-degree-of-freedom Motion

We shall study a two-degree-of-freedom translational motion in the horizontal plane of the manipulator arm described in Chapter 4 of the book (see Figure 4.1.1). A force sensor measuring two horizontal components of the force vector is attached to the lower end of the manipulator arm.

Let us introduce a stationary orthogonal coordinate system OXY in the horizontal plane (manipulator base plane). The translational motion of the manipulator arm along the axes OX and OY involves two respective degrees of freedom. The motion along each axis is actuated by a separate electric motor. Figure 6.1.1 illustrates the schematics of the manipulator.

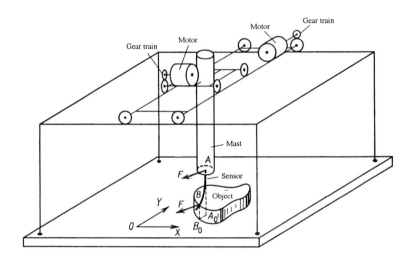

FIGURE 6.1.1
Schematics of a manipulator with two degrees of freedom

Denote by A the point where the sensor is attached to the mast, by A_0, the projection of the point A onto the plane XY (see Figure 6.1.1). Let x, y be the coordinates of the point A_0. By neglecting the inductance of the motor armature, we can write the equations of motions for the two-degree-of-freedom system in hand in the form

$$\mathcal{M}_x \ddot{x} + d_x \dot{x} = \alpha_x u_x + F_x,$$

$$\mathcal{M}_y \ddot{y} + d_y \dot{y} = \alpha_y u_y + F_y \qquad (6.1.1)$$

In the above equations, $\mathcal{M}_x$ and $\mathcal{M}_y$ are equivalent masses of the moving parts (the manipulator arm, motor rotor, gears, etc.) for the respective degrees of freedom; $d_x \dot{x}$ and $d_y \dot{y}$ are combined forces of viscous friction and back e.m.f. in the drives; u_x and u_y are voltages supplied to the motors; $d_x, d_y, \alpha_x, \alpha_y$ are positive constants dependent on the drive characteristics; F_x and F_y are components of the horizontal force $\boldsymbol{F}$ (see Figure 6.1.1) exerted by the force sensor on the manipulator mast.

Each equation in (6.1.1) is analogous to (5.1.8) and can be derived in a similar way.

A sensory probe, a finger sensor, or a wrist sensor may be used for the force measurement. We used sensors of each type in different experiments. Each of the used sensors had two channels with orthogonal sensitivity axes in the horizontal plane. Though we refer to a probe-type force sensor below, the derived equations may also be used for other sensor types.

In this chapter, we shall model the force sensor as an *infinitely thin elastic massless rod* (see Figure 6.1.1) with a uniform bending stiffness in all directions. The rod is vertical when undeformed. As noted above, it is attached to the manipulator arm at the point A.

In the absence of contact between the force sensor and the object, the force acting on the arm is $F = 0$, as the sensor is assumed to be massless. When the contact is established at the point B, the sensor bends (see Figure 6.1.1). Denote the projection of the point B onto the plane XY by B_0. The force exerted by the object on the sensor is applied at the point B. The horizontal component of this force is collinear to the vector $\mathbf{A_0 B_0}$. Since the rod is assumed to be massless, this component is equal to the horizontal force F applied from the sensor to the manipulator at the point A. If the sensor stiffness is high and its deformations are small, then, in accordance with the sensor model adopted in Section 1.2 (see (1.2.4)), we consider the force F to be proportional to the length of the segment $A_0 B_0$:

$$F = k \cdot \mathbf{A_0 B_0} \tag{6.1.2}$$

The coefficient $k = \text{const} > 0$ defines the sensor stiffness (see 5.1.9). Herein we ignore energy dissipation in the sensor.

For simplicity sake, assume that the object is a right cylinder. The object lateral surface is orthogonal to the plane XY, and the top and bottom flanges are parallel to this plane. The sensor contacts the object at the point B of the top flange (see Figure 6.1.1). Let us write an equation for the top flange contour projection onto the plane XY in the parametric form

$$x = x(s), \quad y = y(s), \tag{6.1.3}$$

where s is the length of the contour arc measured from a certain point on it (see Figure 6.1.2).

Let s_B be a parameter (arc length) corresponding to the point B_0. The vector equality (6.1.2) may be rewritten as two scalar ones:

$$F_x = k[x(s_B) - x], \quad F_y = k[y(s_B) - y] \tag{6.1.4}$$

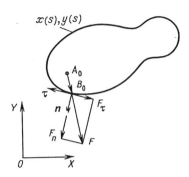

FIGURE 6.1.2
Object contour projection onto the plane XY

The expressions for the tangent $\boldsymbol{\tau}$ and the normal $\boldsymbol{n}$ to the contour (6.1.3) at the point B_0 have the form

$$\boldsymbol{\tau} = \begin{bmatrix} x'(s_B) \\ y'(s_B) \end{bmatrix} = \begin{bmatrix} \cos\psi \\ \sin\psi \end{bmatrix}, \quad \boldsymbol{n} = \begin{bmatrix} -y'(s_B) \\ x'(s_B) \end{bmatrix} = \begin{bmatrix} -\sin\psi \\ \cos\psi \end{bmatrix}, \quad (6.1.5)$$

where $\psi = \psi(s_B)$ is an angle measured in the positive direction between the axis OX and the vector $\boldsymbol{\tau}$ at the point B_0. The prime denotes differentiation with respect to s. It is assumed that the respective derivatives exist. We consider $\boldsymbol{n}$ in (6.1.5) to be the vector of the normal that is external with respect to the object. Clearly, along with the vector $\boldsymbol{\tau}$ in (6.1.5), the vector $-\boldsymbol{\tau}$ is also a unit tangent vector.

The projections F_τ and F_n of the force (6.1.2) onto the tangent and normal to the top contour at the point B_0 (see Figure 6.1.2) are given by

$$F_\tau = \boldsymbol{F} \cdot \boldsymbol{\tau} = k[x'(s_B)(x(s_B) - x) + y'(s_B)(y(s_B) - y)],$$

$$F_n = \boldsymbol{F} \cdot \boldsymbol{n} = k[-y'(s_B)(x(s_B) - x) + x'(s_B)(y(s_B) - y)], \quad (6.1.6)$$

where F_τ is a frictional force between the sensor and the object, and F_n is a force of normal pressure. If the contact point B_0 moves along the contour, i.e., $\dot{s}_B \neq 0$, then the force F_τ is directed against the contouring velocity vector, and the force vector $\boldsymbol{F}$ is on the boundary of the cone of friction:

$$F_\tau = -fF_n\mathrm{sgn}\dot{s}_B \quad (F_n > 0) \tag{6.1.7}$$

In the above equation, $f = \mathrm{const} > 0$ is a Coulomb friction coefficient. Denote a friction angle by β: $\tan\beta = f$. From (6.1.6) and (6.1.7), for

$\dot{s}_B \neq 0$, we obtain

$$x'(s_B)(x(s_B) - x) + y'(s_B)(y(s_B) - y)]$$

$$= f[y'(s_B)(x(s_B) - x) - x'(s_B)(y(s_B) - y)] \operatorname{sgn} \dot{s}_B \qquad (6.1.8)$$

If $\dot{s}_B = 0$, then the force vector $\boldsymbol{F}$ lies in the cone of friction

$$|F_\tau| \leq f F_n \qquad (6.1.9)$$

If the equations for the contour (6.1.3) are given, then, provided the arm coordinates x, y are known, one can find the coordinate s_B of the contact point B from (6.1.8) or (6.1.9). Once s_B is known, one can find the coordinates $x(s_B), y(s_B)$ of the point B from (6.1.3). Then we find the components (6.1.4) of the force $\boldsymbol{F}$, so as to close the system of equations of motion (6.1.1). If the friction coefficient $f = 0$, then the relationships (6.1.8), (6.1.9) are replaced by only one equation

$$x'(s_B)[x(s_B) - x] + y'(s_B)[y(s_B) - y] = 0 \qquad (6.1.10)$$

The equation (6.1.10) is an orthogonality condition for the vectors of contact force $\boldsymbol{F}$ and contour tangent $\boldsymbol{\tau}$.

The relationships (6.1.1), (6.1.3), (6.1.4), (6.1.8), (6.1.9) present a closed form description for a two-degree-of-freedom manipulator in contact with an object. In the absence of contact, one should only use the equations (6.1.1), assuming $F_x = F_y = 0$. These relationships also define moments of establishing and losing contact.

Suppose that control voltages u_x, u_y are given as functions of time and/or phase coordinates. Also, let the initial conditions $x(0)$, $y(0)$, $\dot{x}(0)$, and $\dot{y}(0)$ be given. If for $t = 0$ the sensor contacts the object, then we assume that the coordinates $x(s_B), y(s_B)$ of the contact point B, which satisfy the relationships (6.1.8) or (6.1.9), are known as well. Under these conditions, the solutions to the equations (6.1.1), (6.1.3), (6.1.4), (6.1.8), (6.1.9) describe the manipulator motion for $t > 0$. When solving the equations, in principle, one may obtain several solutions for s_B from (6.1.8), (6.1.9). In this case, to find a unique solution, it is necessary to introduce additional conditions such as the requirement that the function $s_B(t)$ is continuous.

6.2   Control Problem Statement

Let us assume that a force sensor is in contact with an object at the initial time (see Figures 6.1.1, 6.1.2). The object contour is unknown a priori, i.e., the equations (6.1.3) are not given. It is desired to design control of the drive, such that the force sensor keeps contact with an object and moves along its contour, i.e., along a unilateral constraint.

We shall consider a manipulator motion along an object contour as a *superposition* of two *basic* motions – at a *normal* and at a *tangent* to the contour. Accordingly, we define the "desired" (commanded) velocity vector V_d as a *sum* of two terms

$$V_d = V_n + V_\tau = g(F_n - F_d)n + v\tau \tag{6.2.1}$$

In the above equation, $F_d = \text{const} > 0$ is a desired (commanded) value of the normal force component exerted on the force sensor by the object, $g > 0$ is a constant feedback gain, $v = \text{const} > 0$ is a desired (commanded) value of the manipulator velocity along a tangent to the object contour (the desired contour velocity).

The first term V_n in (6.2.1) is a vector directed towards the object if $F_n < F_d$, and from the object if $F_n > F_d$. (Note that n is the vector of the external normal to the object).

One can consider the expression (6.2.1) for the commanded velocity as an extension of (5.3.13). For $v = 0$, (6.2.1) is an extension of (5.3.13) for the case of a two-degree-of-freedom manipulator keeping contact with an object. In the case of $v \neq 0$, (6.2.1) extends (5.3.13) for a manipulator keeping contact with an object and simultaneously moving along its contour.

If the manipulator is required to move in the opposite direction along the contour, one should substitute $-\tau$ for τ in (6.2.1).

We design the manipulator control as a linear feedback of the velocity

$$U = C(V_d - V), \tag{6.2.2}$$

where

$$U = \begin{bmatrix} U_x \\ U_y \end{bmatrix}, \quad C = \begin{bmatrix} c_x & 0 \\ 0 & c_y \end{bmatrix}, \quad V = \begin{bmatrix} \dot{x} \\ \dot{y} \end{bmatrix} = \begin{bmatrix} V_x \\ V_y \end{bmatrix}, \quad V_d = \begin{bmatrix} V_{xd} \\ V_{yd} \end{bmatrix}, \tag{6.2.3}$$

and c_x, c_y are positive constants.

The formula (6.2.2) is an extension of (5.3.12) for the case of two-degree-of-freedom manipulator motion.

In the absence of friction, i.e., for $f = \beta = 0$,

$$F = F \cdot n, \quad F = F_n = [F_x^2 + F_x^2]^{1/2} \tag{6.2.4}$$

For $f = 0$, one can find F_n and the vector of normal n from (6.2.4) by using a force sensor to measure the force vector F. By setting the desired direction of the manipulator motion, we can uniquely define the vector τ, for example, by rotating the vector n clockwise through an angle $\pi/2$. Thus, we can compute all terms in (6.2.1). Therefore, in the absence of friction, the control laws (6.2.2), (6.2.3) can be implemented by using the information from force sensors and velocity sensors measuring the components $\dot{x}$ and $\dot{y}$ of the vector V.

In reality, friction is always present and $\beta \neq 0$. Let a force sensor move in contact with an object. Then the force vector F is on the boundary of the friction cone (see Figure 6.2.1).

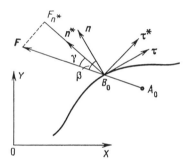

FIGURE 6.2.1
Force vector F on the boundary of the friction cone

If the friction angle β is known, then one can determine the vectors τ, n, and $F_n = F \cos\beta$ from the force sensor readings, and implement the control law (6.2.1), (6.2.2). If the angle β is unknown, then the force information is insufficient to compute the vectors τ, n, and F_n. However, it is possible to guess the friction coefficient. Denote the assumed friction angle by γ as distinct from β, the real friction angle. We can choose γ within the typical range of friction angles for the given materials. Since $0 < \beta < \pi/2$, then $0 < \gamma < \pi/2$. Let us set the velocity vector V_d according to (6.2.1) based on the assumed friction angle

$$V_d = g(F_{n*} - F_d)n^* + v\tau^*, \tag{6.2.5}$$

where n^* is a unit vector making an angle γ with the force vector F measured by the force sensor. A unit vector τ^* is obtained by rotating the vector

n^* clockwise through the angle $\pi/2$, $F_{n^*} = F\cos\gamma$ (see Figure 6.2.1). If γ is the assumed friction angle, then n^* and τ^* are the assumed vectors of normal and tangent to the object contour. For $\gamma = 0$, n^* is directed along the measured force vector F. If $\gamma = \beta$, then $n^* = n$, $\tau^* = \tau$, and the expressions (6.2.1) and (6.2.5) coincide. We will assume that for $\beta \neq 0$, the commanded velocity vector (6.2.1) is computed approximately, as defined by (6.2.5).

The vector (6.2.1) may be approximately found in a different way. By obtaining from position sensors the coordinates x, y of the manipulator arm following the contour, one can find a secant to the object contour and estimate the tangent τ and normal n.

It is desired to find parameters v, F_d, g, c_x, c_y, γ such that the control laws (6.2.2), (6.2.3), (6.2.5) perform well for all friction angles β within the given range.

The paper [89] considers a somewhat different method for computing the commanded velocity as compared to (6.2.1), (6.2.5). The method of [89] performs well in the experiments and is quite acceptable. The control law proposed in this section is based on the *systematic* approach to the design of control for robotic systems with mechanical constraints. This approach is further generalized in Chapter 7. The problem in question may be considered as a problem of control design for a manipulator performing an in plane motion along a constraint. The problem of maintaining contact solved in Chapter 5 may be thought of as design of a controller keeping a manipulator on a constraint. The problem considered in this chapter seems to be the simplest problem of the constrained manipulator motion control. Yet this is also a key problem, since the control algorithms designed for it are used to solve other problems.

6.3   Following a Linear Contour

Let the equations (6.1.3) describe a straight line. Without the loss of generality, it may be assumed that this straight line passes through the coordinate origin

$$x = s \cdot \cos\psi, \quad y = s \cdot \sin\psi, \tag{6.3.1}$$

where $\psi = $ const is an angle between the straight line and the axis OX, the contour path length s is counted from the coordinate origin.

Let the contact point B move so that $\dot{s}_B > 0$. We can find s_B by using (6.1.8), and then find the force vector components F_x and F_y from (6.1.3),

(6.1.4). By substituting F_x and F_y into (6.1.1), we obtain the following equations

$$M_x\ddot{x} + d_x\dot{x} + \frac{k\sin(\psi + \beta)}{\cos\beta}(x\sin\psi - y\cos\psi) = \alpha_x u_x,$$

$$M_y\ddot{y} + d_y\dot{y} + \frac{k\cos(\psi + \beta)}{\cos\beta}(x\sin\psi - y\cos\psi) = \alpha_y u_y \qquad (6.3.2)$$

One can find the vectors $\boldsymbol{\tau}^*$, $\boldsymbol{n}^*$ from (6.1.5), (6.2.5), and then find the components of the control vector U in (6.2.2), (6.2.3), (6.2.5)

$$u_x = c_x\left\{-g\left[\frac{k\cos\gamma}{\cos\beta}(x\sin\psi - y\cos\psi) - F_d\right]\sin(\psi + \beta - \gamma)\right.$$

$$\left. +v\cos(\psi + \beta - \gamma) - \dot{x}\right\},$$

$$u_y = c_y\left\{g\left[\frac{k\cos\gamma}{\cos\beta}(x\sin\psi - y\cos\psi) - F_d\right]\cos(\psi + \beta - \gamma)\right.$$

$$\left. +v\sin(\psi + \beta - \gamma) - \dot{y}\right\} \qquad (6.3.3)$$

The equations (6.3.2), (6.3.3) have a cyclic coordinate. After the coordinate transformation

$$\xi = x\cos\psi + y\sin\psi, \quad \eta = x\sin\psi - y\cos\psi, \qquad (6.3.4)$$

the coordinate ξ appears in (6.3.2), (6.3.3) only through its first $\dot{\xi}$ and second $\ddot{\xi}$ derivatives. Using the variables ξ and η, the equations for the straight line (6.3.1) take the form $\xi = s$, $\eta = 0$.

We look for a stationary regime for the system (6.3.2), (6.3.3), which corresponds to a manipulator motion with a constant velocity and constant contact force:

$$\dot{x} = \dot{s}_B\cos\psi, \quad \dot{y} = \dot{s}_B\sin\psi, \quad \dot{s}_B = \dot{\xi} = V = v + \Delta v = \text{const},$$

$$F_{n^*} = \frac{k\cos\gamma}{\cos\beta}(x\sin\psi - y\cos\psi) = F_d + \Delta F = \text{const} \qquad (6.3.5)$$

By substituting (6.3.5) into the differential equations (6.3.2), (6.3.3), we obtain linear algebraic equations for determining the stationary regime parameters $\Delta V, \Delta F$

$$(d_x + \sigma_x) \cos \psi \Delta v + \left[\frac{\sin(\psi + \beta)}{\cos \gamma} + \sigma_x g \sin(\psi + \beta - \gamma) \right] \Delta F$$

$$= -F_d \frac{\sin(\psi + \beta)}{\cos \gamma} + v \left[\sigma_x \cos(\psi + \beta - \gamma) - (d_x + \sigma_x) \cos \psi \right],$$

$$(d_y + \sigma_y) \sin \psi \Delta v - \left[\frac{\cos(\psi + \beta)}{\cos \gamma} + \sigma_y g \cos(\psi + \beta - \gamma) \right] \Delta F$$

$$= F_d \frac{\cos(\psi + \beta)}{\cos \gamma} + v \left[\sigma_y \sin(\psi + \beta - \gamma) - (d_y + \sigma_y) \sin \psi \right], \quad (6.3.6)$$

where $\sigma_x = \alpha_x c_x, \ \sigma_y = \alpha_y c_y$.

The solutions of (6.3.6) with respect to $\Delta V, \Delta F$ are rather cumbersome expressions, so we only write them for the case where

$$d_x = d_y = d, \quad \sigma_x = \sigma_y = \sigma, \qquad (6.3.7)$$

which means that the degrees of freedom are "partially" identical:

$$\Delta v = \frac{\sigma}{d + \sigma} \frac{v(1 + \sigma g) \cos \gamma - F_d g \sin \gamma}{\cos \beta + \sigma g \cos(\gamma - \beta) \cos \gamma} - v,$$

$$\Delta F = \frac{v \sigma \sin(\gamma - \beta) \cos \gamma - F_d \cos \beta}{\cos \beta + \sigma g \cos(\gamma - \beta) \cos \gamma} \qquad (6.3.8)$$

The expressions (6.3.8) do not depend on the angle ψ between the straight line (6.3.1) and the axis OX, which could be expected under the identity conditions (6.3.7). The parameters (6.3.8) should satisfy the inequalities

$$v + \Delta v > 0, \quad F_d + \Delta F > 0, \qquad (6.3.9)$$

since they are obtained for the motion in contact with an object for $\dot{s}_B > 0$.

For $\gamma \neq \beta$, the tangential velocity vector $v\tau^*$ is not directed along the straight line (6.3.1), but rather along the straight line making the angle $\psi + \beta - \gamma$ with the axis OX. This introduces a systematic error into tracking of the commanded motion velocity and contact force. These errors remain for $\sigma \to \infty$ as well:

$$\Delta v = v \left(\frac{1}{\cos(\gamma - \beta)} - 1 \right), \quad \Delta F = \frac{v}{g} \tan(\gamma - \beta) \qquad (6.3.10)$$

The limit transition for $\sigma \to \infty$ $(c_x, c_y \to \infty)$ in the control law may be considered as a transition to a switching control system [263], since in reality the absolute values of the voltages u_x, u_y are bounded.

The expressions (6.3.10) give the solution of the system (6.3.6) for σ_x, $\sigma_y \to \infty$, even if the conditions (6.3.7) do not hold.

Let us consider some special cases.

For $\beta = \gamma = 0$, we obtain from (6.3.8)

$$\Delta v = -\frac{d}{d+\sigma} v, \quad \Delta F = -\frac{F_d}{1+\sigma g} \qquad (6.3.11)$$

For $\beta = \gamma$ and $\beta \geq 0$, ΔF coincides with (6.3.11) and

$$\Delta v = -\frac{d}{d+\sigma} v - \frac{\sigma g \tan \beta}{(d+\sigma)(1+\sigma g)} F_d \qquad (6.3.12)$$

Compared to (6.3.11), (6.3.12) has an additional term, which describes a frictional force. If $\sigma \to \infty$, then static errors tend to zero for $\beta = \gamma$: Δv, $\Delta F \to 0$. If $g \to \infty$, then $\Delta F \to 0$.

If $\gamma = 0$, then

$$\Delta v = v \left[\frac{\sigma}{(d+\sigma)\cos \beta} - 1 \right], \quad \Delta F = -\frac{v\sigma \tan \beta + F_d}{(1+\sigma g)} \qquad (6.3.13)$$

To study the stability of the stationary regime (6.3.5), let us write the linear equations (6.3.2), (6.3.3) in the deviations $\Delta x, \Delta y$ from this regime:

$$\Delta \ddot{x} + q_x \Delta \dot{x} + r_x \sin \delta \Delta x - r_x \cos \delta \Delta y = 0,$$

$$\Delta \ddot{y} + q_y \Delta \dot{y} - r_y \sin \delta \Delta x + r_y \cos \delta \Delta y = 0 \qquad (6.3.14)$$

In the above equation,

$$q_x = \frac{d_x + \sigma_x}{M_x}, \quad q_y = \frac{d_y + \sigma_y}{M_y},$$

$$r_x = \frac{k}{M_x \cos \beta}[\sin(\psi + \beta) + \sigma_x g \cos \gamma \sin(\psi + \beta - \gamma)],$$

$$r_y = \frac{k}{M_y \cos \beta}[\cos(\psi + \beta) + \sigma_y g \cos \gamma \cos(\psi + \beta - \gamma)], \qquad (6.3.15)$$

The characteristic equation of the system (6.3.14) has the form

$$p^4 + p^3(q_x + q_y) + p^2(r_x \sin \psi + r_y \cos \psi + q_x q_y)$$

$$+ p(q_x r_y \cos \psi + r_x q_y \sin \psi) = 0 \qquad (6.3.16)$$

The equation (6.3.16) has a zero root, since the coordinate ξ is cyclic (see (6.3.14)). We shall consider the stability of the system (6.3.2), (6.3.3) with respect to the three remaining variables ξ, η, $\dot{\eta}$.

Let us write out the Hurwitz asymptotic stability conditions [135]

$$q_x r_y \cos \psi + r_x q_y \sin \psi > 0,$$

$$q_x r_x \sin \psi + q_y r_y \cos \psi + q_x q_y(q_x + q_y) > 0 \qquad (6.3.17)$$

and complement the conditions (6.3.7) on the system parameters with the equality

$$\mathcal{M}_x = \mathcal{M}_y = \mathcal{M} \qquad (6.3.18)$$

Since q_x, $q_y > 0$ (see (6.3.15)), the two inequalities (6.3.17) reduce to the first one, which under the conditions (6.3.7), (6.3.18) has the form

$$\cos \beta + \sigma g \cos(\gamma - \beta) \cos \gamma > 0 \qquad (6.3.19)$$

Since

$$0 \le \beta < \frac{\pi}{2}, \quad 0 \le \gamma < \frac{\pi}{2}, \qquad (6.3.20)$$

the inequality (6.3.19) holds for $\sigma, q \ge 0$.

Under the condition (6.3.19), the inequality (6.3.9) has the form

$$v(1 + \sigma g) > g F_d \tan \gamma, \quad g F_d > v \tan(\beta - \gamma) \qquad (6.3.21)$$

Figure 6.3.1 shows the domain described by the inequalities (6.3.20), (6.3.21) for the variables β, γ. The stable stationary regime of motion exists in this domain. If β is close to $\frac{\pi}{2}$ and γ to zero, then the stationary regime of motion does not exist, as can be seen from Figure 6.3.1.

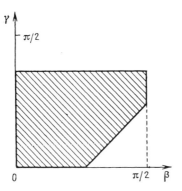

FIGURE 6.3.1
Stability domain in parameter space for following a linear contour

In the absence of feedback, i.e., for $g = 0$, the inequalities (6.3.21) take the form

$$v > 0, \quad \gamma > \beta$$

Given v, F_d, γ and by substituting various β within the typical range of friction angles for the given materials, one can determine coefficients g, σ such that the stationary regime exists.

By using (6.3.8) and the inequalities (6.3.9), one can verify that the motion velocity decreases and the contact force grows with the increase of the angle γ (in the vicinity of $\gamma = \beta$).

Assume now that the degrees of freedom are not identical, i.e., at least one of the equalities (6.3.7), (6.3.18) does not hold. Then we can prove that if the second inequality in (6.3.21) is satisfied, then the conditions (6.3.9) hold for sufficiently large σ_x, σ_y. For each pair of variables β, γ that satisfies the condition (6.3.20), the inequalities (6.3.17) are also satisfied for sufficiently large σ_x, σ_y. In this case, the stationary regime exists and is asymptotically stable.

6.4 Following a Circular Contour

Let the equations (6.1.3) describe a circle of radius R with the center in the coordinate origin O (see Figure 6.4.1).

$$x = -R\cos\varphi, \quad y = R\sin\varphi, \quad s = R\varphi, \tag{6.4.1}$$

where the arc length s is counted clockwise from the point $(-R, 0)$.

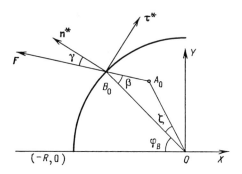

FIGURE 6.4.1
Kinematics for a circular object contour

According to (6.1.5), we obtain from (6.4.1)

$$\boldsymbol{\tau} = [\sin\varphi, \cos\varphi]^T, \quad \boldsymbol{n} = [-\cos\varphi, \sin\varphi]^T$$

Suppose the manipulator is in contact with the circle (6.4.1) and the contact point B moves so that $\dot{s}_B > 0$. Then after a simple trigonometric transformation, the equation (6.1.8) with respect to s_B takes the form

$$x\sin(\varphi_B - \beta) + y\cos(\varphi_B - \beta) = R\sin\beta, \quad \varphi_B R = s_B \tag{6.4.2}$$

One can also obtain (6.4.2) directly by examining Figure 6.4.1. By expanding the expressions for the components F_x, F_y of the force vector $\boldsymbol{F}$, we can present (6.1.1) in the form

$$\mathcal{M}_x\ddot{x} + d_x\dot{x} + F\cos(\varphi_B - \beta) = \alpha_x u_x,$$

$$\mathcal{M}_y\ddot{y} + d_y\dot{y} - F\sin(\varphi_B - \beta) = \alpha_y u_y \tag{6.4.3}$$

By expanding the vectors $\boldsymbol{\tau}^*$, $\boldsymbol{n}^*$ in accordance with (6.2.5), (6.4.2), we obtain the components u_x, u_y of the vector $\boldsymbol{U}$ for the control law (6.2.2), (6.2.3), (6.2.5) as follows:

$$u_x = c_x[-g(F\cos\gamma - F_d)\cos(\varphi_B - \beta + \gamma) - v\sin(\varphi_B - \beta + \gamma) - \dot{x}],$$

$$u_y = c_y[g(F\cos\gamma - F_d)\sin(\varphi_B - \beta + \gamma) + v\cos(\varphi_B - \beta + \gamma) - \dot{y}] \qquad (6.4.4)$$

We apply the Sine Rule to the triangle $A_O O B_O$ (see Figure 6.4.1) and obtain

$$\frac{R}{\sin(\beta + \zeta)} = \frac{r}{\sin\beta} = \frac{F}{k\sin\zeta}, \qquad (6.4.5)$$

where $r^2 = OA_0^2 = x^2 + y^2$, and ζ is the angle $A_0 O B_0$. One can use the equation (6.4.5) to determine the force F and the angle ζ. The first equation in (6.4.5) is equivalent to (6.4.2).

We need to find a stationary regime for the system (6.4.3), (6.4.4) corresponding to the constant displacement velocity and constant contact force:

$$\dot{s}_B = \frac{vR}{r} = \text{const}, \quad r = \text{const} > 0, \quad V = v + \Delta v,$$

$$F\cos\gamma = F_d + \Delta F = \text{const}, \quad \zeta = \text{const}, \quad \varphi_B = \frac{V}{r}t + \text{const},$$

$$x = -r\cos(\varphi_B + \zeta), \quad y = r\sin(\varphi_B + \zeta) \qquad (6.4.6)$$

By substituting (6.4.6) into (6.4.3), (6.4.4), we obtain the algebraic equations for stationary regime parameters:

$$M_x\frac{V^2}{r}\cos(\varphi_B + \zeta) + (d_x + \sigma_x)V\sin(\varphi_B + \zeta) + \frac{F_d + \Delta F}{\cos\gamma}\cos(\varphi_B - \beta)$$

$$= -\sigma_x g\Delta F\cos(\varphi_B - \beta + \gamma) + \sigma_x v\sin(\varphi_B - \beta + \gamma),$$

$$(6.4.7)$$

$$-M_y\frac{V^2}{r}\sin(\varphi_B + \zeta) + (d_y + \sigma_y)V\cos(\varphi_B + \zeta) - \frac{F_d + \Delta F}{\cos\gamma}\sin(\varphi_B - \beta)$$

$$= \sigma_y g\Delta F\sin(\varphi_B - \beta + \gamma) + \sigma_y v\cos(\varphi_B - \beta + \gamma)$$

Unlike (6.3.6), the equations (6.4.7) are nonlinear in the parameters ΔV and ΔF of the stationary regime.

Suppose that the manipulator degrees of freedom are not identical, i.e., at least one of the equations (6.3.7), (6.3.18) does not hold. It is possible to demonstrate that the equations (6.4.7) do not have a solution in this case, and therefore, the stationary regime (6.4.6) does not exist.

Let us assume that the identity conditions (6.3.7), (6.3.18) hold. We add up the first and second equations in (6.4.7) pre-multiplied by $\cos(\varphi_B + \zeta)$ and $-\sin(\varphi_B + \zeta)$, respectively, and obtain the first equation for the transformed system. By adding up the first and second equations pre-multiplied by $\sin(\varphi_B + \zeta)$ and $\cos(\varphi_B + \zeta)$, respectively, we obtain the second equation for the transformed system as

$$\mathcal{M}\frac{V^2}{r} + \Delta F\left[\frac{\cos(\beta + \zeta)}{\cos\gamma} + \sigma g\cos(\beta + \zeta - \gamma)\right]$$

$$= -\sigma v\sin(\beta + \zeta - \gamma) - \frac{F_d}{\cos\gamma}\cos(\beta + \zeta), \tag{6.4.8}$$

$$(d + \sigma)V + \Delta F\left[\frac{\sin(\beta + \zeta)}{\cos\gamma} + \sigma g\sin(\beta + \zeta - \gamma)\right]$$

$$= \sigma v\cos(\beta + \zeta - \gamma) - \frac{F_d}{\cos\gamma}\sin(\beta + \zeta)$$

Clearly, the equations (6.4.8) should be considered jointly with the equalities (6.4.5) that relate the parameters r, ΔF, and ζ.

Let $\beta = \gamma = 0$, and therefore $\zeta = 0$. In this case, we obtain from (6.4.8) (compare to (6.3.11)):

$$\Delta v = -\frac{d}{d + \sigma}v, \quad \Delta F = -\frac{F_d}{1 + \sigma g} - \frac{\mathcal{M}v^2\sigma^2}{(d + \sigma)^2(1 + \sigma g)r} \tag{6.4.9}$$

The first equations in (6.4.9) and (6.3.11) coincide. An additional term in the second expression in (6.4.9) as compared to (6.3.11) introduces a centrifugal force "pushing" the manipulator from an object (for $r > 0$). By substituting the expression $F_d + \Delta F = k(R - r)$ into the second equation in (6.4.9), one can prove that under the condition

$$F_d > \frac{\mathcal{M}v^2\sigma}{(d + \sigma)^2 Rg}, \tag{6.4.10}$$

this equation has the unique solution $r \in (0, R)$.

For $r \to \infty, \zeta \to 0$ and bounded V, the equations (6.4.8) become linear with respect to $\Delta v, \Delta F$. In this case, they coincide with the equations obtained after the respective transformation of (6.3.6) under the identity conditions (6.3.7).

The nonlinear equations (6.4.5), (6.4.8) can be solved in the following way (backwards). Given the velocity V and force F, one can determine the radius r and angle ζ from (6.4.5). Then these parameters can be found from the relationships (6.4.8) that are linear in v and F_d.

By considering the last two equations (6.4.6) as a transformation from the variables x, y to the variables r, φ_B, one can write (6.4.3), (6.4.4) in polar coordinates. Under the identity conditions (6.3.7), (6.3.18), these equations have the form:

$$\mathcal{M}[\ddot{r} - r(\dot{\varphi}_B + \dot{\zeta})^2] + (d + \sigma)\dot{r} - F\cos(\beta + \zeta)$$

$$= \sigma[g(F\cos\gamma - F_d)\cos(\beta + \zeta - \gamma) + v\sin(\beta + \zeta - \gamma)], \qquad (6.4.11)$$

$$\mathcal{M}[r(\ddot{\varphi}_B + \ddot{\zeta}) + 2\dot{r}(\dot{\varphi}_B + \dot{\zeta})] + (d + \sigma)r(\dot{\varphi}_B + \dot{\zeta}) + F\sin(\beta + \zeta)$$

$$= \sigma[-g(F\cos\gamma - F_d)\sin(\beta + \zeta - \gamma) + v\cos(\beta + \zeta - \gamma)]$$

One can determine the angle ζ and force F from (6.4.5). In (6.4.11), the angle φ_B is only present as its first and second derivatives, same as the coordinate ξ in (6.3.2), (6.3.3), i.e., φ_B is a cyclic coordinate.

Clearly, by substituting (6.4.6) into (6.4.11), we obtain the algebraic equations (6.4.8).

To investigate stability of the stationary regime, it is necessary to derive the equations in the variations $\Delta r, \Delta\varphi_B$. In the equations (6.4.11), in addition to the variables r, φ_B, the angle ζ and force F should vary. The variations of the latter may be excluded by using the relationships obtained by varying the equalities (6.4.5). For example, we obtain the relationship between $\Delta\zeta$ and Δr from the first equality in (6.4.5):

$$\Delta\zeta = -\tan(\beta + \zeta)\frac{\Delta r}{r}$$

We omit intermediate computations and present the resulting equations in the variations

$$\Delta\ddot{r} + \left[\frac{d + \sigma}{\mathcal{M}} + \frac{2V}{r}\tan(\beta + \zeta)\right]\Delta\dot{r} + \left[\frac{k}{\mathcal{M}} + \frac{k\sigma g\cos\gamma\cos(\beta + \zeta - \gamma)}{\mathcal{M}\cos(\beta + \zeta)}\right.$$

$$-\left(\frac{V}{r}\right)^2 + \frac{d+\sigma}{M}\frac{V}{r}\tan(\beta+\zeta)\bigg]\Delta r - 2V\Delta\dot{\varphi}_B = 0, \qquad (6.4.12)$$

$$\Delta\ddot{\varphi}_B + \left[\frac{d+\sigma}{M} - \frac{2V}{r}\tan(\beta+\zeta)\right]\Delta\dot{\varphi}_B + \frac{2V}{r^2\cos^2(\beta+\zeta)}\Delta\dot{r}$$

$$+\left[\frac{(d+\sigma)V}{Mr^2\cos^2(\beta+\zeta)} + \frac{k\sigma g}{2Mr}\frac{\sin 2\gamma}{\cos^2(\beta+\zeta)}\right]\Delta r = 0 \qquad (6.4.13)$$

In the equations (6.4.12), (6.4.13), V,ζ,r are the variables in the stationary regime. These equations are derived by using the relationships (6.4.8) for the stationary regime parameters. We obtain the equation (6.4.12) by varying the first equation in (6.4.11). The equation (6.4.13) is produced by a linear composition of (6.4.12) with the variation of the second equation in (6.4.11). Let us consider stability of the system (6.4.12), (6.4.13) with respect to the three variables $\Delta r, \Delta\dot{r}, \Delta\dot{\varphi}_B$. In other words, consider the stability of the trivial solution

$$\Delta r = \Delta\dot{r} = \Delta\dot{\varphi}_B = 0 \qquad (6.4.14)$$

for the system (6.4.12), (6.4.13).

We omit the third-order characteristic equation of the system (6.4.12), (6.4.13), as the expressions for its coefficients are cumbersome. The Hurwitz conditions for this equation are satisfied for

$$\beta + \zeta < \pi/2 \qquad (6.4.15)$$

and a sufficiently large σ. The inequality (6.4.15) may be violated if the friction angle β is close to $\pi/2$, and the stiffness factor k of the force sensor is small.

For $\beta = \gamma = \zeta = 0$, the equations in variations (6.4.12), (6.4.13) are simplified. The third-order characteristic equation for such a simplified system has the form

$$p^3 + \frac{2(d+\sigma)}{M}p^2 + \left[\left(\frac{d+\sigma}{M}\right)^2 + 3\left(\frac{V}{r}\right)^2 + \frac{k}{M}(1+\sigma g)\right]p$$

$$+\frac{(d+\sigma)V^2}{Mr^2} + \frac{k}{M^2}(d+\sigma)(1+\sigma g) = 0$$

For $\sigma, g > 0$, the Hurwitz conditions for the asymptotic stability of the solution (6.4.14) are satisfied.

According to the Lyapunov theorem on stability in the first approximation [29], it follows from the asymptotic stability of the solution (6.4.14) for the system of equations in variations (6.4.12), (6.4.13) that the solution (6.4.6) for the original nonlinear system is asymptotically stable.

Thus, the control (6.2.2), (6.2.5) can provide asymptotic stability for a stationary motion of a manipulator with identical degrees of freedom along a circular contour, i.e., for a motion at a constant velocity with a constant contact force. Hence, provided the initial values of the velocity and contact force do not differ too much from stationary values, the manipulator can follow the circular contour without losing contact with it. In Section 6.3, we proved that an asymptotically stable stationary regime exists for the motion along a straight line even though the identity conditions (6.3.7), (6.3.18) do not hold. For the motion along a circle, a stationary regime does not exist if the identity conditions do not hold. However, since solutions for the equations (6.4.3), (6.4.4) continuously depend on the parameters, the manipulator is still able to track the circle without losing contact with it for "a sufficiently small perturbation" of (6.3.7), (6.3.18). To this end, the initial values of the velocity and contact force should not differ too much from stationary values given by (6.4.8).

For the control (6.2.2), (6.2.5) to be efficient, the set of "acceptable" initial condition should be "large". Since the system (6.4.3), (6.4.4) is nonlinear, it is impossible to estimate the dimension of this set analytically. It can be estimated by solving the system (6.4.3), (6.4.4) numerically, though this presents a challenge, since the order of the phase space is four.

6.5 Experiments in Contour Following

In the preceding sections, we analytically studied the manipulator motion along a straight line and a circle. Such analytical study of a motion for more complex contours is difficult. Yet by using the above-derived equations (6.1.1)–(6.1.9), (6.2.1– (6.2.5), it is possible to conduct a numerical study.

It may be assumed that for a small commanded velocity v, the control (6.2.2), (6.2.5) ensures a manipulator motion without the loss of contact not only along a straight line and a circle, but along smooth contours of a more complex shape as well. It seems that for a small velocity of motion, individual contour sections may be considered as segments of straight lines or arcs of circles. However, the exhaustive proof of the control (6.2.2), (6.2.5) is in the experiments. Studying the equations (6.1.1)–(6.1.9), (6.2.1– (6.2.5), one should not forget that they present a mathematical model and as such, are but an idealization of a real-world system. The experiments can verify if the developed model is adequate for the problem in question.

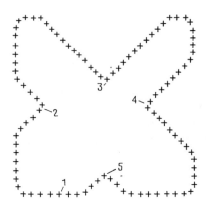

FIGURE 6.5.1
A complex shape contour

The experimental research conducted on the robotic system proves that the control (6.2.2), (6.2.5) allows one to follow object contours of a comparatively complex shape. In the experiments, as the manipulator moves along the contour, position sensors provide the coordinates x, y to the computer. The object contour can be reconstructed from the coordinate data. A finger sensor provides the information on the force F to the computer. Figure 6.5.1 displays an object contour, Figure 6.5.2, the corresponding plot of the force $F(t)$.

In the experiment, $v = 10$ mm/s, $F_d = 1$ N. The points *1, 2, 3, 4, 5* in Figure 6.5.1 correspond to the respectively numbered points in Figure 6.5.2. The contour is concave at these points, and the force time-history exhibits peaks there. These peaks can be reduced by decreasing the commanded contour velocity v. As seen in Figure 6.5.2, the force F does not exceed 4 N. According to (6.1.2), given the sensor stiffness $k = 8 \cdot 10^3$ N/m, the contour tracking error caused by the sensor compliance does not exceed 0.5 mm. If the commanded contour velocity v increases, then the force sensor may lose contact with the contour on the segments with a small curvature radius. The contact loss may be avoided by increasing the commanded contact force F.

For measuring distances and displacements, guiding the manipulator towards an object, and monitoring motion safety, various types of proximity sensors are used, such as inductive, ultrasound, pneumatic, and optical [58, 59, 178, 279]. The paper [105] discusses the possibility of using algo-

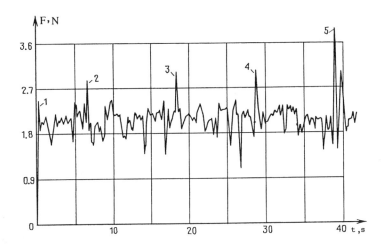

FIGURE 6.5.2
Time history of contact force in the contour following experiments

rithms designed to control manipulators with force sensors for controlling manipulators with proximity sensors. The paper [152] considers a control problem for a gantry manipulator with an optical proximity sensor. The paper [152] shows that control of the form (6.2.1), (6.2.2) enables the manipulator with such proximity sensor to move along an object contour. The paper [152] simulates and analytically studies a motion of the manipulator with an optical proximity sensor along a rectangular and circular object contour. It determines stationary regimes for the motion with a constant velocity at a constant distance from the object, tracking errors for the commanded velocity and desired sensor measurement, and studies the motion stability. The experimental results of the paper [152] demonstrate that control of the form (6.2.1), (6.2.2) enables the manipulator to follow object contours of a rather complex shape.

6.6 Rotating a Steering Wheel

Let us consider a task of rotating a steering wheel by a manipulator instrumented with a force sensor. Such a problem is studied, for example, in [163].

Let the rotation axis of the steering wheel be vertical (see Figure 6.6.1). To control the wheel rotation, a force sensor only needs to measure two

FIGURE 6.6.1
Rotating a steering wheel

horizontal components of the force vector. In our experiments, the force sensor shown in Figure 2.2.7 was attached to the end of the manipulator arm (mast). The lower end of the sensor is inserted (without a clearance) into an opening in the steering wheel rim. A gantry manipulator is used in this task, so that only two translational degrees of freedom can be used to control horizontal motions of the arm. The problem is as follows: design a control enabling the manipulator to rotate the steering wheel.

The position of the sensor tip inserted in the steering wheel is constrained to be on the circumference. To rotate the steering wheel, it is necessary to move the sensor tip along the circumference. In doing so, each point of the mast should move along a curve closely approximating a circle. We assume that the position of the circle center and its radius are a priori unknown. In the problem of contouring an object with the manipulator moving in contact with it, as considered in Sections 6.1– 6.5, the constraint imposed on the force sensor is not binding (is unilateral). Therefore, to avoid the loss of contact between the manipulator and object, the commanded force F_d of contact with the object in (6.2.1) needs to be positive. For the problem we consider in this section, the constraint is binding (bilateral). Hence, using the same control design (6.2.1) to solve this problem, we can take $F_d = 0$:

$$V_d = gF_n n + v\tau \qquad (6.6.1)$$

In the above equation, $V_d = (V_{xd}, V_{yd})$ is a commanded velocity vector for the manipulator motion, n and τ are vectors of normal and tangent to the constraint (the circle), F_n is a projection of the force applied to the sensor onto the normal n, $v = $ const is a commanded velocity of the wheel rotation (contouring velocity), $g > 0$ is a constant gain. The direction of the wheel rotation is determined by the direction of the vector τ and sign of v. The normal n in (6.6.1) can be chosen as a vector directed either from the center or towards the center of the wheel.

If the friction in the wheel axis and the wheel mass are negligible, then the force exerted on the sensor by the wheel is always directed along the radius. By measuring the vector F of this force, we determine the vector of normal $n = F/F$, and therefore, the vector of tangent τ. Then we can rewrite (6.6.1) as follows:

$$V_d = gF + v\tau \qquad (6.6.2)$$

It seems that (6.6.2) may also be used when the force of resistance to the wheel rotation is nonzero, but not too large. If the resistance force of the wheel rotation is large, then (6.6.1) should be used to calculate a commanded velocity vector. To calculate the vectors of normal and tangent to the constraint, one may use a priori information on friction in the wheel axis. These vectors can also be determined by using the position sensors of the manipulator. As the steering wheel rotates, each point of the manipulator arm moves along a curve closely approximating a circle. By measuring consecutive positions of the arm, one can determine a secant to the curve so as to build vectors close to τ and n. In the beginning of the motion, the vectors of normal and tangent should be given. Control voltages applied to the manipulator drives are computed linearly, according to (6.6.2). Thus, the required manipulator control law is designed.

The described control law and its modifications were successfully applied in the experiments.

6.7   Planar Two-link Manipulator

In Chapter 5 and the preceding sections of this chapter, a gantry manipulator motion is studied. In this section, we consider an articulated manipulator (see also Section 7.4).

Consider a two-link articulated manipulator moving in the horizontal plane. Figure 6.7.1 displays the schematics of such a manipulator. A link OC of length l_1 rotates about a stationary point O, where a cylindrical joint with a vertical rotation axis is located. A link CA_1 of length l_2 is attached to the link OC at a point C through a joint with a vertical axis. Each joint is actuated by a separate motor.

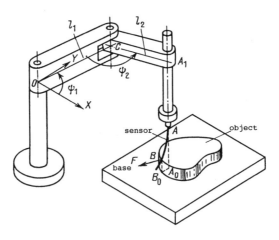

FIGURE 6.7.1
A planar two-link articulated manipulator

In the horizontal plane, we attach a stationary orthogonal coordinate system OXY with an arbitrary but fixed orientation, to the point O. Denote by ψ_1 a counterclockwise rotation angle of the link OC about the vertical axis. Denote by ψ_2 an angle OCA_1 between the links. A force sensor measuring two horizontal force components is mounted on the manipulator at a point A. As in Section 6.1, the sensor is modeled by an infinitely thin massless rod that is vertical when undeformed (see Figure 6.7.1 and 6.1.1).

Consider a cylindrical object that rests on the horizontal base. The lateral surface of the object is orthogonal to the plane XY, the top and bottom surfaces are parallel to it. The point A_0 is a projection of the point A onto the base plane. When the force sensor contacts the object, a horizontal force that may be calculated from (6.1.2) acts on the sensor at the point B. By writing the equation for the object contour in the form (6.1.3), we can use the relationships (6.1.4)– (6.1.10).

Equations of motion for a two-link manipulator are presented in a number of publications (see, for example, [76]). Based on these equations and using the relationships (6.1.2)–(6.1.10), one can design a mathematical model for the manipulator motion in contact with the object.

To enable the force sensor to move along the object contour without a loss of contact, we define the commanded (desired) velocity vector of the manipulator point A in the form (6.2.1) or (6.2.5). The vector projections on the axes OX and OY are V_{xd} and V_{yd} (see (6.2.3)).

The coordinates x, y of the point A may be expressed through the angles ψ_1, ψ_2 as follows:

$$x = l_1 \cos \psi_1 - l_2 \cos(\psi_1 + \psi_2),$$

$$y = l_1 \sin \psi_1 - l_2 \sin(\psi_1 + \psi_2) \tag{6.7.1}$$

Let us differentiate (6.7.1) with respect to time and then solve for the angular velocities $\dot\psi_1, \dot\psi_2$:

$$\dot\psi_1 = \omega_1 = \frac{l}{l_1 \sin \psi_2}[\dot x \cos(\psi_1 + \psi_2) + \dot y \sin(\psi_1 + \psi_2)],$$

$$\dot\psi_2 = \omega_2 = \frac{l}{l_1 l_2 \sin \psi_2} \{\dot x[l_1 \cos \psi_1 - l_2 \cos(\psi_1 + \psi_2)] \tag{6.7.2}$$

$$+ \dot y[l_1 \sin \psi_1 - l_2 \sin(\psi_1 + \psi_2)]\}$$

The equations (6.7.2) are valid for $\psi_2 \neq 0, \pi$. By substituting the commanded values V_{xd}, V_{yd} of the linear velocities $\dot x = V_x$, $\dot y = V_y$ into (6.7.2), we obtain the desired (commanded) values ω_{1d}, ω_{2d} of the angular velocities ω_1, ω_2. The commanded values ω_{1d}, ω_{2d} at a given time depend on the system configuration at that time. To compute ω_{1d}, ω_{2d}, one needs, in addition to a force sensor, sensors for measuring the angles ψ_1, ψ_2. The commanded values ω_{1d}, ω_{2d} are supplied to the velocity servo system controlling the drives in the joints. Then the control voltages u_1, u_2 applied to the electromotors have the form

$$u_i = c_i(\omega_{id} - \omega_i), \quad i = 1, 2, \tag{6.7.3}$$

where c_1, c_2 are constant positive gains, or

$$u_i = u_{i0} \, \text{sgn} \, (\omega_{id} - \omega_i), \quad i = 1, 2, \tag{6.7.4}$$

where u_{10}, u_{20} are maximum possible voltages applied to the motors.

To determine the voltages (6.7.3) or (6.7.4), it is necessary to measure angular velocities ω_1, ω_2 by tachometers.

Chapter 7

Control of Constrained Motion

In this chapter, we develop a mathematical model for a general manipulator system with a force sensor. Manipulator compliance (in the joints, links or base) is not taken into account. We consider the system compliance to be lumped in the force sensor. For a number of tasks, it is appropriate to consider a manipulator motion as a motion with mechanical constraints imposed on a manipulated object. This is the case, for instance, in mating operations such as assembly, or in robotic machining. In such tasks, the manipulator control should make a manipulated object move along the constraints imposed on it. If these constraints are not binding, then the problem of keeping contact with the constraint needs to be addressed. This problem has been discussed in Chapter 5. Chapter 6 considers the manipulator motion along an object contour, which is a special case of motion along (not binding) constraints.

For a robotic system with imposed constraints, the use of information on forces (deformations, strains) acting in the system seems to provide an adequate description of the process. Hence, control based on the force information is an adequate approach to solve the problem of moving a system along constraints.

This chapter proposes a general approach to design of robotic system control for manipulating an object subjected to constraints. This approach is an extension of the method for setting a commanded velocity presented in the preceding chapter. The control is designed by superimposing basic motions along normals and at a tangent to the constraints. The commanded velocity vector is given as a sum of normal and tangential components. A component of the commanded velocity directed along a normal depends on the difference between a projection of the measured force on the normal and a commanded value of this force.

First, the proposed general approach to control of constraint motions is applied to an articulated manipulator with three degrees of freedom tracking an object contour. Then we study control of a motion along a screw constraint. Such motion occurs in automated assembly of threaded joints. One more typical control problem where a constraint is imposed on a manipulator is opening and closing a hatch lid. It is considered in the last section of this chapter.

7.1 General Mathematical Model for a Manipulator with a Force Sensor

Let us consider a robotic manipulator assuming all its links are rigid bodies. We do not take into account compliance in the joints connecting the adjacent links. The manipulator is equipped with a force sensor, which includes an elastic element and transducers measuring deformations or displacements in this element. Such sensor may be placed in the manipulator wrist (immediately preceding the gripper) or in the gripper fingers. Figure 7.1.1 schematically displays a sensor placed in the wrist, Figure 7.1.2, a sensor placed in the gripper fingers.

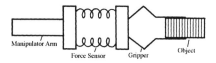

FIGURE 7.1.1
Wrist sensor schematics

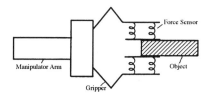

FIGURE 7.1.2
Schematics of a sensor in gripper fingers

Under the assumptions made, we can write the equations of motion for the manipulator links "before" the force sensor in the form of Lagrange's equations [76, 196]:

$$\frac{d}{dt}\left(\frac{\partial T}{\partial \dot{q}_i}\right) - \frac{\partial T}{\partial q_i} = -\frac{\partial \Pi}{\partial q_i} + Q_i, \quad i = 1, \ldots, n \qquad (7.1.1)$$

In the above equation, T and Π are the kinetic and potential energy of the system, q_i and Q_i are generalized coordinates and forces, respectively, n is the number of degrees of freedom.

If the sensor is placed in the gripper fingers, then the vector $q = [q_i]$ defines the gripper coordinates unambiguously. If the sensor is placed in the wrist, then its deformations may lead to the gripper displacement relative to the rest of the arm.

Since we do not take into account compliance in the manipulator structure, the potential energy Π is only defined by the gravity and depends on the manipulator configuration, i.e., on the vector q. If the manipulator is actuated by electric motors, then the kinetic energy T comprises the energy of rotating motor rotors and reduction gear (see, for example, (5.7.5), (5.8.3)).

The motor torque is directly proportional to the electrical current in the rotor armature [65]. By neglecting inductance in the motor circuit, we can write this torque in the form (5.1.6). The control force is defined by the voltage applied to the motor. The back e.m.f. provides a dissipative force. These forces are included in the vector of generalized forces $Q = [Q_i]$.

Let

$$l(q) = [l_j(q)], \quad j = 1, \ldots, \nu; \quad \nu \leq n; \quad \nu \leq 6 \qquad (7.1.2)$$

be a coordinate vector that defines the position and orientation of the arm link immediately preceding the gripper with respect to the stationary coordinate frame. As an example of the forward kinematics relationship (7.1.2), consider (6.7.1), (7.4.4).

Let $U = [u_k]$ be an $r \times 1$ column vector of control signals applied to the manipulator. Such signals may be voltages applied to the motors. Let $\Phi = [\varphi_j]$ be an $m \times 1$ column vector of forces and torques acting from the force sensor on the manipulator, either on the arm "before" the gripper, in case of a wrist sensor, or on the gripper, in case of a finger sensor. Components φ_j $(j = 1, \ldots, m)$ are the projections of force and torque vectors onto the axes of the stationary coordinate frame. We assume that only forces and torques measured by the force sensor act on the manipulator. The number m can assume values from one to six.

The column vector Q of the generalized forces applied to the manipulator can be represented in the form

$$Q = EU + D(q)\Phi + Q', \qquad (7.1.3)$$

where $E, D(q), Q' = [Q_i']$ are $n \times r$, $n \times m$, and $n \times 1$ matrices, respectively. The forces Q_i' comprise viscous and Coulomb friction forces as well as back e.m.f.

Assume that a manipulated object (part, workpiece) is held in the gripper. Let $\Lambda(q)$ be a $p \times 1$ column vector defining the position and orientation of the manipulated object in the stationary coordinate frame. If the sensor does not deform, then the arm link adjacent to it and the manipulated object will form a single rigid body. Hence the dimension p of the vector $\Lambda(q)$ does not exceed the dimension ν of the coordinate vector $l(q)$ of the manipulator link to which the force sensor is attached $(p \leq \nu)$. For example, if the manipulated object is modeled as a massive particle, then $p < \nu$. For $p = \nu$, we assume that $\Lambda(q) = l(q)$. For $p < \nu$, we assume that components of the vector $\Lambda(q)$ coincide with p components of the vector $l(q)$, for instance, with the first p components $l_1(q), \ldots, l_p(q)$. Further, denote by z a column vector defining the position and orientation of the manipulated object relative to the stationary coordinate frame, provided the force sensor is arbitrarily deformed. Assume that the vector z has a dimension p and for an undeformed sensor we have

$$z = \Lambda(q)$$

Assume also that the dimension p of the vector z coincides with the number m of components measured by the force sensor $(p = m)$. We write the vector z in the form

$$z = \Lambda(q) + H(l)\xi \quad (z = [z_j]; \ j = 1, \ldots, m), \tag{7.1.4}$$

where $\xi = [\xi_j]$ $(j = 1, \ldots, m)$ is a vector defining displacements in the elastic element of the sensor. The $m \times m$ matrix H depends on the sensor design and orientation relative to the stationary coordinate frame. If the sensor is in the wrist, then the gripper and the manipulated object it holds can be considered as a single rigid body.

By neglecting the mass of the elastic element, one may consider the sensor as a lumped spring (see Section 1.2). For small deformations of the elastic element, forces acting in it can be assumed proportional to the deformations (Hooke's law) and their derivatives (see (1.2.8)):

$$\Phi' = K\xi + B\dot{\xi} \quad (\Phi' = [\varphi'_j]; \ j = 1, \ldots, m), \tag{7.1.5}$$

In the above equation, Φ' is a column-vector of forces and torques acting from the sensor on the adjacent manipulator link, in projections onto the axes of the coordinate frame attached to this link. K and B are $m \times m$ constant positive-definite matrices, K being the stiffness matrix, B, the damping matrix describing energy dissipation in the sensor. We denote by $A(l)$ the transformation matrix for force and torque components in the transition from the coordinate frame attached to the adjacent arm link to the stationary frame and obtain

$$\Phi = A(l)\Phi' = A(l)[K\xi + B\dot{\xi}] \tag{7.1.6}$$

The relationship $\Phi' = K\xi$ is analogous to (1.2.7) or (1.2.9) and describes the static characteristics of the force sensor. By using the transformation matrix $A(l)$, the vector of forces and torques (vector of "static" forces) calculated from the sensor output can be projected onto the axes of the stationary coordinate frame

$$F = A(l)K\xi \quad (F = [f_i], \quad j = 1, \ldots, \mu) \tag{7.1.7}$$

The equations of motion for the manipulated object (together with the gripper, for a wrist sensor) are equations of motion for a rigid body. Forces $-\varphi_j$ $(j = 1, \ldots, m)$ act on this body. For the sake of simplicity, we assume that the coordinates z_j and forces φ_j $(j = 1, \ldots, m)$ are such that the virtual work through the displacement δz_j is $-\varphi_j \delta z_j$. Then we can write the equations of motion for the rigid body in the form

$$\frac{d}{dt}\left(\frac{\partial T_0}{\partial \dot{z}_j}\right) - \frac{\partial T_0}{\partial z_j} = -\frac{\partial \Pi_0}{\partial z_j} - \varphi_j + R_j, \quad j = 1, \ldots, m \tag{7.1.8}$$

In the above equation, T_0 and Π_0 are the kinetic and potential energy of the object, R_j $(j = 1, \ldots, m)$ are components of the resultant vector of the constraint forces imposed on the object. The potential energy Π_0 only depends on gravity.

Let the constraints imposed on the manipulated object be given in the form

$$\eta_s(z, t) \geq 0, \quad s = 1, \ldots, \mu_1, \quad \eta_s(z, t) = 0, \quad s = \mu_1, \ldots, \mu,$$

$$1 \leq \mu_1 \leq \mu, \quad \mu \leq m - 1 \tag{7.1.9}$$

Constraints are typical for assembly operations where mating of parts is required and for part machining. For example, to insert a peg into a hole or to drive a screw into a nut, the peg or screw should satisfy the constraints given by (7.1.9). A similar situation occurs when a tool (for instance, a grinding wheel) moves along a part surface during machining.

Consider the motion along constraints, i.e., under the conditions

$$\eta = [\eta_s(z, t)] = 0, \quad s = 1, \ldots, \mu, \tag{7.1.10}$$

We assume the constraints (7.1.10) to be ideal (frictionless) and obtain a vector R for the constraint forces acting on the manipulated object in the form

$$R = \sum_{s=1}^{\mu} \zeta_s \, \text{grad} \, \eta_s, \tag{7.1.11}$$

where ζ_s $(s = 1, \ldots, \mu)$ are Lagrange multipliers $(\zeta = [\zeta_s])$. From (7.1.11), we obtain the expressions for components of the vector of constraint forces:

$$R_j = \sum_{s=1}^{\mu} \zeta_s \frac{\partial \eta_s}{\partial z_j}, \quad j = 1, \ldots, m \tag{7.1.12}$$

For a given control U, the relationships (7.1.1)– (7.1.4), (7.1.6), (7.1.8), (7.1.10), (7.1.12) allow one to determine the manipulator motion. Thus, these relationships are equations of motion for the robotic system with a force sensor.

7.2   Discussion on Equations of Motion. Simplified Model

The constraint equations (7.1.10) may get so complicated that it is impossible to solve them analytically in any variables in order to exclude the latter from the equations of motion. In this case, it is possible to study the equations numerically, for instance, as follows.

By differentiating (7.1.10) twice with respect to time, we obtain

$$\sum_{j=1}^{m} \frac{\partial \eta_s}{\partial z_j} \ddot{z}_j + \Psi_s(z, \dot{z}) = 0 \quad s = 1, \ldots, \mu \tag{7.2.1}$$

In (7.2.1)

$$\Psi_s(z, \dot{z}) = \sum_{j=1}^{m} \sum_{k=1}^{m} \frac{\partial^2 \eta_s}{\partial z_j \partial z_k} \dot{z}_j \dot{z}_k + 2 \sum_{j=1}^{m} \frac{\partial^2 \eta_s}{\partial z_j \partial t} \dot{z}_j + \frac{\partial^2 \eta_s}{\partial t^2}$$

By integrating the equations of motion numerically at each calculation step, one can solve $n + m + \mu$ algebraic equations (7.1.1), (7.1.8), (7.2.1) (first, the relationships (7.1.2)–(7.1.4), (7.1.6), (7.1.12) for $n + m + \mu$ components of columns $\ddot{q}$, $\ddot{z}$, ζ should be substituted into them). In so doing and using the relationships (7.1.11) for constraints, one can obtain the right hand sides of the differential equations.

If some of the constraints imposed on the object are not binding (see the inequalities in (7.1.9)), then after solving the problem with binding constraints, one should verify that

$$R \cdot \operatorname{grad} \eta_s \geq 0, \quad s = 1, \ldots, \mu_1 \qquad (7.2.2)$$

For "orthogonal" constraints, i.e., if

$$\operatorname{grad} \eta_{s_1} \cdot \operatorname{grad} \eta_{s_2} = 0, \quad s_1, s_2 = 1, \ldots, \mu; \quad s_1 \neq s_2, \qquad (7.2.3)$$

the inequalities (7.2.2) are equivalent to $\zeta_s \geq 0$ $(s = 1, \ldots, \mu_1)$.

Let the manipulator arm link adjacent to the force sensor be stationary, i.e., $l \equiv \text{const}$. Then the equations (7.1.8) describe oscillations of the manipulated object with a deformation of the force sensor (such oscillations are considered in Sections 3.3 and 3.4). If the constraints are scleronomous, i.e. $\frac{\partial \eta_s}{\partial t} \equiv 0$ $(s = 1, \ldots, \mu)$, then these oscillations die out because of the dissipation. When the oscillations die out, the stationary regime is achieved such that

$$f_i = -\frac{\partial \Pi_0}{\partial z_j} + R_j, \quad j = 1, \ldots, m \qquad (7.2.4)$$

In the absence of gravity, $\frac{\partial \Pi_0}{\partial z_j} = 0$ $(j = 1, \ldots, m)$, so the expression (7.2.4) is simplified to become

$$f_i = R_j, \quad j = 1, \ldots, m \quad (F = R) \qquad (7.2.5)$$

If we assume the object and, for a wrist sensor, the gripper to be massless and neglect damping forces in the sensor $(B = 0)$, then the equations (7.1.8) transform into (7.2.5).

Provided the oscillation decay time in the force sensor is negligibly small compared to characteristic times of change for the vectors l and n, we can substitute approximate (quasistatic) relationships (7.2.4) or (7.2.5) for (7.1.8). The approximate relationships (7.2.5) explain why the terms force sensor, force and torque sensor, and force sensing transducer are used for elastic mechanical elements: the deformations of these elements allow determination of the forces acting between the manipulator and constraints.

If the vector l is known, then we may consider the relationships (7.1.4), (7.1.7), (7.1.10), (7.1.11), (7.2.5) as $4m+\mu$ algebraic equations with respect to $4m+\mu$ unknowns, i.e., components of the vectors z, F, ξ, ζ, and R. These components are functions of the vector l and time. By substituting these functions into (7.1.3), we obtain the system (7.1.1)–(7.1.3) that enables us to determine the manipulator motion for a given control U. Therefore, by substituting algebraic equations (7.2.4) or (7.2.5) for differential equations (7.1.8), we obtain a simplified model for the manipulator with a force sensor.

In the two preceding chapters, the one- and two-degree-of-freedom equations of motion for the manipulator with a force sensor are derived in the context of the simplified model.

The above mathematical models for the manipulator with a force sensor have an affinity to the elastic manipulator model developed in the book [28].

7.3   Control of Manipulator Motion Along Constraints

In this section, we assume that the constraints (7.1.9), (7.1.10) are scleronomous, i.e., the vector η does not depend on time: $\eta = \eta(z)$.

Consider an m-dimensional space $\Re^m$ of the manipulator coordinate vectors z. Let

$$n_s = \frac{\operatorname{grad} \eta_s}{|\operatorname{grad} \eta_s|}, \quad s = 1, \ldots, \mu, \tag{7.3.1}$$

be unit vectors of normals to the surfaces (7.1.10); $|\operatorname{grad} \eta_s|$ is the norm of the vector $\operatorname{grad} \eta_s$. We assume that the vectors (7.3.1) differ depending on the point $z \in P$, where P is the intersection of the surfaces (7.1.10). In other words, we assume that P is an $(m - \mu)$-dimensional manifold.

Let τ be a unit vector from the tangent space at the point $z \in P$:

$$\tau \cdot n_s = 0, \quad s = 1, \ldots, \mu \tag{7.3.2}$$

If $\mu = m$, then the manifold P is a point, and (7.3.2) defines the vector $\tau = 0$. If $\mu = m - 1$, then P is a one-dimensional manifold in $\Re^m$ and the vector τ may be defined at each point $z \in P$ up to the sign. In this case, let us choose the direction of the vector τ to be the same as the direction τ_0 of the desired motion for the manipulated object. If $\mu < m - 1$, then (7.3.2) defines a set of unit vectors τ at each point $z \in P$. Let us assume that by analyzing each individual operation, we can select from this set a vector τ_0 that defines the desired motion of the manipulated object along the constraint. In the control problem of following an object contour with the manipulator considered in Chapter 6 and Section 7.4, $\mu = m - 1$, and the vector τ_0 is uniquely defined by the desired direction of the manipulator motion. In the problem of moving along a screw constraint in the assembly of a threaded connection considered in Section 7.5, $\mu = m - 1$ as well; in this case, the vector τ_0 is uniquely defined if a screw is to be driven in. If "unscrewing" is required, then the vector τ_0 is opposite in sign.

Let us design a manipulator control such that a manipulated object moves along constraints. By generalizing the approach to design of manipulator control for following an object contour described in Section 6.2, we shall design the motion along constraints as a *superposition* of $\mu+1$ basic motions – *along normals and a tangent to constraints*. Accordingly, we define the "commanded" value $V_{z,d}$ of the velocity vector $\dot{z}$ as a sum of $\mu + 1$ items

$$V_{z,d} = V_{n_1} + \ldots + V_{n_\mu} + V_\tau$$

$$= g_1(R_{n_1} - R_{n_1,d})n_1 + \ldots + g_\mu(R_{n_\mu} - R_{n_\mu,d})n_\mu + v\tau_0, \quad (7.3.3)$$

where $R_{n_s} = R \cdot n_s$ is a projection of the vector of constraint forces and torques onto the normal n_s; $R_{n_s,d}$ is a desired (commanded) value of the projection R_{n_s}; v is a given (commanded) value for the manipulated object velocity of motion along the desired direction τ_0, g_s are constant gains to be chosen $(s = 1, \ldots, \mu)$. If $g_s > 0$, then the component V_{n_s} of the vector $V_{z,d}$ is directed towards the increase of η_s for $R_{n_s} > R_{n_s,d}$, and in the opposite direction for $R_{n_s} < R_{n_s,d}$.

Keeping in mind that the deformation vector ξ is small, according to (7.1.4), we obtain $z \approx \Lambda$, $\dot{z} \approx \dot{\Lambda}$. Hence, instead of the commanded value $V_{z,d}$ of the velocity $\dot{z}$, we define the commanded value $V_{\Lambda,d}$ of the velocity $\dot{\Lambda}$ in the form (7.3.3)

$$V_{\Lambda,d} = V_{n_1} + \ldots + V_{n_\mu} + V_\tau$$

$$= g_1(R_{n_1} - R_{n_1,d})n_1 + \ldots + g_\mu(R_{n_\mu} - R_{n_\mu,d})n_\mu + v\tau_0 \quad (7.3.4)$$

Finally, by using the approximate equalities (7.2.5), we change the commanded velocity (7.3.4) to the following:

$$V_{\Lambda,d} = V_{n_1} + \ldots + V_{n_\mu} + V_\tau$$

$$= g_1(F_{n_1} - F_{n_1,d})n_1 + \ldots + g_\mu(F_{n_\mu} - F_{n_\mu,d})n_\mu + v\tau_0, \quad (7.3.5)$$

where $F_{n_s} = F \cdot n_s$ is a projection of the force vector F measured by the sensor onto the normal to the constraint n_s; $F_{n_s,d}$ is a desired (commanded) value of the projection F_{n_s} which is equal to $R_{n_s,d}$ $(s = 1, \ldots, \mu)$. To take gravity into account, the approximate equalities (7.2.4) should be used instead of (7.2.5). In so doing, the expression (7.3.5) changes in the obvious way.

If it is possible to determine the vectors τ_0 and n_s during the system operation or a priori, then all the parameters in (7.3.5) become known. Vectors of tangent and normals may be determined in a variety of ways. These vectors are sometimes known a priori. Such is the case, for example, for the problem of moving along a screw constraint with the known pitch of the screw, where these vectors are constant (see Section 7.5). Also, tangents and normals are known a priori in grinding (see Chapter 9). Vectors of tangent and normals may be computed directly during a grinding operation from the output of force sensors or manipulator position sensors. This is the way these vectors are determined for the control problems of a manipulator contouring an object (see Chapter 6).

In a general case, the vector z of manipulated object coordinates comprises both linear coordinates defining the position of some object point and angular coordinates defining its orientation. In this case, prior to composing the expression (7.3.5) for the commanded velocity, we should transform linear or angular variables so that all components of the vector z have the same physical dimension [163]. A similar "adjustment" of physical dimensions is performed to solve the problems in Sections 7.5 and 9.3.

Note that the same mechanical constraints may be represented by different equations depending on the coordinate frames and dimensions that define an object position. Hence, vectors of normals to constraints and, consequently, the complete vector (7.3.5) cannot always be defined uniquely. For each problem, the arising ambiguity can be handled individually by using additional considerations.

The expression (7.3.5) for the commanded velocity may be considered as an extension of the formula (6.2.1). For $v = 0$, (7.3.5) is an extension of (5.3.13) for the case of maintaining contact under several constraints.

For $g_1 = \ldots = g_\mu = g$, (7.3.5) has the form

$$V_{\Lambda,d} = g(F_n - F_{n,d}) + v\tau_0, \qquad (7.3.6)$$

where F_n is a sum of vectors obtained by projecting the vector F onto normals n_s $(s = 1, \ldots, \mu)$, and $F_{n,d}$ is a commanded value of this sum. On substituting some notation, (6.2.1) coincides with (7.3.6).

Provided the vector F_n differs little from the vector F, and $F_{n,d} = 0$, we can write instead of (7.3.6)

$$V_{\Lambda,d} = gF + v\tau_0 \qquad (7.3.7)$$

Since the vectors of normals n_s $(s = 1, \ldots, \mu)$ do not appear in (7.3.7), it is convenient to use this formula, if normals are not known.

In (7.3.5)–(7.3.7), v may be either a constant or a function of time and phase coordinates. The vector $v\tau_0$ can be represented as a weighted sum of unit vectors in the tangent manifold. These weights may be chosen based on various considerations. For instance, the weights can be set depending on the difference between the current and desired (commanded) position of the manipulator arm.

It is not always possible to define vectors n_s, τ_0 exactly for specific problems. However, one can solve a specific problem, provided the given vectors do not differ "too much" from vectors of normals and tangent to constraints. Such is the case for problems considered in Sections 6.3–6.6, 7.6, 8.6, 9.3, 10.2, 11.3. To define vectors of normals and tangent approximately, considerations intrinsic to the problem in hand can be applied. So, for the problem of opening a hatch lid (Section 7.6), it is important to set up the initial motion (opening). In so doing, one can use approximations for vectors of normals and a tangent to constraints that are valid only at the beginning of the motion. In Section 10.2, the final stage of inserting a peg into a hole is the peg motion inside the hole. The vector of tangent to constraint chosen in Section 10.2 is really a tangent vector only in the final part of the motion. When solving the problem of contouring an object in Section 8.6, we have chosen vectors directed "along" each degree of freedom of the gantry manipulator, instead of vectors of normals and tangent. This choice is justified if the object contour is elongated in one of these directions.

Let us now design control for the manipulator. By differentiating the column Λ with respect to time, we obtain

$$\dot{\Lambda} = W(q)\dot{q}, \tag{7.3.8}$$

where $W(q)$ is an $m \times n$ matrix. If $m = n$ and $\det W(q) \neq 0$, then we obtain from (7.3.8):

$$\dot{q} = W^{-1}(q)\dot{\Lambda} \tag{7.3.9}$$

By using (7.3.9), we can unambiguously find the joint velocity $\dot{q}$ from the given Cartesian velocity $\dot{\Lambda}$ [294], and, in particular, the commanded value $V_{q,d}$ of the velocity $\dot{q}$ from the commanded value $V_{\Lambda,d}$ of the velocity $\dot{\Lambda}$. If $m < n$, then (7.3.8) does not have a unique solution in $\dot{q}$. To find a unique solution, we need to add $n - m$ relationships to the system (7.3.8), by using some additional considerations.

We choose the control for the manipulator drives to be linear with respect to velocity:

$$EU = C(V_{q,d} - V_q) = C(V_{q,d} - \dot{q}), \tag{7.3.10}$$

where C is an $n \times n$ matrix to be chosen. The number of drives in robotic systems typically coincides with the number of degrees of freedom $(r = n)$. In this case, it is advisable to choose the matrix C as a diagonal matrix, so that each drive "tracks" the commanded velocity for its "own" degree of freedom. The control laws (6.2.2), (6.7.3), (7.4.6), (7.5.15), and (7.6.2) are designed in this way, as well as the control laws for all the experiments described in the book.

A control other than linear may be used. For example, the control laws (5.4.3), (6.7.4), (7.4.7) define switching control laws.

If $r = n$ and $\det E \neq 0$, then (7.3.10) is solved with respect to the column matrix U as follows:

$$U = E^{-1}C(V_{q,d} - \dot{q}) \tag{7.3.11}$$

By substituting (7.3.9) into (7.3.11), we obtain:

$$U = E^{-1}C[W^{-1}(q)V_{\Lambda,d} - \dot{q}] \tag{7.3.12}$$

Generally, to implement control laws (7.3.5), (7.3.12), it is necessary to measure coordinates q_i and velocities $\dot{q}_i$ $(i = 1, \ldots, n)$. To do this, one needs to use position and velocity sensors in addition to force sensors.

The relationships (7.3.5)–(7.3.8), (7.3.12) present a general algorithm for designing control of a robotic system motion along constraints on the basis of force information. These relationships can be modified for a particular problem. Any component of the commanded velocity (7.3.5) normal to the constraint may not depend on the corresponding force projection linearly. Drive control may also be nonlinear. To maintain a given contact force, dynamical regulators can be used, as in Chapter 8. As mentioned above, instead of vectors of normals and tangent to constraints, it is possible to use vectors closely approximating them. This important fact is used to solve a number of problems, such as the applied problems in Part III of the book.

7.4 Articulated Manipulator Motion in Contact With an Object

Consider an articulated manipulator with three degrees of freedom shown in Figure 7.4.1. The manipulator base link of length l_0 rotates about a vertical axis ON. The link NC of length l_1 is connected to the base link at a point N through a cylindrical joint. The link CA of length l_2 is connected to the link NC at a point C through another cylindrical joint. The joint axes at the attachment points N and C are horizontal. Each of the three manipulator links is actuated by its own drive.

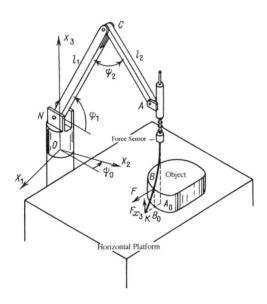

FIGURE 7.4.1
An articulated manipulator with three degrees of freedom

Let O be an intersection point of the rotation axis of the base link and the base supporting the manipulator. We attach a stationary orthogonal coordinate system $OX_1X_2X_3$ to the point O. The axis OX_3 of this system coincides with the straight line ON and is directed vertically upwards. The axes OX_1, OX_2 lie in the horizontal plane, and their orientation is arbitrary, but fixed. Denote by ψ_0 a counterclockwise rotation angle of the base link about the axis OX_3. Denote by ψ_1 an angle between the link NC and the horizontal plane, by ψ_2, an angle between the links NC and CA.

A three-component force sensor is attached to the tip A of the manipulator to measure a vertical and two horizontal force components. The sensor is modeled as a massless lumped spring and an infinitely thin rod. We assume that the spring axis is vertical for any manipulator configuration. The spring models a single-component force sensor measuring a vertical force component. The rod is also vertical when undeformed, and coincides with the axis of the spring. As in Sections 6.1, 6.7, it models a two-component force sensor module measuring a horizontal force component.

An object having a cylindrical lateral surface with a vertical generator rests on the horizontal platform. The upper surface of the object is in the horizontal plane

$$x_3 = c \quad (c = \text{const}) \tag{7.4.1}$$

The rod modeling a two-component force sensor contacts the object at the point B of the upper surface and bends. The tip K of the rod presses against the platform causing the spring to compress.

Denote by A_0 a point of intersection of the spring axis with the surface of the platform. When the force sensor contacts the object and the platform, a horizontal component of the force exerted on the sensor can be computed from the formula (6.1.2). A vertical component F_{x_3} of this force is given by the formula (see (5.1.9)):

$$F_{x_3} = \begin{cases} 0, & \text{if } x_3 \geq a + l, \\ -k_{x_3}(x_3 - a - l), & \text{if } x_3 \leq a + l, \end{cases} \tag{7.4.2}$$

where x_3 is a z-coordinate of the point at which the upper end of the spring is attached (it can be assumed that this z-coordinate coincides with the z-coordinate of the point A); a is a z-coordinate of the lower end of the spring (it is constant if we neglect the deflection of the rod); l is a length of the undeformed spring, k_{x_3} is a force sensor stiffness along the vertical axis (the spring stiffness coefficient).

By deriving the equations of motion for the three-link manipulator described above and using the relationships (6.1.2)–(6.1.10), (7.4.2), we can develop a mathematical model for the manipulator moving in contact with an object.

To make the manipulator follow the edge of the object upper surface without losing contact with it, we shall define a vector of commanded velocity for the manipulator tip A. To this end, we use the relationships (7.3.5). The manipulator needs to follow the object edge formed by the intersection of the horizontal plane (7.4.1) and the cylindrical surface. The vector n_1 of normal to the platform H and the plane (7.4.1) is $[0, 0, 1]^T$ (it is external relative to the platform H and the object J). The vectors n_2 of normal to the cylindrical lateral surface with the vertical generator belong to the horizontal plane. The tangent vector τ to the contour also belongs to the horizontal plane. Hence, according to (7.3.5), the vertical component $V_{x_3,d}$ of the commanded velocity vector is given by

$$V_{x_3,d} = g_1(F_{x_3} - F_{x_3,d}), \tag{7.4.3}$$

where $F_{x_3,d} = \text{const} > 0$ is a commanded vertical component of the force acting on the sensor from the object, $g_1 = \text{const} > 0$ is a feedback gain. The horizontal component of the commanded velocity vector, according to (7.3.5), is given by the expression (6.2.1). The expressions (6.2.1), (7.4.3) define all three components $V_{x_1,d}$, $V_{x_2,d}$, $V_{x_3,d}$ of the commanded velocity vector of the manipulator tip A.

The coordinates x_1, x_2, x_3 of the point A are related to the angles ψ_0, ψ_1, ψ_2 by the following forward kinematic relationships:

$$x_1 = [l_1 \cos \psi_1 - l_2 \cos(\psi_1 + \psi_2)] \cos \psi_0,$$

$$x_2 = [l_1 \cos \psi_1 - l_2 \cos(\psi_1 + \psi_2)] \sin \psi_0,$$

$$x_3 = l_0 + l_1 \sin \psi_1 - l_2 \sin(\psi_1 + \psi_2) \tag{7.4.4}$$

By differentiating (7.4.4) with respect to time and then solving for the angular velocities $\dot{\psi}_0$, $\dot{\psi}_1$, $\dot{\psi}_2$, we obtain the inverse velocity kinematics as

$$\dot{\psi}_0 = \omega_0 = \frac{-\dot{x}_1 \sin \psi_0 + \dot{x}_2 \cos \psi_0}{l_1 \cos \psi_1 - l_2 \cos(\psi_1 + \psi_2)},$$

$$\dot{\psi}_1 = \omega_1 = \frac{1}{l_1 \sin \psi_2} [(\dot{x}_1 \cos \psi_0 + \dot{x}_2 \sin \psi_0) \cos(\psi_1 + \psi_2)$$

$$+ \dot{x}_3 \sin(\psi_1 + \psi_2)], \tag{7.4.5}$$

$$\dot{\psi}_2 = \omega_2 = \frac{1}{l_1 l_2 \sin \psi_2} \{(\dot{x}_1 \cos \psi_0 + \dot{x}_2 \sin \psi_0)$$

$$\times [l_1 \cos \psi_1 - l_2 \cos(\psi_1 + \psi_2)] + \dot{x}_3 [l_1 \sin \psi_1 - l_2 \sin(\psi_1 + \psi_2)]\}$$

The equations (7.4.5) are valid for

$$\psi_2 \neq \{0, \pi\}, \quad l_1 \cos \psi_1 - l_2 \cos(\psi_1 + \psi_2) \neq 0$$

By substituting into (7.4.5) the commanded values $V_{x_1,d}$, $V_{x_2,d}$, $V_{x_3,d}$ for the linear velocities $\dot{x}_1 = V_{x_1}$, $\dot{x}_2 = V_{x_2}$, $\dot{x}_3 = V_{x_3}$ computed from (6.2.1), (7.4.3), we obtain the commanded values ω_{0d}, ω_{1d}, ω_{2d} for the angular velocities ω_0, ω_1, ω_2. To determine these commanded values at any given time, in addition to the force sensor outputs, one needs measurements of angles ψ_0, ψ_1, ψ_2. The commanded values of the angular velocities are applied to the servo systems controlling the drives in the joints. Then the voltages u_0, u_1, u_2 applied to the electromotors have the form

$$u_i = c_i(\omega_{id} - \omega_i), \quad i = 0, 1, 2, \tag{7.4.6}$$

where c_0, c_1, c_2 are constant positive gains, or

$$u_i = u_{i0} \operatorname{sgn}(\omega_{id} - \omega_i), \quad i = 0, 1, 2, \tag{7.4.7}$$

where u_{i0} is the maximum possible voltage applied to the drive i.

In the above-stated problem of the manipulator motion along a contour in the horizontal plane, the desired motion of the manipulator tip is a two-degree-of-freedom motion. Yet, unlike the gantry manipulator considered in Chapter 6 where only two degrees of freedom were used, this motion involves all three degrees of freedom of the articulated manipulator. This is why a three-component force sensor is required to solve the problem.

7.5 Motion Along a Screw Constraint

We shall consider the arm of the gantry manipulator moving along three degrees of freedom (see Figure 4.1.1). The lower end of the manipulator arm has a wrist enabling the end-effector to rotate about the vertical axis. The end-effector includes a four-component (wrist) force sensor and a gripper (see Figure 4.3.3).

We introduce a stationary orthogonal coordinate system $OX_1X_2X_3$. As the arm moves translationally along the first, the second, or the third degree of freedom, it moves along the axis OX_1, OX_2, or OX_3. The axes OX_1 and OX_2 belong to the horizontal plane, the axis OX_3 is directed vertically upwards. The manipulator schematics are shown in Figure 7.5.1 (see also Figure 6.1.1). The screw is to be inserted into a stationary threaded hole (a Nut). We assume the axes of the screw and the threaded hole to be parallel.

We place the origin of the stationary coordinate system $OX_1X_2X_3$ at some point on the axis of the threaded hole (the Nut). The motion along

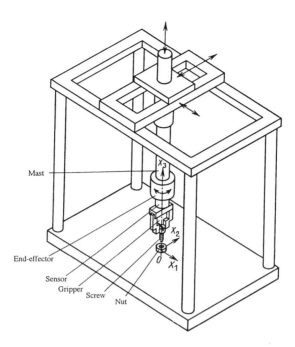

FIGURE 7.5.1
A gantry manipulator assembling a threaded joint

each of the four degrees of freedom is actuated by an individual electric motor.

Denote by x_1, x_2, x_3 the coordinates of some point on the end-effector axis, and by ψ, its counterclockwise rotation angle counted from the axis OX_1. Denote by h the screw pitch, and by σ, the ratio $h/2\pi$. To derive the equations of motion, we use, in addition to the angle ψ, a variable $x_4 = \sigma\psi$ measured in units of length.

By neglecting inductance in the rotor circuit of the motors [65], we can write the equations of motion for the four-degree-of-freedom system in hand (excluding the force sensor) in the form

$$\mathcal{M}_i \ddot{x}_i + d_i \dot{x}_i = e_i u_i + \varphi_i \quad (i = 1, 2),$$

$$\mathcal{M}_3 \ddot{x}_3 + d_3 \dot{x}_3 = e_3 u_3 + \varphi_3 - m_3 g,$$

$$\mathcal{M}_4 \ddot{x}_4 + d_4 \dot{x}_4 = e_4 u_4 + \varphi_4 \qquad (7.5.1)$$

In the above equations, $\mathcal{M}_i$ $(i = 1, \ldots, 4)$ are equivalent masses of all the moving parts (the manipulator mast, motor rotors, gears, etc.) for the respective degrees of freedom $(\mathcal{M}_4\sigma^2$ is the equivalent moment of inertia for the end-effector); m_3 is the mass of the mast; g is the acceleration of gravity; $d_i\dot{x}_i$ $(i = 1, \ldots, 4)$ are combined forces of viscous friction and back e.m.f. in the drives; u_i $(i = 1, \ldots, 4)$ are voltages applied to the motors; e_i, d_i $(i = 1, \ldots, 4)$ are positive constants dependent on the drive characteristics; φ_i $(i = 1, \ldots, 4)$ are forces exerted by the four-component force sensor on the manipulator mast and end-effector. The last equation in (7.5.1) is composed from the equations of moments of forces for the end-effector; $\sigma\varphi_4$ is a moment of forces applied by the force sensor; $d_4\sigma^2\dot{\psi}$ is a moment of forces produced by forces of viscous friction and back e.m.f. The equations (7.5.1) are analogous to (5.1.8) and (6.1.1). They are the equations of motion for the "part" of the manipulator "before" the force sensor, same as (7.1.1). The variables x_i $(i = 1, \ldots, 4)$ represent the components vectors q, l, and Λ simultaneously, because $n = v = m = 4$ for the problem in hand.

The screw is held in the gripper jaws (see Figure 7.5.1). Since a wrist force sensor is used, the screw together with the gripper can be considered a single rigid body. When the sensor is undeformed, the screw axis coincides with the end-effector axis. On the axis of the end-effector and the screw axis, we choose some points A and A', which coincide when the force sensor is undeformed. We shall assume that x_1, x_2, x_3 in (7.5.1) are the coordinates of the point A.

The four-component force sensor measures two horizontal force components, a vertical force component and a torque about the vertical axis. Sensors of this type are described in Chapter 2 (see Figure 2.3.9). More precisely, the sensor measures (and experiences) the corresponding deformations. As the sensor deforms, the gripper with the screw moves in the horizontal and vertical directions relative to the end-effector and rotates about the vertical axis. Let us attach an orthogonal coordinate system to the upper flange of the wrist force sensor. To this end, we translate the system $OX_1X_2X_3$ and rotate it about the axis OX_3 through the angle ψ. Let ξ_1, ξ_2 be linear deformations of the sensor along the two horizontal axes of the attached system, ξ_3, a vertical component of the deformation, and ε, an angular deformation $(\xi_4 = \sigma\varepsilon)$. It is assumed that the screw axis remains vertical through all force sensor deformations. Denote by z_1, z_2, and z_3 horizontal and vertical coordinates of the point A' of the screw, counted along the stationary axes OX_1, OX_2, and OX_3, respectively. Denote by γ a counterclockwise rotation angle of the screw about its vertical axis, counted from the axis OX_1. Let $z_4 = \sigma\gamma = h\gamma/(2\pi)$.

It follows from the above that we can write the matrix relationship (7.1.4) for the considered problem in the form of the four scalar equations:

$$z_1 = x_1 + \xi_1 \cos\psi - \xi_2 \sin\psi, \quad z_2 = x_2 + \xi_1 \sin\psi - \xi_2 \cos\psi,$$

$$z_3 = x_3 + \xi_3, \quad z_4 = x_4 + \xi_4 \quad (\gamma = \psi + \varepsilon) \tag{7.5.2}$$

The equations of motion for the rigid body with four degrees of freedom, consisting of the gripper and the screw, can be written in the form (see (7.1.8))

$$m\ddot{z}_i = -\varphi_i + R_i \quad (i = 1, 2),$$

$$m\ddot{z}_3 = -\varphi_3 + R_3 - mg, \tag{7.5.3}$$

$$J\ddot{z}_4 = -\varphi_4 + R_4$$

In this equation, m is the mass of the body, $J\sigma^2$ is its moment of inertia relative to the screw axis; R_i $(i = 1, \ldots, 4)$ are components of the constraint forces imposed on the screw. The last equation in (7.5.3) is derived from the equations of moments of forces for the screw axis, $R_4\sigma$ is a moment of constraint forces about this axis. To derive this equation, we omit the torque $-\varphi_1'\xi_2 + \varphi_2'\xi_1$ of the forces $-\varphi_1'$ and $-\varphi_2'$ with respect to the screw axis. For large stiffness coefficients of the sensor and small deformations (see (7.5.4), (7.5.5)), this torque is negligibly small compared to the torque $\varphi_4\sigma$. If forces of viscous friction are absent in the force sensor and its stiffness coefficients along the horizontal axes are the same (see (7.5.4)), then the torque $-\varphi_1'\xi_2 + \varphi_2'\xi_1$ is zero, and the equations (7.5.3) are exact.

According to (7.1.5), forces straining the force sensor are proportional to the deformations and their derivatives. We assume the stiffness and damping matrices to be diagonal for the considered four-component sensor, i.e. $K = \text{diag}[k_i]$, $B = \text{diag}[\beta_i]$. Hence the formula (7.1.5) has the following scalar form:

$$\varphi_i' = k_i\xi_i + \beta_i\dot{\xi}_i \quad (i = 1, \ldots, 4) \tag{7.5.4}$$

The values φ_i' $(i = 1, \ldots, 4)$ are force and torque projections onto the coordinate system attached to the top flange of the force sensor.

By projecting the forces onto the axes of the stationary coordinate system $OX_1X_2X_3$, we obtain (see (7.1.6))

$$\varphi_1 = \varphi_1' \cos\psi - \varphi_2' \sin\psi = k_1\xi_1 \cos\psi - k_2\xi_2 \sin\psi$$

$$+\beta_1 \dot{\xi}_1 \cos \psi - \beta_2 \dot{\xi}_2 \sin \psi,$$

$$\varphi_2 = \varphi_1' \sin \psi + \varphi_2' \cos \psi = k_1 \xi_1 \sin \psi + k_2 \xi_2 \cos \psi \qquad (7.5.5)$$

$$+\beta_1 \dot{\xi}_1 \sin \psi + \beta_2 \dot{\xi}_2 \cos \psi,$$

$$\varphi_i = \varphi_i' = k_i \xi_i + \beta_i \dot{\xi}_i \qquad (i = 3, 4)$$

Constraints imposed on the screw motion inside the threaded hole (see (7.1.10)) restrain the screw displacements in the horizontal plane and relate the translational motion along the vertical axis OX_3 to the rotation about this axis:

$$z_i = 0, \quad i = 1, 2, \quad z_3/h = \gamma/(2\pi) \quad (z_3 = z_4) \qquad (7.5.6)$$

Recall that $z_4 = h\gamma/(2\pi) = \sigma\gamma$. The last relationship in (7.5.6) describing the screw constraint [196] corresponds to a right-hand thread. For a left-hand thread, $-\gamma$ should be substituted for the angle γ in (7.5.6), and $-z_4$, for z_4, (or $-z_3$ for z_3). For given control signals u_i $(i = 1, \ldots, 4)$, the relationships (7.5.1)–(7.5.3), (7.5.5), (7.5.6) represent a closed system of equations of motion for a manipulator with a threaded part in the gripper.

We proceed now, in accordance with the methods of Section 7.3, to design control enabling the manipulator to drive the screw into the stationary threaded hole, i.e, to move it along the screw constraint. The projections f_i $(i = 1, \ldots, 4)$ of the force vector F, measured by the sensor, onto the axes of the stationary coordinate system $OX_1X_2X_3$ (see (7.1.7)) are given by the formulae (7.5.5), provided that the terms dependent on deformation velocities are eliminated:

$$f_1 = k_1 \xi_1 \cos \psi - k_2 \xi_2 \sin \psi, \quad f_2 = k_1 \xi_1 \sin \psi + k_2 \xi_2 \cos \psi,$$

$$f_i = k_i \xi_i \quad (i = 3, 4) \qquad (7.5.7)$$

We set the commanded values of constraint forces $R_{n_s,d} = 0$ $(s = 1, 2, 3)$ and define the vector $V_{\Lambda,d}$ of the commanded velocity in variables x_i $(i = 1, \ldots, 4)$

$$V_{\Lambda,d} = [V_{x_i,d}] \quad (i = 1, \ldots, 4), \qquad (7.5.8)$$

according to (7.3.4), in the form

$$V_{\Lambda,d} = \sum_{s=1}^{3} g_s R_{n_s} n_s + v\tau_0 \quad (v = \text{const} > 0) \tag{7.5.9}$$

It follows from (7.5.3) that the expressions (7.2.4) for the stationary regime can be written for the problem in hand in the form

$$f_j = R_j \quad (j = 1, 2, 4), \quad f_3 = R_3 - mg \tag{7.5.10}$$

Vectors of normals and tangent to the surfaces (7.5.6) have the form

$$n_1 = \begin{bmatrix} 1 \\ 0 \\ 0 \\ 0 \end{bmatrix}, \quad n_2 = \begin{bmatrix} 0 \\ 1 \\ 0 \\ 0 \end{bmatrix}, \quad n_3 = \frac{1}{\sqrt{2}} \begin{bmatrix} 0 \\ 0 \\ 1 \\ -1 \end{bmatrix}, \quad \tau = \pm\frac{1}{\sqrt{2}} \begin{bmatrix} 0 \\ 0 \\ 1 \\ 1 \end{bmatrix} \tag{7.5.11}$$

When the screw is driven into the hole, it goes downwards, so the vector τ_0 in (7.5.9) should be chosen as

$$\tau_0 = -\frac{1}{\sqrt{2}} \begin{bmatrix} 0 & 0 & 1 & 1 \end{bmatrix}^T \tag{7.5.12}$$

If the screw should be driven out of the hole, then we achieve the corresponding control by substituting the vector $-\tau_0$ into (7.5.9).

For this problem, we know a priori the equations for constraints along which the motion is performed. Hence the vectors of tangent and normals to constraints are known a priori. Note that these vectors are constants.

From (7.5.10), (7.5.11), we obtain

$$R_{n_1} = f_1, \quad R_{n_2} = f_2, \quad R_{n_3} = \frac{1}{\sqrt{2}}(f_3 + mg - f_4) \tag{7.5.13}$$

By substituting (7.5.7), (7.5.8), (7.5.11)– (7.5.13) into (7.5.9), we obtain

$$V_{x_1,d} = g_1 f_1 = g_1(k_1\xi_1 \cos\psi - k_2\xi_2 \sin\psi),$$

$$V_{x_2,d} = g_2 f_2 = g_2(k_1\xi_1 \sin\psi - k_2\xi_2 \cos\psi),$$

$$V_{x_3,d} = \frac{1}{2}g_3(f_3 - f_4 + mg) - \frac{1}{\sqrt{2}}v = \frac{1}{2}g_3(k_3\xi_3 - k_4\xi_4 + mg) - \frac{1}{\sqrt{2}}v,$$

$$(7.5.14)$$

$$V_{x_4,d} = -\frac{1}{2}g_3(f_3 - f_4 + mg) - \frac{1}{\sqrt{2}}v = -\frac{1}{2}g_3(k_3\xi_3 - k_4\xi_4 + mg) - \frac{1}{\sqrt{2}}v$$

In Section 7.3, we used the relationships (7.2.5) to pass from (7.3.4) to (7.3.5). In the present section, for a similar transition from (7.5.9) to (7.5.14), we use the relationships (7.2.4) that have the form (7.5.10) for this problem.

Note that by substituting $g'_3 = g_3/2$ and $v' = v/\sqrt{2}$ for g_3 and v, we can simplify the last two equations in (7.5.14).

In order to go back from x_4 to the angular variable ψ and find the expression for the commanded value ω_d of the angular velocity $\dot\psi$, we should divide the last expression in (7.5.14) by $\sigma = h/(2\pi)$. In so doing, we also substitute $2\pi M/h$ for f_4, where M is the torque measured by the force sensor, and obtain

$$\omega_d = -\pi g_3(f_3 - 2\pi M/h + mg)/h - \sqrt{2}\pi v/h$$

The last term in the above expression and in that for $V_{x_3,d}$ describes the motion at a tangent to the screw constraint. Their ratio is the relationship between velocities of the rotational and translational screw motions.

Let the matrix C in the control law (7.3.10) be diagonal. Then

$$e_i u_i = c_i(V_{x_i,d} - V_{x_i}) = c_i(V_{x_i,d} - \dot x_i)$$

$$(c_i = \text{const} > 0, \quad i = 1,\dots,4) \qquad (7.5.15)$$

According to (7.1.11), (7.1.12), (7.5.11), we obtain expressions for components of the constraint forces:

$$R_1 = \zeta_1, \quad R_2 = \zeta_2, \quad R_3 = \zeta_3, \quad R_4 = -\zeta_3, \qquad (7.5.16)$$

It follows from the last two equations in (7.5.16) that

$$R_4 = -R_3 \qquad (7.5.17)$$

Assume that

$$k_1 = k_2 = k, \quad \beta_1 = \beta_2 = \beta \qquad (7.5.18)$$

Then for the control (7.5.14), (7.5.15), taking into account the relationships (7.5.2), (7.5.5), (7.5.6), we can rewrite the first and second equations in (7.5.1) as follows:

$$\mathcal{M}_1\ddot{x}_1 + (d_1 + c_1 + \beta)\dot{x}_1 + (1 + c_1 g_1)kx_1 + \frac{\beta}{\sigma}\dot{x}_4 x_2 = 0,$$

$$\mathcal{M}_2\ddot{x}_2 + (d_2 + c_2 + \beta)\dot{x}_2 + (1 + c_2 g_2)kx_2 - \frac{\beta}{\sigma}\dot{x}_4 x_1 = 0 \quad (7.5.19)$$

The third and fourth equations in (7.5.1) take the form

$$\mathcal{M}_3\ddot{x}_3 + (d_3 + c_3)\dot{x}_3 + \frac{1}{2}c_3 g_3(k_4\xi_4 - k_3\xi_3) - k_3\xi_3 - \beta_3\dot{\xi}_3$$

$$= \frac{1}{2}c_3 g_3 mg - \frac{1}{\sqrt{2}}c_3 v - m_3 g, \quad (7.5.20)$$

$$\mathcal{M}_4\ddot{x}_4 + (d_4 + c_4)\dot{x}_4 + \frac{1}{2}c_4 g_3(k_3\xi_3 - k_4\xi_4) - k_4\xi_4 - \beta_4\dot{\xi}_4$$

$$= -\frac{1}{2}c_4 g_3 mg - \frac{1}{\sqrt{2}}c_4 v$$

The first and second equations in (7.5.3) under the condition (7.5.18), by using the relationships (7.5.2), (7.5.5), (7.5.6), can be transformed to yield

$$kx_1 + \beta\dot{x}_1 + \frac{\beta}{\sigma}\dot{x}_4 x_2 + R_1 = 0, \quad kx_2 + \beta\dot{x}_2 - \frac{\beta}{\sigma}\dot{x}_4 x_1 + R_2 = 0 \quad (7.5.21)$$

In view of the relationships (7.5.2), (7.5.5), (7.5.17), we can write the third and fourth equations in (7.5.3) as follows:

$$m(\ddot{x}_3 + \ddot{\xi}_3) + \beta_3\dot{\xi}_3 + k_3\xi_3 = R_3 - mg,$$

$$J(\ddot{x}_4 + \ddot{\xi}_4) + \beta_4\dot{\xi}_4 + k_4\xi_4 = -R_3 \quad (7.5.22)$$

To close the system of equations (7.5.19)–(7.5.22), we should write the last condition in (7.5.6):

$$x_3 + \xi_3 = x_4 + \xi_4 \quad (7.5.23)$$

We shall search for a stationary regime for the system (7.5.19)– (7.5.23) corresponding to the manipulator motion with a constant rotation velocity of the end-effector, constant force sensor deformations, and constant constraint forces

$$x_1 = x_2 = 0, \quad R_1 = R_2 = 0, \quad \dot{x}_3 = \text{const}, \quad \dot{x}_4 = \text{const},$$

$$\xi_3 = \text{const}, \quad \xi_4 = \text{const}, \quad R_3 = \text{const} \tag{7.5.24}$$

The functions $x_1 = x_2 = 0$, $R_1 = R_2 = 0$ satisfy the equations (7.5.19), (7.5.21) for any velocity $\dot{x}_4$.

We use the relationships (7.5.20), (7.5.22), (7.5.23) to find the stationary parameters (7.5.24) and, after some rearrangement, obtain the linear algebraic equations

$$(d_3 + c_3)\dot{x}_3 + (1 + c_3 g_3)k_4\xi_4 = -\frac{1}{\sqrt{2}}c_3 v - (m + m_3)g,$$

$$(d_4 + c_4)\dot{x}_3 - (1 + c_4 g_3)k_4\xi_4 = -\frac{1}{\sqrt{2}}c_4 v, \tag{7.5.25}$$

$$k_3\xi_3 + k_4\xi_4 = -mg, \quad \dot{x}_3 = \dot{x}_4 \quad R_3 = -k_4\xi_4$$

Under the simplifying assumptions

$$d_4 = d_3 = d, \quad c_4 = c_3 = c \tag{7.5.26}$$

the solution of these equations has the form

$$\dot{x}_3 = \dot{x}_4 = -\frac{cv}{\sqrt{2}(d + c)} - \frac{(m + m_3)g}{2(d + c)}, \quad R_3 = \frac{(m + m_3)g}{2(1 + cg_3)},$$

$$\xi_3 = -\frac{mg}{k_3} - \frac{(m + m_3)g}{2(1 + cg_3)}, \quad \xi_4 = -\frac{(m + m_3)g}{2k_4(1 + cg_3)} \tag{7.5.27}$$

For $c \to \infty$, the variables (7.5.27) assume their limit values

$$\dot{x}_3 = \dot{x}_4 = -\frac{v}{\sqrt{2}}, \quad R_3 = 0, \quad \xi_3 = -\frac{mg}{k_3}, \quad \xi_4 = 0 \tag{7.5.28}$$

It follows from (7.5.25) that the equations (7.5.28) hold for $c_3, c_4 \to \infty$ even without the assumptions (7.5.26).

Under the conditions (7.5.24), (7.5.27), $z_1 = z_2 = 0$, $\dot{z}_3 = \dot{x}_3$, $\dot{z}_4 = \dot{x}_4$, and the screw is driven into the threaded hole. Therefore, the above-formulated problem of motion along a screw constraint is solved by the stationary regime (7.5.27).

Let us now study stability of the solution (7.5.24), (7.5.27). Denote variations of the variables x_i $(i = 1, \ldots, 4)$, ξ_3, ξ_4, R_j $(j = 1, 2, 3)$ by Δx_i, $\Delta \xi_1$, $\Delta \xi_2$, ΔR_j. Then the equations in variations corresponding to the system (7.5.19), (7.5.21) have the form

$$M_1 \Delta \ddot{x}_1 + (d_1 + c_1 + \beta)\Delta \dot{x}_1 + (1 + c_1 g_1)k\Delta x_1 + \frac{\beta}{\sigma}\dot{x}_4 \Delta x_2 = 0,$$

$$M_2 \Delta \ddot{x}_2 + (d_2 + c_2 + \beta)\Delta \dot{x}_2 + (1 + c_2 g_2)k\Delta x_2 - \frac{\beta}{\sigma}\dot{x}_4 \Delta x_1 = 0$$

$$k\Delta x_1 + \beta \Delta \dot{x}_1 + \frac{\beta}{\sigma}\dot{x}_4 \Delta x_2 + \Delta R_1 = 0, \qquad (7.5.29)$$

$$k\Delta x_2 + \beta \Delta \dot{x}_2 - \frac{\beta}{\sigma}\dot{x}_4 \Delta x_1 + \Delta R_2 = 0$$

Now let us write the equations (7.5.20), (7.5.22), (7.5.23) in deviations from the stationary regime

$$M_3 \Delta \ddot{x}_3 + (d_3 + c_3)\Delta \dot{x}_3 + \frac{1}{2}c_3 g_3 (k_4 \Delta \xi_4 - k_3 \Delta \xi_3) - k_3 \Delta \xi_3 - \beta_3 \Delta \dot{\xi}_3 = 0,$$

$$M_4 \Delta \ddot{x}_4 + (d_4 + c_4)\Delta \dot{x}_4 + \frac{1}{2}c_4 g_3 (k_3 \Delta \xi_3 - k_4 \Delta \xi_4) - k_4 \Delta \xi_4 - \beta_4 \Delta \dot{\xi}_4 = 0,$$

$$(7.5.30)$$

$$m(\Delta \ddot{x}_3 + \Delta \ddot{\xi}_3) + \beta_3 \Delta \dot{\xi}_3 + k_3 \Delta \xi_3 - \Delta R_3 = 0,$$

$$J(\Delta \ddot{x}_4 + \Delta \ddot{\xi}_4) + \beta_4 \Delta \dot{\xi}_4 + k_4 \Delta \xi_4 + \Delta R_3 = 0,$$

$$\Delta x_3 + \Delta \xi_3 = \Delta x_4 + \Delta \xi_4$$

For $\dot{x}_4 = 0$, the first two equations in (7.5.29) do not depend on each other. In this case, the asymptotic stability conditions are the inequalities

$$d_i + c_i + \beta > 0, \quad 1 + c_i g_i > 0, \quad (i = 1, 2) \qquad (7.5.31)$$

Now assume that $\dot{x}_4 \neq 0$. The first two equations in (7.5.29) form an independent system of the fourth order. It is easy to demonstrate that the necessary and sufficient asymptotic stability condition for a trivial solution of this system includes the inequalities (7.5.31) and the condition that the leading Hurwitz determinant is positive. The latter condition has the form (see (5.9.16))

$$a_3(a_1a_2 - a_0a_3) - a_1^2a_4 > 0, \qquad (7.5.32)$$

where a_i $(i = 0, \ldots, 4)$ are coefficients of a characteristic polynomial. The velocity $\dot{x}_4$ only appears in the constant term a_4 of the polynomial:

$$a_4 = (1 + c_1g_1)(1 + c_2g_2)k^2 + \left(\frac{\beta}{\sigma}\dot{x}_4\right)^2 \qquad (7.5.33)$$

It follows from (7.5.33) that the inequality (7.5.32) limits from above the velocity $\dot{x}_4$, i.e., the angular rotation velocity $\dot{\psi}$ in the stationary regime. The value of this velocity is given by the first equality in (7.5.27).

The system (7.5.30) has a closed form. Its stability is determined by a characteristic polynomial obtained by expanding the determinant

$$\begin{vmatrix} \mathcal{M}_3p^2 + (d_3 + c_3)p & 0 & -\beta_4p - (1 + \frac{1}{2}c_3g_3)k_3 & \frac{1}{2}c_3g_3k_4 \\ 0 & \mathcal{M}_4p^2 + (d_4 + c_4)p & \frac{1}{2}c_4g_3k_3 & -\beta_4p - (1 + \frac{1}{2}c_4g_3)k_4 \\ mp^2 & Jp^2 & mp^2 + \beta_3p + k_3 & Jp^2 + \beta_4p + k_4 \\ 1 & -1 & 1 & -1 \end{vmatrix}$$

$$(7.5.34)$$

This characteristic polynomial is of the sixth order. It can be seen that $p = 0$ is one of the polynomial zeros. This is explained by the fact that the equations (7.5.20), (7.5.22), (7.5.23) have one cyclic coordinate. We shall further study the problem of asymptotic stability for the five remaining phase coordinates, which means dealing with a fifth-order polynomial.

To make the study as simple as possible, we assume the equalities (7.5.26) and (7.5.35) to hold:

$$\mathcal{M}_3 = \mathcal{M}_4 = \mathcal{M}, \quad k_3 = k_4 = k', \quad \beta_3 = \beta_4 = \beta' \qquad (7.5.35)$$

Under the conditions (7.5.26) and (7.5.35), the above fifth-order polynomial reduces to a product of polynomials of the second order

$$\mathcal{M}p^2 + (d + c + \beta')p + (1 + cg_3)k' \qquad (7.5.36)$$

and third order

$$\mathcal{M}(m + J)p^3 + [(m + J)(d + c + \beta') + 2\mathcal{M}\beta']p^2$$

$$+[2(d + c)\beta' + (2\mathcal{M} + J + m)k']p + 2(d + c)k' \qquad (7.5.37)$$

The zeros of the polynomial (7.5.36) are in the left half-plane of the complex plane provided that the following conditions of the form (7.5.31) hold:

$$d + c + \beta' > 0, \quad 1 + cg_3 > 0 \qquad (7.5.38)$$

It follows from the Hurwitz inequalities that zeros of the polynomial (7.5.37) are to the left of the imaginary axis for all sufficiently large values of the coefficient c. Hence the stationary regime (7.5.27) for driving the screw into the hole can be made asymptotically stable by choosing feedback gains in the control laws (7.5.14), (7.5.15).

Thus, in this section we have demonstrated how to design a control law of the form (7.3.5), (7.3.10) for the task of motion along a screw constraint.

Chapter 10 (see Sections 10.3, 10.4) studies both theoretically and experimentally the assembly of a threaded connection. The control law designed in this section is used there for driving screws in.

7.6 Opening a Hatch Lid

Let us consider a task where a manipulator with a force sensor is used to open or close a hatch lid.

Let the rotational axis of the lid be horizontal (see Figure 7.6.1). We assume that the lid initial position is horizontal as well. In the beginning, the manipulator arm is held above the handle on the lid. The handle can be pulled by a hook attached to the lower end of a force sensor. The upper end of the sensor is rigidly fixed in the arm. The sensor measures three force components.

FIGURE 7.6.1
The manipulator opening a hatch lid

The manipulator is required to raise and turn the hatch lid by pulling the handle with the hook. To solve this problem with a gantry manipulator, we need to use its three translational degrees of freedom. Same as in Chapter 4, we assume that the axes OX, OY of the orthogonal coordinate system $OXYZ$ are horizontal, and the axis OZ is directed vertically upwards. Each degree of freedom moves the manipulator arm along one of the axes.

It is desired to design control enabling the manipulator to open and close the hatch lid.

The hook pulling the handle is mechanically constrained to stay on a circumference during the lid rotation. This circumference is in a vertical plane P orthogonal to the axis of the lid rotation. The center of the circumference is on the lid rotation axis, its radius is the distance between the axis and the handle. To open or close the lid, it is necessary to move the hook along this circumference. We assume that the location of the center and the radius of the circumference are not known in advance. The hook is attached to the manipulator arm through the force sensor. Each point of the manipulator arm should move along a curve closely approximating a circle. The circumference along which the hook should move in space can be defined by two equations. One of the equations defines the plane P, the other should define a sphere or a cylindrical surface intersecting this plane. In this case, the expression (7.3.5) or the three-dimensional vector V_d of the commanded velocity of the manipulator arm takes the form

$$V_d = g_1 F_{n_1} n_1 + g_2 F_{n_2} n_2 + v\tau \qquad (7.6.1)$$

In this equation, n_1 is a vector of normal to the plane P, n_2 is a vector of normal to the sphere or cylinder, τ is a vector of tangent to the circumference, F_{n_1} and F_{n_2} are respective projections of the force F applied to the force sensor onto the normals n_1 and n_2, $v = \text{const}$ is a desired velocity of opening or closing the hatch (a velocity of the handle). In this case, the commanded values $F_{n_1,d}$ and $F_{n_2,d}$ of the projections F_{n_1} and F_{n_2} in (7.3.5) are assumed to be zero.

If the orientation of the lid rotation axis is known a priori, then the plane P, and therefore, the vector n_1 are also known. Provided that the mass of the lid and frictional forces in the lid rotation axis and at the contact point between the hook and the handle are negligibly small, the force vector F measured by the force sensor is normal to the circumference. In this case, the direction of the vector n_2 can be determined by projecting the force vector F onto the plane P. The tangent vector τ is orthogonal to the vectors n_1 and n_2. Thus, one can determine all parameters in (7.6.1) by using a force sensor. Then we substitute the vector V_d into the formula of the form (7.3.10) and obtain

$$U = C(V_d - V), \qquad (7.6.2)$$

where U is a three-dimensional vector of voltages applied to the electric motors for the three degrees of freedom of the manipulator, C is a diagonal gain matrix, $V = [V_x, V_y, V_z]^T$ is an arm velocity vector measured by tachometers.

The control based on the formulae (7.6.1), (7.6.2) works well, provided the lid mass and frictional forces are small.

If $g_1 = g_2 = g$, then (7.6.1) takes the form (see (6.6.1), (7.3.6))

$$V_d = g F_n + v\tau, \qquad (7.6.3)$$

where F_n is a projection of the force vector F onto the plane that is normal to the circumference.

If the orientation of the lid rotation axis is known a priori, then the task of opening (closing) a hatch is analogous to the task of rotating a steering wheel considered in Section 6.6. The problem gets more complicated if the orientation of the hatch lid rotation axis is not known a priori. In the latter case, the vectors of normals n_1, n_2, and the vector of tangent τ cannot be found from the force sensor output, and therefore, the expressions (7.6.1), (7.6.3) may not be used. If in the beginning of the operation, the hatch lid position is horizontal and the manipulator arm is above the handle, then at the initial instant of time the vector τ of tangent to the circumference is vertical, i.e., equals $(0,0,1)^T$ and the vector F_n is horizontal. Let us write the vector relationship (7.6.3) for the initial time instant in the form of three scalar ones:

$$V_{xd} = gF_x, \quad V_{yd} = gF_y, \quad V_{zd} = v \qquad (7.6.4)$$

The expression (7.6.4) for the commanded velocity can be used in the lid opening process as long as the tangent vector τ does not differ too much from the vector $(0,0,1)^T$.

In the experiments, the commanded velocity was defined by the expressions (7.6.4) throughout the entire lid opening operation. We used the force sensor shown in Figure 2.2.7. The information on two horizontal components of the force vector was used for control. The experiments demonstrate that under the control (7.6.2), (7.6.4), the manipulator is able to lift the lid to an almost vertical position. The control (7.6.2), (7.6.4) does not allow rotation of the lid through 180°. Yet using the control (7.6.2), (7.6.4), it is possible to determine the orientation of the rotation axis in the process of lifting the lid. To this end, one needs to determine manipulator arm coordinates using position sensors. By measuring arm coordinates several times during the lid lifting process, one can approximately find the plane P, or more precisely, the vector n_1 of normal to it. After that, one can use the expressions (7.6.1) or (7.6.3) for the commanded velocity. To use (7.6.1), (7.6.3), it is necessary to measure all three force vector components. By computing the commanded velocity from (7.6.1) or (7.6.3), one can rotate the lid through any angle.

Chapter 8

Discrete-Time Manipulator Control Design

Robotic systems are controlled by digital computers. This usually means that a control output is a zero-order hold function of time. Chapters 5 to 7 consider continuous-time models of control. The present chapter analyzes dynamics of a manipulator with a force sensor in a discrete-time control setting. We design several discrete-time controllers, among them those that take into account compliance in the manipulator structure.

Manipulator control problems of this chapter differ from those studied in Chapters 5 – 7 not only because they consider discrete-time control. The present chapter is also different in that it formulates control problems with incomplete state information. Therefore, the control design includes estimations of the system state from available measurements. By using state estimators (observers), it is possible to design regulators that, in addition to oscillations in a nonrigid manipulator structure, compensate for the motion of a point of contact with an object.

In this chapter, we make a number of simplifying assumptions. We only study a one-degree-of-freedom manipulator motion. We consider a cascade servo system that has a stiff inner velocity feedback loop. In other words, we consider an asymptotic case of a very large velocity feedback gain. For rheological properties of compliance in the manipulator structure, a linear model is accepted.

The models used for control designs in this chapter are more complex and, therefore, less general than those of Chapter 5. At the same time, the designed controllers yield transient processes of a somewhat better quality in the experiments.

8.1 Problems of Keeping Contact With a Stationary and Moving Object

Recall the problem of keeping a manipulator in contact with a stationary object formulated in Section 5.2. Similar to Section 5.2, herein we assume two measurements to be available. These are a force sensor output F defining the manipulator arm motion with respect to the object, and a manipulator velocity $\dot{x}$ measured by a tachometer. Sections 5.2–5.3 consider a few control laws for keeping a manipulator in contact with an object. The experiments show that these control laws do not always ensure a satisfactory quality of transient processes as the manipulator establishes contact with an object. The transient processes observed in the experiments cannot be fully explained within the simplest model of motion of a one-degree-of-freedom manipulator with a force sensor developed in Section 5.1. Sections 5.6–5.8 examine some possible causes why the closed-loop behavior differs from what was predicted by the simplest model. These causes include a control signal delay and elastic compliance in the manipulator structure (lumped compliance in the manipulator base or arm). This section attempts to design control taking into account the above-mentioned factors in order to improve the quality of transient processes.

In Chapter 5, we stated one more important problem: keeping contact with a moving object (see Section 5.9). This problem may be formulated as a regulation problem in which a given contact force is maintained despite the disturbances acting on the system. In this case, we consider the object velocity as a disturbance variable. Control theory recommends approaches based on the estimation of a disturbance variable for such regulation problems. This kind of approach is actually employed in Section 5.9, which proposes various techniques to estimate the velocity of a moving object. The goal of the present chapter is to develop a systematic method for taking into account the motion of an object contact point by using an internal dynamical model for such motion. Once the dynamical model of the disturbance is available, the design of a regulator compensating for the disturbances becomes a standard procedure [141, 250].

Note that the problem of maintaining a given contact force while the contact point moves is not only relevant for a moving manipulated object. A manipulator motion with imposed mechanical constraints is probably the most important case where this problem statement is applicable. Chapters 6 and 7 demonstrate how to control a manipulator motion with imposed constraints by using a superposition of two basic motions. The first motion maintains a given constraint force by moving normally to the constraints, while the second motion is directed tangentially to the constraints. As a result, the contact point image in the configuration space moves along the constraint. Inaccuracies in normals to constrains and dynamical factors influencing the motion of the contact point act as disturbances on the first basic motion. The control should render this motion as invariant to the disturbances as possible.

In this chapter we shall consider a model problem of a two-degree-of-freedom motion along a constraint (an object contour), which is similar to the problem of keeping contact with a moving object. It resembles the tracking problem considered in Sections 6.1–6.5.

Let us now introduce mathematical models to be used in this chapter. We shall model manipulator structure compliance as a lumped compliance in the manipulator frame or gear train in the same way as in Sections 5.7 and 5.8. For a delay in the control loop, however, we use a model that differs from those studied in Section 5.6.

Consider a gantry manipulator such as one used in the experimental robotic system (see Chapter 4). The drive control voltage is computed by a hardware velocity servo-system and a computer that sets a commanded (desired) velocity v_d. We shall consider the commanded velocity to be a control parameter. It takes some time to compute the velocity v_d, which causes a delay in the control loop. A typical sequence of the control operation is shown in Figure 8.1.1.

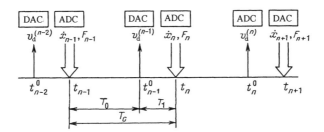

FIGURE 8.1.1
Digital control sequence diagram

At each control cycle, at a time instant t_n the control program samples data from Analog-Digital Converters (ADCs) to obtain the information $\dot{x}_n = \dot{x}(t_n)$ from the velocity sensor (tachometer), and $F_n = F_{t_n}$ from the force sensor. This data is used to compute the next commanded velocity $v_d^{(n)}$, which is passed to the Digital-Analog Converter (DAC) at a time instant $t_n^0 > t_n$. The computed voltage is applied to the hardware servo system. The voltage is kept by the DAC till the computer generates the next output value. At the time instant $t_{n+1} > t_n^0$, the control program samples sensor outputs again, and the next control cycle begins. Time parameters for all control cycles are usually the same. Denote

$$T_c = t_{n+1} - t_n, \quad T_1 = t_{n+1} - t_n^0 \tag{8.1.1}$$

The time parameters are defined by the computation load of the digital computer, in particular, by the algorithm complexity. Independent of the algorithm, in all the experiments described in this chapter, the times T_c and T_1 are the same, i.e., for all algorithms $T_c = 17.3$ ms, $T_1 = 2.8$ ms.

8.2   The Simplest Discrete-Time Model for One-degree-of-freedom Motion

To describe a one-degree-of-freedom manipulator motion, we shall use the model (5.1.8). We neglect the frictional force, assuming that the only external force applied to the manipulator is the one acting from the sensor, i.e., $F_a = -F$, where F is given by (5.1.9). The control voltage u at the motor is supplied by the analog velocity servo system. We shall assume that the control voltage is given by (5.3.12). Thus, (5.1.8), (5.1.9), and (5.3.12) yield

$$\mathcal{M}\ddot{x} + d_2\dot{x} + F = d_1 g_2(v_d - \dot{x}), \tag{8.2.1}$$

where v_d is a commanded velocity set by the digital computer, F is a force measured by the force sensor,

$$F = \begin{cases} 0 & , \text{ if } x < x_s, \\ k(x - x_s) & , \text{ if } x \geq x_s, \end{cases} \tag{8.2.2}$$

where x_s is a coordinate at which the manipulator establishes contact with the object.

Let us first consider the case where the force F applied to the manipulator from the force sensor can be neglected in the left-hand side of (8.2.1). For the robotic system described in Chapter 4, $|F| \leq F_\theta$, where $F_\theta \cong 10$ N is the measurement range of the force sensor. The frictional force ignored in (8.2.1) has the same order of magnitude. The maximum control force $d_1 u_0$ acting on the manipulator is an order of magnitude greater and is in the range of 100 N. Considering the feedback gain in the velocity servo system to be large, we shall neglect F.

By introducing the notation

$$\gamma_1 = (d_1 g_2)/\mathcal{M}, \quad \gamma = (d_1 g_2 + d_2)/\mathcal{M} \tag{8.2.3}$$

and neglecting F, we obtain from (8.2.1)

$$\ddot{x} + \gamma\dot{x} = \gamma_1 v_d \tag{8.2.4}$$

As noted above, the commanded velocity v_d of the manipulator remains constant at the time interval (t_n^0, t_{n+1}^0). Let us consider the motion of the system (8.2.4) for $v_d = \text{const}$. By solving (8.2.4), we obtain

$$x(t) = x(t_0) + \frac{1}{\gamma}\left[1 - e^{-\gamma(t-t_0)}\right]\dot{x}(t_0)$$

$$+ \left[t - t_0 + \frac{1}{\gamma}(e^{-\gamma(t-t_0)} - 1)\right]\frac{\gamma_1}{\gamma}v_d, \tag{8.2.5}$$

$$\dot{x}(t) = \dot{x}(t_0)e^{-\gamma(t-t_0)} + \left[1 - e^{-\gamma(t-t_0)}\right]\frac{\gamma_1}{\gamma}v_d$$

We introduce a 2×2 fundamental matrix $\Psi(t)$ of the system (8.2.4) and a 2×1 column vector χ

$$\Psi(t) = \begin{bmatrix} 1 & \frac{1}{\gamma}(1 - e^{-\gamma t}) \\ 0 & e^{-\gamma t} \end{bmatrix}, \quad \chi(t) = \begin{bmatrix} \left(t + \frac{1}{\gamma}(e^{-\gamma t} - 1)\right)\frac{\gamma_1}{\gamma} \\ (1 - e^{-\gamma t})\frac{\gamma_1}{\gamma} \end{bmatrix} \tag{8.2.6}$$

Let us now derive difference equations to relate the state coordinates of the system at two subsequent control steps. Recall the control time sequence shown in Figure 8.1.1. At time t_n, the system state coordinates are $x(t_n)$ and $\dot{x}(t_n)$. On the time interval (t_n, t_n^0) the system moves at a constant commanded velocity $v_d^{(n-1)}$ set at time t_{n-1}^0. Thus, at time t_n^0, according to (8.2.5) and (8.2.6), the state coordinates are given by

$$\begin{bmatrix} x(t_n^0) \\ \dot{x}(t_n^0) \end{bmatrix} = \Psi(T_0)\begin{bmatrix} x(t_n) \\ \dot{x}(t_n) \end{bmatrix} + \chi(T_0)v_d^{(n-1)}, \tag{8.2.7}$$

where $T_0 = t_n^0 - t_n = T_c - T_1$.

At time t_n^0 the control program produces the next value of the commanded velocity $v_d^{(n)}$, and in the interval (t_n^0, t_{n+1}^0) the system moves at

$v_d = v_d^{(n)}$. By analogy with (8.2.7), we can write the state at the time instant t_{n+1} in the form

$$
\begin{bmatrix} x(t_{n+1}) \\ \dot{x}(t_{n+1}) \end{bmatrix} = \Psi(T_1) \begin{bmatrix} x(t_n^0) \\ \dot{x}(t_n^0) \end{bmatrix} + \chi(T_1) v_d^{(n)}
\tag{8.2.8}
$$

By substituting (8.2.8) into the right side of (8.2.7), we obtain a relation between the state coordinates through the entire control cycle. We shall write this relation as a system of first-order difference equations. To this end, we introduce a delayed input variable

$$
u_n = v_d^{(n-1)}
\tag{8.2.9}
$$

and an extended state vector $\psi^{(n)}$ of the considered discrete-time system

$$
\psi^{(n)} = \begin{bmatrix} x(t_n) & \dot{x}(t_n) & u_n \end{bmatrix}^T
\tag{8.2.10}
$$

According to (8.2.7)–(8.2.9), we can write the following first-order linear difference equation with respect to $\psi^{(n)}$:

$$
\psi^{(n+1)} = \begin{bmatrix} \Psi(T_c) & \Psi(T_1)\chi(T_0) \\ 0 & 0 \end{bmatrix} \psi^{(n)} + \begin{bmatrix} \chi(T_1) \\ 1 \end{bmatrix} v_d^{(n)}
\tag{8.2.11}
$$

Zero block matrices in the right-hand side of (8.2.11) have dimensions 2×1 and 1×1. To derive (8.2.11), we used the identity $\Psi(T_1)\Psi(T_0) = \Psi(T_1 + T_0) = \Psi(T_c)$.

The difference equation (8.2.11) gives a linear discrete-time model of the controlled system – a manipulator with a force sensor. The commanded velocity $v_d^{(n)}$ is the discrete-time control input to the system. When a manipulator is in contact with an object, all state coordinates of the system are directly accessible. Thus, according to (8.2.2) and (8.2.10), $\psi_1^{(n)} = x(t_n) = x_s + F(t_n)/k$, where $F(t_n) = F_n$ is a force sensor reading at the measurement instant t_n at the n-th control step, $\psi_2^{(n)} = \dot{x}(t_n) = \dot{x}_n$ is a tachometer reading at the n-th measurement. According to (8.2.9), $\psi_3^{(n)} = u_n = v_d^{(n-1)}$ is the control input computed at the previous control step.

For the problem of establishing contact between a manipulator and an object, the goal of control is to achieve a stationary regime

$$
\psi^{(n)} = \psi_d = \begin{bmatrix} x_s + F_d/k & 0 & 0 \end{bmatrix}^T
$$

such that a constant given contact force F_d is maintained. The steady-state ψ_d is reachable, since the system (8.2.11) is controllable. The controllability can be verified by using the rank criterion [250]. Since the system (8.2.11) has rank 3, and the control vector has rank 1, the desired stationary regime can be reached in three control steps [250].

To explore the difference system (8.2.11) further, we make one more simplifying assumption. We assume the feedback velocity gain g_2 in (8.2.1) to be large and consider asymptotics for $g_2 \to \infty$. For the experimental system considered in this book, g_2 was large.

It follows from (8.2.3) that for $g_2 \to \infty$

$$\gamma \to \infty, \quad \gamma_1/\gamma \to 1$$

According to (8.2.6), we obtain from (8.2.11), in the limit,

$$\psi^{(n+1)} = \begin{bmatrix} 1 & 0 & T_0 \\ 0 & 0 & 0 \\ 0 & 0 & 0 \end{bmatrix} \psi^{(n)} + \begin{bmatrix} T_1 \\ 1 \\ 1 \end{bmatrix} v_d^{(n)} \tag{8.2.12}$$

The equations of motion for the second and third components of the vector $\psi^{(n)}$ coincide. Hence these components are equal and, according to (8.2.9), (8.2.10), $\dot{x}(t_n) = v_d^{(n-1)}$. Thus, the system motion can be described by two state variables instead of three. Let us introduce a state vector

$$q^{(n)} = \begin{bmatrix} x(t_n) - x_s - F_d/k & \dot{x}(t_n) \end{bmatrix}^T, \tag{8.2.13}$$

where F_d is a commanded (desired) contact force applied by the manipulator to the object. The vector q is zero in the desired steady-state. We rewrite (8.2.12) as follows:

$$q^{(n+1)} = \Phi q^{(n)} + b v_d^{(n)}, \tag{8.2.14}$$

$$\Phi = \begin{bmatrix} 1 & T_0 \\ 0 & 0 \end{bmatrix}, \quad b = \begin{bmatrix} T_1 \\ 1 \end{bmatrix}$$

The equation (8.2.14) can be obtained not only from (8.2.4), but also from (8.2.1) for $g_2 \to \infty$. For the latter derivation, one should consider the system (8.2.1) for $g_2 \gg 1$ to be singularly perturbed and separate slow and fast motions of this system. In Section 8.2, difference equations of motion taking into account compliance in the manipulator structure are derived in such a way.

Let us now study control laws for the system (8.2.14). The state vector q is defined by (8.2.13) so that the coordinate origin, according to (8.2.2), corresponds to the stationary regime $F(t_n) = F_n \equiv F_d$. We consider a manipulator motion in contact with an object. For such motion, as noted above, all state coordinates of the system are accessible to measurement. In particular, it follows from (8.2.2) and (8.2.13) that

$$F_n = F(t_n) = kq_1^{(n)} + F_d \qquad (8.2.15)$$

We start with the control law (5.3.13) for the commanded velocity considered in Chapter 5. Since the commanded velocity v_d is given only at discrete time instants, we can rewrite this control law in the form

$$v_d^{(n)} = g[F_d - F(t_n)] \qquad (8.2.16)$$

We substitute (8.2.16) into (8.2.14) and obtain, in view of (8.2.15), an equation for the closed-loop dynamics

$$q^{(n+1)} = \Phi_f q^{(n)}, \quad \Phi_f = \begin{bmatrix} 1 - T_1 kg & T_0 \\ -kg & 0 \end{bmatrix} \qquad (8.2.17)$$

If the eigenvalues of the matrix Φ_f are inside the unit circle in the complex plane, then the closed system (8.2.17) is stable and converges asymptotically to the desired steady-state, namely, to the coordinate origin. The eigenvalues z_1 and z_2 of the matrix Φ_f satisfy the characteristic equation

$$\det(\Phi_f - Iz) = z^2 + (T_1 kg - 1)z + T_0 kg = 0, \qquad (8.2.18)$$

where I is a 2×2 identity matrix.

The conditions for the roots of (8.2.18) to be inside the unit circle have the form [250]

$$T_c gk > 0, \quad T_0 gk < 1, \quad (T_1 - T_0)kg < 2 \qquad (8.2.19)$$

Recall that $T_c = T_0 + T_1$. The first inequality in (8.2.19) means that the feedback should be "negative". The third inequality in (8.2.19) tends to be satisfied for all $g > 0$, since, as a rule, $T_0 > T_1$. This is due to the fact that T_0 is the time between the measurement and control input instants, which is taken by control computations, while T_1 is the time between the control input and the next measurement instants which does not depend on control computations. The second inequality in (8.2.19) coincides with the inequality (5.6.7), if we set the control loop delay $\theta = T_c$ and $g_1/g_2 = g$, according to (5.3.13). Thus, the inequalities in (8.2.19) derived independently of those in (5.6.7) yield the same estimate for a limiting feedback gain in the case of control loop delay.

The linear discrete-time system (8.2.14) is controllable (which can be easily checked), so it can be brought to a desired stable state in a finite number of control steps (in two steps) [141, 250]. To this end, we use a linear state feedback system of the form

$$v_d^{(n)} = - \begin{bmatrix} gk & h \end{bmatrix} q^{(n)}, \tag{8.2.20}$$

where g and h are feedback gains. In view of (8.2.13) and (8.2.15), we can rewrite the control law (8.2.20) in the form

$$v_d^{(n)} = g[F_d - F(t_n)] - h\dot{x}(t_n) \tag{8.2.21}$$

The motion of the system (8.2.14) under the control (8.2.20) is given by a difference equation

$$q^{(n+1)} = \Phi_c q^{(n)}, \quad \Phi_c = \begin{bmatrix} 1 - T_1 kg & T_0 - T_1 h \\ -kg & -h \end{bmatrix} \tag{8.2.22}$$

The characteristic equation of the system (8.2.22) has the form

$$z^2 + (T_1 kg + h - 1)z + T_0 kg - h = 0, \tag{8.2.23}$$

Two zero roots of the characteristic equation correspond to the desired "dead-beat" controller bringing the system to the steady-state in two steps. By setting the free and linear terms to zero in (8.2.23), we find coefficients for the dead-beat controller

$$g = 1/(T_c k), \quad h = T_0/T_c \tag{8.2.24}$$

The control law (8.2.21), unlike (8.2.16), includes a *discrete-time velocity feedback*. Recall that we already have a *continuous-time velocity feedback* in place implemented by the analog servo system (see (5.3.12)). Note also that as far as the continuous-time model of the manipulator motion is concerned, the controller configuration is not affected by one more velocity feedback loop closed via the computer. The benefit of a velocity feedback closed through the computer becomes apparent only after taking into account the discreteness of control and building the corresponding manipulator motion model. The experiments show that the quality of transient processes is somewhat better with the control law (8.2.21) than with the control of the form (8.2.16) considered in Chapter 5.

8.3 Discrete-Time Model for Manipulator with Transmission Compliance

The simplest continuous-time model for dynamics of a manipulator with force feedback considered in Sections 5.1–5.5 does not explain some peculiarities of the robotic system performance in the experiments, such as sustained oscillations that appear for high feedback gains.

The discrete-time model, built on the basis of the continuous-time model in the previous section, accounts for the loss of stability due to the increase of force feedback gain. Still, this model fails to describe the experimentally observed transient processes adequately.

This section and the following one consider discrete-time models of motion for a compliant manipulator (see Sections 5.7 and 5.8 for the continuous-time models). In the present section, we assume a lumped compliance in the gear train. We shall use the dynamical model for the manipulator with a force sensor and a compliant gear train developed in Section 5.7. By using (5.7.2)–(5.7.4), we obtain the equations of motion for the manipulator in contact with an object:

$$\mathcal{M}\ddot{x} + d_2\dot{x} - \beta_c\dot{\xi} - k_c\xi = d_1 u, \qquad (8.3.1)$$

$$m_1\ddot{x} + \beta\dot{x} + k(x - x_s) + m_1\ddot{\xi} + (\beta + \beta_c)\dot{\xi} + (k + k_c)\xi = 0$$

In the above equations, x is a coordinate of the manipulator arm "before compliance", ξ is a displacement in the visco-elastic element which models compliance in the gear train. The control voltage u is computed by the velocity servo system according to (5.3.12). The other parameters in (8.3.1) are explained in Section 5.7.

Let us assume that the velocity feedback gain in the servo system is large. In order to apply the methods of time-scale separation, we represent (8.3.1), (5.3.12) in the form of a first-order system of differential equations

$$\frac{dx}{dt} = \dot{x},$$

$$\frac{d\dot{x}}{dt} = \frac{\beta_c}{M}(\dot{x} + \dot{\xi}) - \frac{\beta_c + d_2}{M}\dot{x} + k_c\xi + \frac{d_1 g_2}{M}(v_d - \dot{x}),$$

$$\frac{d\xi}{dt} = (\dot{x} + \dot{\xi}) - \dot{x}, \tag{8.3.2}$$

$$\frac{d}{dt}(\dot{x} + \dot{\xi}) = -\frac{(\beta + \beta_c)}{m_1}(\dot{x} + \dot{\xi}) + \frac{\beta_c}{m_1}\dot{x} - k(x - x_s) - (k + k_c)\xi$$

In (8.3.2), we consider the feedback gain to be a large parameter

$$g_2 \gg 1, \quad \varepsilon = 1/g_2 \ll 1, \tag{8.3.3}$$

where ε is a small parameter.

Under the condition (8.3.3), we can consider (8.3.2) as a singularly perturbed system. In (8.3.2), the velocity $\dot{x}$ is a fast variable, while x, ξ, $\dot{x} + \dot{\xi}$ are slow variables. The fast system describing the dynamics of the fast variable in the limit has the form

$$\varepsilon\frac{d\dot{x}}{dt} = \frac{d_1}{M}(v_d - \dot{x}) \tag{8.3.4}$$

Since the equation (8.3.4) is linear and its solution $\dot{x} = v_d$ is asymptotically stable for $v_d = \text{const}$, the conditions of the Tikhonov's theorem are satisfied [174, 259]. Hence, outside of the boundary layer, the motion of the system (8.3.2), (8.3.3) is close to that of the slow system, which is obtained by setting the right-hand side of (8.3.4) equal to zero:

$$\frac{dx}{dt} = v_d, \quad \frac{d\xi}{dt} = \dot{\xi}, \tag{8.3.5}$$

$$\frac{d}{dt}(\dot{x} + \dot{\xi}) = \frac{dv_d}{dt} + \frac{d\dot{\xi}}{dt} = -\frac{(\beta + \beta_c)}{m_1}\dot{\xi} - \frac{\beta}{m_1}v_d - k(x - x_s) - (k + k_c)\xi$$

The commanded velocity v_d in (8.3.2) is a control parameter. The control goal is to reach a steady-state such that a constant given (commanded) contact force F_d is maintained between the manipulator and the object. According to (5.9.21), in this steady-state the coordinates x and ξ have the values

$$x_0 = x_s + \frac{F_d}{k} + \frac{F_d}{k_c}, \quad \xi_0 = -\frac{F_d}{k_c} \tag{8.3.6}$$

The commanded velocity v_d is set by the control computer and is a piecewise constant function of time. For the motion with a constant commanded velocity, according to (8.3.4), $\dot{x} = v_d = \text{const}$, and we can rewrite (8.3.5) in the form

$$\frac{dx}{dt} = \dot{x}, \qquad \frac{d\dot{x}}{dt} = 0 \qquad \frac{d\xi}{dt} = \dot{\xi}, \tag{8.3.7}$$

$$\frac{d\dot{\xi}}{dt} = -\frac{k}{m_1}(x - x_s) - \frac{\beta}{m_1}\dot{x} - \Omega_c^2 \xi - \gamma_c \dot{\xi},$$

where $\Omega_c^2 = (k_c + k)/m_1$, $\gamma_c = (\beta + \beta_c)/m_1$.

For $x \equiv 0$, the last two equations in (8.3.7) describe damped oscillations. We assume the damping in the structure to be subcritical ($\gamma_c < 2\Omega_c$), and denote the oscillation period by Θ:

$$\Theta = (\Omega_c^2 - \gamma_c^2/4)^{-1/2} \tag{8.3.8}$$

Let us introduce a state vector q for state coordinate deviations from the desired steady-state (8.3.6)

$$q = \begin{bmatrix} x - x_0 & \Theta\dot{x} & \xi - \xi_0 & \Theta\dot{\xi} \end{bmatrix}^T \tag{8.3.9}$$

By using the notation of (8.3.8), (8.3.9), we rewrite the slow system (8.3.7) in the form

$$\dot{q} = \frac{1}{\Theta}Aq, \quad A = \begin{bmatrix} 0 & 1 & 0 & 0 \\ 0 & 0 & 0 & 0 \\ 0 & 0 & 0 & 1 \\ -p & -r & -(1+\delta^2) & -2\delta \end{bmatrix} \tag{8.3.10}$$

In the above equations, the nondimensional parameters are $p = k\Theta^2/m_1$, $r = \beta\Theta/m_1$, $\delta = \gamma_c\Theta/2$, and according to (8.3.8), $\Omega_c^2\Theta^2 = 1 + \delta^2$. Denote by $\Psi(t)$ the fundamental matrix of the system (8.3.10):

$$\Psi(t) = e^{At/\Theta} \tag{8.3.11}$$

Let us now proceed with the derivation of difference equations relating the system state at two subsequent control steps. Similar to Section 8.2, we shall follow the time sequence of the control program shown in Figure 8.1.1. We start with the instant t_n of the n-th measurement. Denote by $q^{(n)} =$

$q(t_n)$ a state vector of the discrete-time system in hand. On the time interval (t_n, t_n^0) between the measurement and control input instants, the system moves with a constant velocity equal to the commanded value. We assume that by the measurement instant t_n, the transient process in the fast system caused by a recurrent commanded velocity value $v_d^{(n-1)}$ has died out at an instant t_{n-1}^0, and $\dot{x}(t_n) = v_d^{(n-1)}$. The system motion on the interval (t_n, t_n^0) can be described by the slow system (8.3.10). The system state at the time instant t_n^0, when the commanded velocity changes, is given by

$$q(t_n^0) = \Psi(T_0) = q^{(n)}, \qquad (8.3.12)$$

where $T_0 = t_n^0 - t_n$. Note that according to (8.3.10), $\dot{x}(t_n^0)$ equals $v_d^{(n-1)}$, same as at the time instant t_n.

After the servo system receives the next commanded velocity value $v_d^{(n)}$ at time t_n^0, a transient process is initiated, which is described by the fast system (8.3.4). In this transient process of duration ε, the fast variable $\dot{x}$ changes its value from $\dot{x}(t_n^0) = v_d^{(n-1)}$ to $v_d^{(n)}$, while the slow variables x, ξ, $\dot{x} + \dot{\xi}$ do not change. Since the variable $\dot{x} + \dot{\xi}$ does not change, the arm motion velocity $\dot{\xi}$ changes to $-(v_d^{(n)} - v_d^{(n-1)}) = -v_d^{(n)} + \dot{x}(t_n^0)$. Let us write the state vector change during the transient process (in the boundary layer) in the form

$$q[t_n^0 + O(\varepsilon)] = (I + \Psi_0)q(t_n^0) + b_0 v_d^{(n)} + O(\varepsilon), \qquad (8.3.13)$$

$$b_0 = \begin{bmatrix} 0 & 1 & 0 & -1 \end{bmatrix}^T, \qquad \Psi_0 = \begin{bmatrix} 0 & -b_0 & 0 & 0 \end{bmatrix},$$

where I is a 4×4 identity matrix.

After the transient process settles down, the system motion coincides with the motion of the slow system (8.3.10) again. By neglecting the boundary layer transients (assuming ε to be infinitely small), we obtain, in view of (8.1.1), the state at time t_{n+1} of the next measurement

$$q^{(n+1)} = q(t_{n+1}) = \Psi(T_1)q(t_n^0) \qquad (8.3.14)$$

By combining the relationships (8.3.12)–(8.3.14), we obtain the expression for the system state variation in the period between consecutive measurements

$$q^{(n+1)} = \Phi q^{(n)} + b v_d^{(n)}, \qquad (8.3.15)$$

where $\Phi = \Psi(T_1)(I + \Psi_0)\Psi(T_0) = \Psi(T_c) + \Psi(T_c)\Psi_0\Psi(T_0)$ is a 4×4 matrix, $b = \Psi(T_1)b_0$ is a 4×1 matrix.

Two parameters are accessible for measurement in the system: the velocity $\dot{x}(t_n)$, which is given by the tachometer reading at the observation time t_n, and the force sensor output $F(t_n)$, which relates to the coordinates x and ξ by (5.7.5). At the same time, (8.3.15) is a fourth-order system, so we shall introduce an observation equation relating measured and state variables. We introduce an observation vector

$$y^{(n)} = \begin{bmatrix} \dot{x}(t_n) & F(t_n) - F_d \end{bmatrix}^T, \tag{8.3.16}$$

which is zero in the desired steady-state, similar to the vector q. According to (5.7.8), (8.3.6), and (8.3.9), we obtain

$$y^{(n)} = Cq^{(n)}, \tag{8.3.17}$$

$$C = \begin{bmatrix} 0 & 1/\Theta & 0 & 0 \\ k & 0 & k & 0 \end{bmatrix}$$

The relationships (8.3.15)–(8.3.17) represent a dynamical model of a controlled system comprising a manipulator with a compliant gear train contacting a stationary object, and an analog velocity servo system of the manipulator.

8.4 Manipulator Model with Compliant Base

In this section, we shall continue the development of a discrete-time dynamical model for a manipulator with structural compliance. Unlike the previous section, we now consider a lumped compliance in the manipulator frame. The previous section makes use of the model developed in Section 5.7. Correspondingly, in this section, we shall employ the model proposed in Section 5.8.

When the manipulator contacts the object, the coordinate x set by the drive and displacement η in the visco-elastic element modeling the frame compliance satisfy the differential equations (5.8.8). We assume that the velocity servo system applies the control voltage u to the motor according to (5.3.12) and obtain from (5.8.8)

$$\mathcal{M}\ddot{x} + m\ddot{\eta} + d_2\dot{x} + \beta(\dot{x} + \dot{\eta}) + k(x + \eta - x_s) = d_1g_2(v_d - \dot{x}),$$

$$m\ddot{x} + \mathcal{M}\ddot{\eta} + \beta_b\dot{\eta} + \beta(\dot{x} + \dot{\eta}) + k_b\eta + k(x + \eta - x_s) = 0 \qquad (8.4.1)$$

In the above equation, v_d is a commanded motion velocity set by computer, g_2 is a feedback gain in the velocity servo system, which we consider to be large, according to (8.3.3), same as in Section 8.3.

We represent (8.4.1) as a system of first-order differential equations:

$$\frac{dx}{dt} = \dot{x}, \quad \frac{d\dot{x}}{dt} = \frac{\mathcal{M}\mathcal{M}_1 d_1 g_2}{\mathcal{M}\mathcal{M}_1 - m^2}(v_d - \dot{x}) + \dots, \quad \frac{d\eta}{dt} = \dot{\eta},$$

$$\frac{d}{dt}(\mu\dot{x} + \dot{\eta}) = -\frac{\beta_b + \beta}{\mathcal{M}_1}(\mu\dot{x} + \dot{\eta}) - \frac{\beta(1 - \mu) - \mu\beta_b}{\mathcal{M}_1}\dot{x} \qquad (8.4.2)$$

$$+ k_b\eta + k(x + \eta - x_s),$$

where $\mu = m/\mathcal{M}_1$ is a nondimensional parameter, and the ellipsis denotes the terms not relevant for the subsequent derivation (according to (5.8.7), $0 < \mu < 1$).

Similar to (8.3.2), under the condition (8.3.3), we may consider (8.4.2) as a singularly perturbed system. In (8.4.2), the velocity $\dot{x}$ is a fast variable, while x, η, $\mu\dot{x} + \dot{\eta}$ are slow variables. The fast system describing the dynamics of the fast variable in asymptotics has the form

$$\varepsilon\frac{d\dot{x}}{dt} = \frac{\mathcal{M}\mathcal{M}_1 d_1}{\mathcal{M}\mathcal{M}_1 - m^2}(v_d - \dot{x}) \qquad (8.4.3)$$

Similar to (8.3.4), the equation (8.4.3) is linear and its solution $\dot{x} = v_d$ is asymptotically stable for $v_d = $ const. We obtain the slow system from (8.4.2) by equating the right-hand side of (8.4.3) to zero:

$$\frac{dx}{dt} = v_d, \quad \frac{d\eta}{dt} = \dot{\eta},$$

$$\frac{d}{dt}(\mu\dot{x} + \dot{\eta}) = \mu\frac{dv_d}{dt} + \frac{d\dot{\eta}}{dt} = -\frac{(\beta + \beta_b)}{\mathcal{M}_1}\dot{\eta} - \frac{\beta}{\mathcal{M}_1}v_d \qquad (8.4.4)$$

$$-k(x - x_s) - (k + k_b)\eta$$

The commanded velocity v_d is a zero-order hold function of time. Since (8.4.3) yields $\dot{x} = v_d$, we can rewrite (8.4.4) for $v_d = $ const in the form similar to (8.3.7):

$$\frac{dx}{dt} = \dot{x}, \quad \frac{d\dot{x}}{dt} = 0 \quad \frac{d\eta}{dt} = \dot{\eta}, \tag{8.4.5}$$

$$\frac{d\dot{\eta}}{dt} = -\frac{k}{\mathcal{M}_1}(x - x_s) - \frac{\beta}{\mathcal{M}_1}\dot{x} - \Omega_b^2\eta - \gamma_b\dot{\eta},$$

where $\gamma_b = (\beta_b + \beta)/\mathcal{M}_1$, $\Omega_b^2 = (k_b + k)/\mathcal{M}_1$. By analogy with (8.3.6) and (8.3.8), we introduce the notation

$$\Theta = [\Omega_b^2 - \gamma_b^2/4]^{-1/2}, \tag{8.4.6}$$

$$x_0 = x_s + F_d/k + F_d/k_b, \quad \eta_0 = -F_d/k_b,$$

where x_0 and η_0 are the coordinates corresponding to the desired stationary regime (5.8.11) with a constant contact force $F \equiv F_d$. We introduce a state vector

$$q = \begin{bmatrix} x - x_0 & \Theta\dot{x} & \eta - \eta_0 & \Theta\dot{\eta} \end{bmatrix}^T \tag{8.4.7}$$

For the desired stationary regime, the state vector q is zero. We can rewrite (8.4.5) in the form coinciding with (8.3.10), though the nondimesional parameters δ, r in the matrix A will have a different meaning than in (8.3.10):

$$p = k\Theta^2/\mathcal{M}_1, \quad r = \beta\Theta/\mathcal{M}_1, \quad \delta = \gamma_b\Theta/2, \quad 1 + \delta^2 = \Omega_b^2\Theta^2$$

The derivation of difference equations relating the system state (8.4.7) at two subsequent measurement instants t_{n+1} and t_n coincides with the similar derivation made in Section 8.3 up to the notation. The only difference is that the vector b_0 in the equation analogous to (8.3.13) has a somewhat different form in this case. Let us find this vector. After the next commanded velocity value $v_d^{(n)}$ is applied at the time instant t_n^0, completing the transient process of duration ε, the fast variable $\dot{x}$ changes from $\dot{x}(t_n) = v_d^{(n-1)}$ to $v_d^{(n)}$. The slow variables x, η, $\mu\dot{x} + \dot{\eta}$ do not vary. Since the variable $\mu\dot{x} + \dot{\eta}$ does not change, the variation $\dot{\eta}$ equals $-\mu(v_d^{(n)} - v_d^{(n-1)})$. Thus, instead of (8.3.13), we obtain

$$q[t_n^0 + O(\varepsilon)] = (I + \Psi_0)q(t_n^0 - 0) + b_0\Theta v_d^{(n)} + O(\varepsilon), \tag{8.4.8}$$

$$b_0 = \begin{bmatrix} 0 & 1 & 0 & -\mu \end{bmatrix}^T, \quad \Psi_0 = \begin{bmatrix} 0 & -b_0 & 0 & 0 \end{bmatrix}$$

The observation equation in this case coincides with (8.3.17).

For the case in question, difference equations of motion have the form (8.3.15). The expressions for the matrices Ψ in Sections 8.3 and 8.4 coincide, and the meaning of the parameters Θ, δ, p, and r, on which these matrices depend, are close. Comparing the expressions for the matrices b_0 and Ψ_0 in (8.3.13) and (8.4.8), one can see that they coincide if we take $\mu = 1$ in (8.4.8). Recall that $\mu = m/\mathcal{M}_1$ and $0 < \mu < 1$, according to (5.8.7).

Therefore, we can consider that the equations (8.3.15)– (8.3.17), where the matrices Φ and b are given by (8.3.10), (8.4.8), describe the motion of a manipulator with a force sensor and a lumped compliance in its structure. For $0 < \mu < 1$ in (8.4.8), the compliance is lumped in the frame, and for $\mu = 1$, it is lumped in the gear train.

Similar to Section 8.2, we consider in more detail the case where the force acting on the manipulator from the force sensor is negligible. In this case, we should take $k = 0$ and $\beta = 0$ in the equations of motion (8.3.1) and (8.4.1), and $p = 0$, $r = 0$, in (8.3.10). Then we can write the expression for the fundamental matrix Ψ in (8.3.11) in the explicit form:

$$\Psi(t) = \begin{bmatrix} 1 & t/\Theta & 0 & 0 \\ 0 & 1 & 0 & 0 \\ 0 & 0 & \psi_{33}(t/\Theta) & \psi_{34}(t/\Theta) \\ 0 & 0 & \psi_{43}(t/\Theta) & \psi_{44}(t/\Theta) \end{bmatrix},$$

$$\psi_{33}(\Theta) = e^{-\delta\theta}(\cos\theta + \delta\sin\theta), \quad \psi_{34}(\Theta) = e^{-\delta\theta}\sin\theta,$$

$$\psi_{43}(\Theta) = -e^{-\delta\theta}(1 + \delta^2)\sin\theta, \quad \psi_{44}(\Theta) = -e^{-\delta\theta}(\cos\theta - \delta\sin\theta)$$

Then, by using the expressions (8.4.8) for the matrices b_0 and Ψ_0 and taking into account (8.3.15), we obtain the following expressions for the matrices Φ and b:

$$\Phi = \begin{bmatrix} 1 & \tau - \tau_1 & 0 & 0 \\ 0 & 0 & 0 & 0 \\ 0 & \mu\psi_{34}(\tau_1) & \psi_{33}(\tau) & \psi_{34}(\tau) \\ 0 & \mu\psi_{44}(\tau_1) & \psi_{33}(\tau) & \psi_{44}(\tau) \end{bmatrix}, \quad b = \begin{bmatrix} \tau_1 \\ 1 \\ -\mu\psi_{34}(\tau) \\ -\mu\psi_{44}(\tau) \end{bmatrix} \quad (8.4.9)$$

In the above expressions, we used the notation

$$\tau = T_c/\Theta, \quad \tau_1 = T_1/\Theta \quad (8.4.10)$$

Hence $T_0/\Theta = (T_c - T_1)/\Theta = \tau - \tau_1$. To use the designed models for the high-performance controller design, it is necessary to estimate the parameters that appear in the models.

In all the experiments considered in the present chapter, the arm of the manipulator (described in Chapter 4) moves horizontally. The force sensor employed in the experiments is displayed in Figure 4.3.3. The sensor stiffness $k = 3.35 \cdot 10^3$ N/m in the horizontal direction is much less than the stiffness of the manipulator structure. Hence it is possible to neglect the force acting on the manipulator from the sensor.

The matrices A, B, and C in (8.4.8), (8.4.9) depend on the three unknown parameters: μ that gives a ratio of masses in the system, δ that gives an oscillation decrement in the structure, and τ that defines a period (frequency) of the oscillations. All the parameters were estimated experimentally. In doing so, we assumed that the case $0 < \mu < 1$ corresponds to the compliance in the frame.

The parameters τ and τ_1 from (8.4.9) are related to Θ by (8.4.10). The time parameters T_0 and T_1 from (8.1.1) that appear in (8.4.10) can be measured directly. In all the experiments described in this chapter, these parameters were kept the same: $T_0 = 17.3$ ms and $T_1 = 2.8$ ms.

The parameters μ, δ, and Θ were estimated simultaneously by applying a PRBS (Pseudo-Random Binary Sequence) signal to the manipulator in contact with the object and fitting the model response against the logged data. In all the experiments $\mu < 0.7$ was obtained, so it may be assumed that the compliance is lumped in the frame. The oscillations of the frame were also observed visually. The identification yielded the following set of system parameters:

$$\Theta = 0.138, \qquad \mu = 0.55, \qquad \delta = 0.6 \qquad\qquad (8.4.11)$$

8.5 Discrete Control of Contact Transition

We continue solving the first of the two problems stated in Section 8.1. The problem is to design control that would improve transient processes in a manipulator establishing contact with an object. We have started solving this problem in Section 8.2. The control law (8.2.21) takes into account a delay in control computations, but assumes that the manipulator structure is rigid. In this section, we take into account structure compliance when designing the control. We present experimental results for a manipulator with a force sensor. In the experiments, the manipulator establishes contact with an object, with controllers that take or do not take into account structure compliance.

Let us state the control design problem for a manipulator with a structure compliance. In the system (8.3.15), (8.3.17), (8.4.9), only two of the four components of the vector $q^{(n)}$ are directly accessible. Hence we shall design a linear dynamic regulator with a state observer for the system. Using such a regulator, the closed-loop eigenvalues (poles) of the system (8.3.15), (8.3.17), (8.4.9) may be placed arbitrarily [141, 250].

We shall use a standard linear quadratic gaussian (LQG) regulator design (see [141, 250]). It has a number of attractive properties, such as a low sensitivity to the deviation of a real controlled system dynamics from the model [51, 143]. The LQG regulator includes a full-order state observer (a canonical Kalman filter) that for the system in hand has the form

$$\hat{q}^{(n+1)} = \Phi\hat{q}^{(n)} + \Theta b v_d^{(n)} + L(y^{(n)} - G\hat{q}^{(n)}), \qquad (8.5.1)$$

$$v_d^{(n)} = -C\hat{q}^{(n)}), \qquad (8.5.2)$$

where $\hat{q}^{(n)}$ is an estimate of the state vector $q^{(n)}$, $y^{(n)}$ is the observation vector. The regulator (8.5.1), (8.5.2) allows to obtain a recurrent estimate of the state $\hat{q}^{(n)}$ and a control input $v_d^{(n)}$ at the n-th control step by the time t_n^0 on the basis of the previous estimate $\hat{q}^{(n-1)}$, the control input $v_d^{(n-1)}$ at the time t_{n-1}^0, and the vector $y^{(n-1)}$ sampled at the time t_{n-1}.

The matrices of the observer L feedback gains (8.5.1) and state regulator G (8.5.2) are obtained by solving the two respective matrix algebraic Riccati equations using standard methods [141, 250]. For the numerical design of the LQG regulator, diagonal noise covariances were assumed. The quadratic state penalty was imposed primarily on a force deviation from the commanded force (the sum of the first and the third components of the vector $q^{(n)}$). One of the experimentally tested state feedback gain matrices (8.5.2) was

$$G = \Theta^{-1} \begin{bmatrix} g_1 & g_2 & g_3 & g_4 \end{bmatrix},$$

where

$$g_1 = 3.27, \quad g_2 = 0.89, \quad g_3 = 2.54, \quad g_4 = 1.13 \qquad (8.5.3)$$

The LQG regulator (8.5.1), (8.5.2) was implemented on the control computer for the experimental robotic system. As noted in Section 8.1, the time T_c of the program cycle was 17.3 ms.

Let us now describe the results of the experiments in which the manipulator establishes contact with an object. In each experiment, the arm of the research manipulator described in Chapter 4 moved horizontally along the axis OX with a constant velocity v_0. The force sensor shown in Figure 4.3.3 is mounted on the end of the arm. A rigid rod is attached to the sensor flange to contact an object. Once the contact is detected, the control which ensures a constant given contact force is applied. Note that the hardware used in the experiments differs from that used in the experiments of Chapter 5. Accordingly, the results obtained for comparable control laws in this chapter somewhat differ from those presented in Chapter 5.

We start with the linear force feedback control law (8.2.16) studied in Chapter 5. The experimental results for the manipulator establishing contact with an object under the control (8.2.16) are illustrated in Figure 8.5.1 (the curves 1). For this case, $F_d = 6$ N, $gk\Theta = 3.5$, $v_0 = 7.5$ mm/s. Oscillations in the transient process are caused by the compliance in the structure (frame) of the manipulator.

The control law (8.2.21) includes a discrete-time feedback of both measurements. As noted in Section 8.2, an additional velocity feedback loop is not justified within the continuous-time models considered in Chapter 5. However, the simplest discrete-time model studied in Section 8.2 predicts that a discrete-time velocity feedback $\dot{x}(t_n)$ should enhance the quality of transient processes. The experimental results for the same force feedback gain $gk\Theta = 3.5$ and a discrete-time velocity feedback gain $h = 0.5$ correspond to the curves 2 in Figure 8.5.1. The transient processes significantly improve compared to the curves 1.

An increase of the force feedback gain g in the control law (8.2.16) leads to increased oscillations in the system in comparison with the case of $gk\Theta = 3.5$ (see the curves 1).

For $gk\Theta \approx 7$, sustained oscillations develop in the experiment. Yet, the poles (eigenvalues) of the closed-loop system (8.2.16), (8.3.15), (8.4.9) for the parameters (8.4.11), only leave the unit circle for $gk\Theta \approx 19$. This difference between the modeling and experiment can be attributed to the fact that the friction in the real system is Coulomb rather than viscous. Also, when a system with Coulomb friction is linearized, the equivalent viscous friction coefficient decreases with the growth of the oscillation amplitude.

The experimental results for the regulator (8.5.1), (8.5.2) for feedback gains (8.5.3) correspond to the curves 3 in Figure 8.5.1. The performance is much better than for the pure force feedback (see the curve 1). At the same time, only small improvement is obtained compared to the discrete-time force and velocity feedback (see the curves 2). This can be explained by the imperfection of the dynamical manipulator model (8.3.15), (8.3.17), (8.4.9). For instance, the assumed linear friction and compliance does not exactly hold in practice. Also, in the beginning of the transient process the variation of the commanded velocity from one control cycle to another are large and the high-gain analog velocity feedback may saturate.

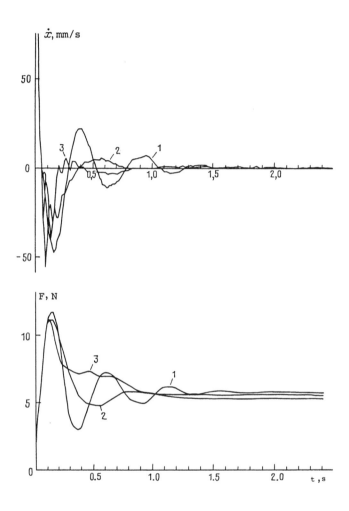

FIGURE 8.5.1
Contact transition in the experiments. Upper graph – manipulator velocity, lower – contact force

8.6   Maintaining Contact Force with a Moving Object

Let us now turn to the second problem stated in Section 8.1: design of a regulator that keeps a given contact force in the presence of disturbances caused by the motion of the contact point.

We set the following model problem related to the control of the contouring motion considered in Sections 6.1– 6.5. Let a force sensor be modeled by a massless elastic rod (same as in Chapter 6), be in contact with an object lying on the manipulator table (see Figure 6.1.1). The first degree of freedom moves the manipulator arm in the direction of the horizontal axis OX and allows contact with the lateral surface of the object to be maintained. We assume that the object is elongated along the horizontal axis OY. The second degree of freedom of the manipulator moves the arm in the direction of the axis OY. As a result, the manipulator arm with the sensor can be held against various points on the object contour and contact them.

We denote by x_s an arm coordinate where the force sensor contacts the object, $x_s = x_s(y)$. It is assumed that the manipulator arm displacement along the axis OY is defined by a given time function $y = y(t)$, which may be unknown in the control design. Thus, $x_s[y(t)] = x_s(t)$.

Accurately speaking, $x_s = x_s[y_s(t)]$, where y_s is a coordinate projection of the contact point onto the axis OY. The coordinate y_s differs from the arm coordinate y by the deformation of the force sensor (see Section 6.1). However, we assume the deformation to be small compared to the arm displacement y and neglect the difference.

We shall control the motion along the axis OX. The control goal is to maintain the component F_x of the force of contact with the object equal to the given value F_d. In doing so, we consider the arm motion in the direction of the axis OY that causes the coordinate x_s to vary as a disturbance. The above-formulated control design problem is similar to the one-dimensional problem of keeping contact with a moving object considered in Section 5.9. Unlike Section 5.9, herein we shall only consider the manipulator arm motion in contact with an object.

The considered approach to control can be obtained using the "compliant joint" concept [197, 199] for the motion of a gantry manipulator along an object contour. According to this concept discussed in Section 1.3, the degree of freedom moving the arm along the axis OX is chosen to be "compliant" (controlled with respect to force). This degree of freedom is most involved in the motion at normal to the contour. The problem of maintaining a given contact force using the compliant joint control can be formulated for a manipulator with a different kinematic scheme such as an articulated manipulator moving along an imposed constraint. The contact force is maintained by controlling certain degrees of freedom and the force regulation disturbances are caused by the motion of other degrees.

The same formulation is applicable to the contour following control problem in Chapter 6. For the latter problem, we can consider as disturbances the curvature of the contour along which the manipulator moves and inaccurately known normal to the contour. By analogy, for the general problem of motion with imposed constraints (see Chapter 7), the motion along the constraint disturbs the primitive motion maintaining a given force of contact.

We shall use the simplest discrete-time model of a one-dimensional manipulator motion developed in Section 8.2. This model is based on relationships (8.2.1), (8.2.2). In this section, we assume that $x_s = x_s(t)$ in (8.2.2). Similar to Section 8.2, we shall use the assumption that the analog velocity servo system is "rigid". Under this assumption, the force acting on the manipulator from the force sensor does not appear in the equations of motion. Hence the equations of motion do not depend on the coordinate x_s. At the same time, the coordinate x_s appears in the definition (8.2.13) of the state vector q, since $q_1^{(n)} = x(t_n) - x_s - F_d/k$. The equation of motion for the component $q_1^{(n)}$ in (8.2.14) is obtained assuming $x_s = \text{const}$. Let us rewrite (8.2.14) in the form

$$x(t_{n+1}) - F_d/k = x(t_n) - F_d/k + T_0 q_2^{(n)} + T_1 v_d^{(n)} \qquad (8.6.1)$$

By subtracting $x_s(t_{n+1})$ from the left-hand side of (8.6.1), and $x_s(t_n) + [x_s(t_{n+1}) - x_s(t_n)]$ from its right-hand side, we rewrite (8.6.1) as

$$q_1^{(n+1)} = q_1^{(n)} + T_0 q_2^{(n)} + T_1 v_d^{(n)} - T_c v^{(n)}, \qquad (8.6.2)$$

where $v^{(n)} = [x_s(t_{n+1}) - x_s(t_n)]/T_c$ is an average velocity of x_s on the interval (t_n, t_{n+1}). The expression (8.6.2) preserves the definition (8.2.13) of the state vector q and its components, yet takes into account that $x_s = x_s(t)$. In view of the modified equation of motion (8.6.2), the difference equation (8.2.14) for the first component of the vector q has the form

$$q^{(n+1)} = \Phi q^{(n)} + d v^{(n)} + b v_d^{(n)}, \quad d = \begin{bmatrix} -T_c & 0 \end{bmatrix}^T, \qquad (8.6.3)$$

where Φ and b are the same matrices as in (8.2.14).

Suppose that a priori information of the contact point motion can be represented as an internal dynamical model with known parameters. Such representation allows us to use the standard regulator theory for designing the control $v_d^{(n)}$, which makes the system (8.6.3) invariant to the disturbance input $v^{(n)}$. Consider the simplest disturbance model. We assume that the contact point moves with a constant, but unknown velocity. It corresponds to the uniform motion of the arm in the direction of the axis OY, while keeping contact with a rectilinear contour of unknown orientation. The dynamical model is

$$v^{(n+1)} = v^{(n)} \tag{8.6.4}$$

Let us now design the regulator. By introducing an extended state vector

$$\psi^{(n)} = \begin{bmatrix} q^{(n)} \\ v^{(n)} \end{bmatrix}, \tag{8.6.5}$$

the equations (8.6.3), (8.6.4) can be written in the form

$$\psi^{(n+1)} = F\psi^{(n)} + Dv_d^{(n)}, \quad F = \begin{bmatrix} \Phi & d \\ 0 & 1 \end{bmatrix}, \quad D = \begin{bmatrix} b \\ 0 \end{bmatrix} \tag{8.6.6}$$

The system (8.6.6) is not controllable, since the evolution of the disturbance coordinate $\psi_3^{(n)} = v^{(n)}$ is uncontrollable. The control goal is to reach a stationary regime such that a desired constant contact force F_d is maintained. In this regime

$$\psi_2^{(n)} - \psi_3^{(n)} = \dot{x}(t_n) - v^{(n)} = 0$$

and

$$\psi_2^{(n)} = x(t_n) - x_s(t_n) - F_d/k = 0$$

The model (8.6.6) of the dynamics of a manipulator keeping contact with an object while following a linear contour should be complemented by an observation equation. The component $\psi_3^{(n)} = v^{(n)}$ of the extended state vector is not accessible to measurement. The components of the vector $q^{(n)}$ can be measured, as stated in Section 8.2. We introduce an observation vector $y^{(n)}$ using (8.3.16). According to (8.2.13), (8.2.15), (8.6.5), and (8.3.16), the observation equation has the form

$$y^{(n)} = C\psi^{(n)}, \quad C = \begin{bmatrix} 0 & 1 & 0 \\ k & 0 & 0 \end{bmatrix} \tag{8.6.7}$$

To achieve the stated control goal, we need to reconstruct the state of the system (8.6.6) using the observations (8.6.7). Similar to Section 8.5, we can employ a full-order state observer for it [141, 250]. It is also possible to estimate the state of the system (8.6.6) by using a Luenberger reduced-order observer [250]. For the system in question, the first two components of the vector $\psi^{(n)}$ can be measured directly and constitute the vector $q^{(n)}$. Thus, the reduced-order observer has the form [250]

$$\hat{v}^{(n+1)} = \hat{v}^{(n)} + l[q^{(n+1)} - \Phi q^{(n)} - bv_d^{(n)} - d\hat{v}^{(n)}] \qquad (8.6.8)$$

In this equation, l is a 1×2 observer gain matrix. According to the expression for the vector d (see (8.6.3)), the second element of the matrix l does not affect dynamics of the observer (8.6.8). Hence we assume $l = [l_1 \quad 0]$. In view of (8.2.14) and (8.6.3), we can write (8.6.8) in the coordinate form as

$$\hat{v}^{(n+1)} = \hat{v}^{(n)} + l_1[q_1^{(n+1)} - q_1^{(n)} - T_0 q_2^{(n)} - T_1 v_d^{(n)} + T_c \hat{v}^{(n)}]$$

We express the components of the vector $q^{(n)}$ in terms of measured parameters, according to (8.2.13) and (8.2.15), and rewrite the previous equation in the form

$$\hat{v}^{(n+1)} = (1 + l_1 T_c)\hat{v}^{(n)} + l_1[(F(t_{n+1}) - F(t_n))/k$$

$$-T_0\dot{x}(t_n) - T_1 v_d^{(n)}] \qquad (8.6.9)$$

The gain $l_1 = -1/T_c$ provides a dead-beat (single-step) estimate convergence for the observer (8.6.9). Since the system is subjected to various disturbances, it is advisable to define the model inaccuracy as $-1/T_c < l_1 < 0$.

We can use the state estimate obtained by a full-order observer or the reduced-order observer (8.6.9) in the feedback regulator of the form

$$v_d^{(n)} = -G\hat{\psi}^{(n)} \qquad (8.6.10)$$

For the reduced-order observer, we assume that $\hat{\psi}^{(n)} = [q^{(n)^T} \quad \hat{v}^{(n)}]^T$. In accordance with the "separation principle" [141, 250], the control (8.6.10) provides the same closed-loop poles as the control

$$v_d^{(n)} = -G\psi^{(n)} \qquad (8.6.11)$$

To design the regulator (8.6.11), let us change variables in (8.6.3) and denote

$$q_v^{(n)} = \begin{bmatrix} x(t_n) - x_s(t_n) - F_d/k \\ \dot{x}(t_n) - v^{(n)} \end{bmatrix} = q^{(n)} - \begin{bmatrix} 0 \\ v^{(n)} \end{bmatrix},$$

$$v_v^{(n)} = v_d^{(n)} - v^{(n)} \tag{8.6.12}$$

By using (8.2.14) and (8.6.4), we can write the system (8.6.3) in variables (8.6.12) as

$$q_v^{(n+1)} = \Phi q_v^{(n)} + b v_v^{(n)} \tag{8.6.13}$$

The equations (8.6.13) have the same form as (8.2.14). The change of variables in (8.6.12) means that we change to a coordinate system moving with the velocity of the contact point, and for a "rigid" velocity servo system, such coordinate system is equivalent to the stationary one.

The control goal is to bring the system (8.6.13) to a stationary regime that corresponds to maintaining a constant contact force F_d. In view of (8.6.12) and (8.2.10), this stationary regime corresponds to the vector $q_v^{(n)} = 0$. Section 8.2 demonstrates that the control law (8.2.20) solves a similar problem for the system (8.2.14), which coincides in form with (8.6.13). For the system in hand, such control law has the form

$$v_v^{(n)} = -\begin{bmatrix} gk & h \end{bmatrix} q_v^{(n)}$$

or, taking into account (8.6.12),

$$v_d^{(n)} = -\begin{bmatrix} gk & h & 1+h \end{bmatrix} \cdot \begin{bmatrix} q^{(n)} \\ v^{(n)} \end{bmatrix} \tag{8.6.14}$$

By using the notation (8.6.5), we can rewrite (8.6.14) in the form (8.6.11), where

$$G = \begin{bmatrix} gk & h & 1+h \end{bmatrix} \tag{8.6.15}$$

Thus, the formulated problem can be solved by the regulator (8.6.10), (8.6.15) using the state estimate (8.6.9), provided the coefficients g and h are chosen such as described in Section 8.2, for instance, by the formula (8.2.24).

Using the reduced-order observer (8.6.8), $\hat{\psi}^{(n)} = [q^{(n)} \ \hat{v}^{(n)}]^T$, and in view of (8.2.13), (8.2.10), we can rewrite the relationship (8.6.14) in the form

$$v_d^{(n)} = -g[F(t_n) - F_d] - h\dot{x}(t_n) + (1+h)\hat{v}^{(n)} \qquad (8.6.16)$$

The control (8.6.16) only differs from (8.2.21) in that it includes a direct feedback of the object velocity estimate. For $h = 0$, this control law has the form (5.9.18). However, the contact point velocity v is estimated in Section 5.9 differently than in the present section.

8.7 Control of Manipulator with Structural Compliance for a Moving Contact Point

We shall continue with the regulator design started in Section 8.6. The regulator should ensure that the manipulator maintains a given force of contact with an object independent of the contact point motion. In Section 8.6, the manipulator is assumed to be rigid, hence the designed observers interpret oscillations caused by the structure compliance as oscillations of the contact point motion velocity. The error is particularly large if the real motion of the contact point is also oscillatory.

In this section, we shall use the dynamical model (8.3.15)– (8.3.17) that takes into account the manipulator structure compliance. We derive the equations of motion for a moving contact point in the same way as in Section 8.6. The equations (8.3.15), (8.4.9), as well as (8.2.14) remain valid, provided the force acting on the manipulator from the force sensor may be neglected. Thus, with the change of the coordinate x_s, only the first component $q_1^{(n)}$ of the vector $q^{(n)}$ varies.

According to (8.4.7), (8.4.6), $q_1^{(n)} = x(t_n) = x_s - F_d/k - F_d/k_b$. The first equation in (8.3.15), (8.4.9) assumes that $x_s = \text{const}$, so we can rewrite it in the form

$$x(t_{n+1}) - F_d/k - F_d/k_b = x(t_n) - F_d/k - F_d/k_b$$

$$+(\tau - \tau_1)q_2^{(n)} + \Theta\tau_1 v_d^{(n)} \qquad (8.7.1)$$

By subtracting $x_s(t_{n+1})$ from the left-hand side of (8.7.1), and $x_s(t_n) + [x_s(t_{n+1}) - x_s(t_n)]$ from its right-hand side, we obtain a difference equation for the first component q_1 of the vector (8.4.7), which is valid for $x_s = x_s(t)$:

$$q_1^{(n+1)} = q_1^{(n)} + (\tau - \tau_1)q_2^{(n)} + \Theta\tau_1 v_d^{(n)} - T_c v^{(n)} \qquad (8.7.2)$$

In this equation, $v^{(n)} = [x_s(t_{n+1}) - x_s(t_n)]/T_c$ is an average velocity of the contact point on the interval (t_n, t_{n+1}). In view of (8.7.2) and (8.4.10), we rewrite the equations of motion (8.3.15) in the form

$$q^{(n+1)} = \Phi q^{(n)} + \Theta d v^{(n)} + \Theta b v_d^{(n)},$$

$$d = \begin{bmatrix} -\tau & 0 & 0 & 0 \end{bmatrix}^T, \qquad (8.7.3)$$

where Φ and b are the same matrices as in (8.4.9), and the vector q has the form (8.4.7).

We shall complement the equations of motion (8.7.3) by relationships that define the dynamics of the contact point velocity $v^{(n)}$. Let us start with the simplest first-order dynamical model.

Aperiodic motion of the contact point.

A general first-order dynamical model has the form

$$v^{(n+1)} = \alpha v^{(n)} \qquad (8.7.4)$$

Suppose that $0 < \alpha \le 1$ in (8.7.4). For $\alpha = 1$, the dynamical model (8.7.4) coincides with (8.6.4).

To design a regulator and estimate the state of the system, we introduce an extended state vector

$$\psi^{(n)} = \begin{bmatrix} q^{(n)} \\ \Theta v^{(n)} \end{bmatrix} \qquad (8.7.5)$$

We shall use the notation (8.7.5) and write the equations (8.7.3), (8.7.4) in the form analogous to (8.6.6):

$$\psi^{(n+1)} = F_1 \psi^{(n)} + \Theta b_1 v_d^{(n)}, \quad F_1 = \begin{bmatrix} \Phi & d \\ 0 & \alpha \end{bmatrix}, \quad b_1 = \begin{bmatrix} b \\ 0 \end{bmatrix}, \quad (8.7.6)$$

where zero matrices are of appropriate sizes. By using (8.7.5) and (8.3.16), we write the observation equation in the form

$$y^{(n)} = C_1 \psi^{(n)}, \quad C_1 = \begin{bmatrix} 0 & 1/\Theta & 0 & 0 & 0 \\ k & 0 & k & 0 & 0 \end{bmatrix} \qquad (8.7.7)$$

In the above equation, the 2×5 matrix C_1 can be represented as $C_1 = [C \ 0]$, where C is given by (8.3.17), and 0 is a 2×1 zero matrix.

To reconstruct the five components of the extended state vector $\psi^{(n)}$ using two observations $y^{(n)}$, we employ a full-order state observer. Compared to a reduced-order observer, this observer has a simpler structure and better filtering properties, since it does not assume that measurements are absolutely accurate. For the considered system (8.7.6), (8.7.7), the full-order observer has the form

$$\hat{\psi}^{(n+1)} = F_1\hat{\psi}^{(n)} + \Theta b_1 v_d^{(n)} + L(y^{(n)} - C_1\hat{\psi}^{(n)}) \qquad (8.7.8)$$

In this equation, $\hat{\psi}^{(n)}$ is a state estimate at the n-th sample, L is a 5×2 observer gain matrix. The state estimate error $\psi^{(n)} - \hat{\psi}^{(n)}$ asymptotically tends to zero, provided the eigenvalues of the matrix $F_1 - LC_1$ are inside the unit circle.

Let us now design a regulator maintaining a given (commanded) contact force F_d. According to the definition of the vector $\psi^{(n)}$ in (8.4.7), (8.7.5), in the desired stationary regime the system state belongs to the manifold $\psi_1^{(n)} = \psi_3^{(n)} = \psi_4^{(n)} = 0$, $\psi_2^{(n)} = \psi_5^{(n)} = \Theta[\dot{x}(t_n) - v^{(n)}] = 0$. This stationary regime can be stabilized by the control law

$$v_d^{(n)} = G\hat{\psi}^{(n)} \qquad (8.7.9)$$

In the above equation, G is a 1×5 matrix of feedback gains chosen so that the 5×5 matrix $F - \Theta DG$ has four eigenvalues inside the unit circle and one eigenvalue α (α may be unity). The latter eigenvalue corresponds to the motion along the uncontrolled coordinate $\psi_5^{(n)} = v^{(n)}$. In this case, the spectrum of the closed-loop system coincides with the combined spectra of the matrices $F_1 - LC_1$ and $F_1 - \Theta b_1 G$ [141, 250].

Note that under the control law

$$v_d^{(n)} = -G_1\hat{\psi}^{(n)} - G_2 y^{(n)}, \qquad (8.7.10)$$

where $G_1 + G_2 C_1 = G$, the closed-loop system has the same set of eigenvalues as with the control (8.7.9). This can be proved in the same way as the separation theorem [141, 250].

The feedback matrices L and G can be found similarly to Section 8.5 by solving the two Riccati equations defining the LQG regulator. Note that as the coordinate $\psi_5^{(n)} = v^{(n)}$ is uncontrollable, its deviation from zero should not be penalized in the LQG design.

Oscillatory motion of the contact point.

Oscillatory motion of the contact point presents the most challenge in the case where the oscillation period of the contact point is comparable to that of the manipulator structure. Then, to design a dynamic regulator, we need to solve two problems. We should distinguish between the two oscillation types in measurements and control the manipulator to make it follow oscillations of the contact point without exciting oscillations in the structure. To design such a regulator, we shall use a dynamical model describing the oscillatory motion of the contact point. This model has a second order.

Let the manipulator move in contact with an object that has a wavy contour. Then we assume that $x_s(y) = a\sin(2\pi y/\lambda)e^{-y/\Delta}$, where y is a coordinate of the point where the force sensor contacts the object, λ is a wavelength of the contour, Δ is a characteristic damping length. The manipulator moves uniformly with a velocity v_y in the direction of the axis OY. Hence the coordinate x_s depends on time as

$$x_s(t) = ae^{-\alpha t}\sin(\Omega t), \qquad (8.7.11)$$

where $\alpha = -v_y/\Delta$, $\Omega = 2\pi v_y/\lambda$. The velocity of the contact point is

$$v(t) = \dot{x}_s(t) = ae^{-\alpha t}(\Omega\cos\Omega t - \alpha\sin\Omega t) \qquad (8.7.12)$$

The dependencies (8.7.11), (8.7.12) are generated by the following dynamical system:

$$\begin{bmatrix} \dot{v} \\ \dot{x}_s \end{bmatrix} = A_0 \begin{bmatrix} v \\ x_s \end{bmatrix}, \qquad A_0 = \begin{bmatrix} -2\alpha & -(\Omega^2 + \alpha^2) \\ 1 & 1 \end{bmatrix} \qquad (8.7.13)$$

We shall use a discrete-time contour model to relate the variables v and x_s at times t_n and t_{n+1}. By integrating (8.7.13) on the interval t_n, t_{n+1} and denoting $v^{(n)} = v(t_n)$, $x_s^{(n)} = x_s(t_n)$, we obtain the required model as

$$\begin{bmatrix} v^{(n+1)} \\ x_s^{(n+1)} \end{bmatrix} = \Phi_0 \begin{bmatrix} v^{(n)} \\ x_s^{(n)} \end{bmatrix}, \qquad \Phi_0 = \exp(A_0 T_c)$$

$$= \begin{bmatrix} \gamma(\cos\Omega T_c - \frac{\alpha}{\Omega}\sin\Omega T_c) & -\gamma(\Omega + \frac{\alpha^2}{\Omega})\sin\Omega T_c \\ \frac{1}{\Omega}\gamma\sin\Omega T_c & \gamma(\frac{\alpha}{\Omega}\sin\Omega T_c + \cos\Omega T_c) \end{bmatrix} \qquad (8.7.14)$$

$$(\gamma = \exp(-\alpha T_c))$$

We introduce a six-component extended state vector

$$\psi^{(n)} = \begin{bmatrix} q^{(n)} \\ \Theta v^{(n)} \\ \Theta x_s^{(n)} \end{bmatrix} \tag{8.7.15}$$

In view of (8.7.15), we write the equations (8.7.3), (8.7.14) in the form

$$\psi^{(n+1)} = F_2 \psi^{(n)} + \Theta b_2 v_d^{(n)},$$

$$F_2 = \begin{bmatrix} \Phi & d_2 \\ 0 & \Phi_0 \end{bmatrix}, \quad d_2 = \begin{bmatrix} d & 0 \end{bmatrix}, \quad b_2 = \begin{bmatrix} b \\ 0 \\ 0 \end{bmatrix} \tag{8.7.16}$$

We obtain an observation equation from (8.3.16), (8.3.17) as

$$y^{(n)} = C_2 \psi^{(n)}, \quad C_2 = \begin{bmatrix} C & 0 & 0 \end{bmatrix} \tag{8.7.17}$$

In (8.7.16), (8.7.17) zero matrices have appropriate orders. The dynamical regulator should stabilize the regime

$$\psi_1^{(n)} = \psi_3^{(n)} = \psi_4^{(n)} = 0, \quad \psi_2^{(n)} - \psi_5^{(n)} = 0$$

and has the form analogous to (8.7.8), (8.7.9). The feedback matrix G of the regulator and the observer (filter) gain matrix L can be obtained by the LQG design procedure.

8.8 Experiments in Maintaining Contact Force for a Moving Contact Point

In the experiments, two degrees of freedom were used to move the arm of the gantry manipulator described in Chapter 4 in the horizontal plane. We used one channel of the force sensor mounted on the end of the arm as shown in Figure 4.3.3. A metal rod attached to the sensor established contact with an object.

In each of the experiments, the manipulator moves along the axis OX and establishes contact with the object placed on the manipulator table. Upon establishing contact, the control is applied to maintain a given contact force projection onto the axis OX (the sensitivity axis of the force sensor). After some time, the arm starts moving in the direction of the axis OY with a constant velocity. The contact with the object is maintained by moving the arm in the direction of the axis OX. At each control step, the control program samples the output $\dot{x}(t_n)$ of the tachometer, $F(t_n)$ of the force sensor, and $x_0(t_n)$ of the potentiometric transducer that defines the position of the arm. The dependence $x_0(t)$ defines a displacement of the contact point. The experimental results are subsequently plotted.

We assume that the manipulator structure compliance does not change as the arm moves horizontally in the direction of the axis OX. Hence the motion of the arm with the force sensor in the direction of the axis OX, regardless of its displacement, can be described by the equations (8.3.15)–(8.3.17), (8.4.9) with the parameters (8.4.11).

Uniform motion of the contact point.

For a uniform motion of the contact point, the system dynamics are described by the relationships (8.7.6), (8.7.7) for $\alpha = 1$. When designing the state observer (8.7.8), we consider the noise covariance matrices to be diagonal. The covariances and penalty weights in the LQG design were chosen to ensure a high quality of transient processes in the experiment. Let us denote the feedback gains in (8.7.9) by

$$G = \Theta^{-1}\begin{bmatrix} g_1 & g_2 & g_3 & g_4 & g_5 \end{bmatrix}$$

In the experiments, the feedback gains in the regulator were as follows:

$$g_1 = 3.40, \quad g_2 = 0.84, \quad g_3 = 2.88, \quad g_4 = 1.29, \quad g_5 = -1.80 \quad (8.8.1)$$

In the experiments, the manipulator establishes contact with a bar placed on the manipulator table. The lateral surface plane of the bar makes a 30° angle with the axis OY. Some time after establishing contact with the object, the manipulator starts moving in the direction of the axis OY. In the process, the velocity projection onto the axis OX of the point of contact between the manipulator and the bar increases stepwise from zero to the constant $\dot{x}_s = 40\text{mm/s}$. The curve 1 in Figure 8.8.1 corresponds to the experimental results using the dynamical regulator (8.7.8), (8.7.9). Even though the object motion velocity is rather high, a small error of tracking contact force is maintained, except in the beginning of the motion.

For comparison, Figure 8.8.1 (the curves 2) displays the results of a similar experiment where only the measured force feedback (8.2.16) is used ($gk\Theta = 3$). One can see a considerable steady-state error and, in addition, significant oscillations in the beginning of the motion.

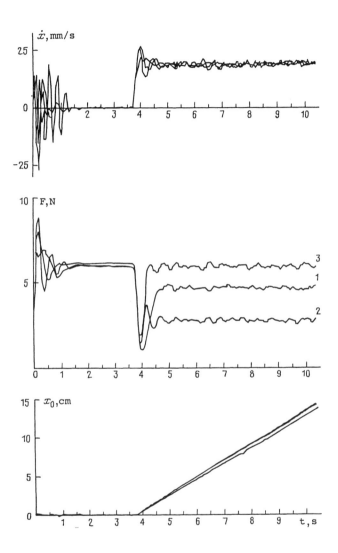

FIGURE 8.8.1
Experimental results for uniform motion of the contact point.
Upper plot – manipulator velocity; middle plot – contact force;
lower plot – contact coordinate

The feedback gains (8.8.1) are such that g_1 and g_3 are close and comparatively large. Equal gains g_1 and g_3 would correspond to the force feedback, since $\psi_1^{(n)} + \psi_3^{(n)} = (F_n - F_d)/k$. Therefore, we also tried a control law of the form (8.7.10) that includes measurement and object velocity estimate feedback. Such control has the same form as (8.6.16):

$$v_d^{(n)} = -g[F(t_n) - F_d] - h\dot{x}(t_n) - g_0\hat{v}^{(n)} \tag{8.8.2}$$

In this case, the velocity estimate $\hat{v}^{(n)}$ is obtained using the observer (8.7.8). The feedback gains in (8.8.2) have been chosen as close as possible to those in (8.8.1): $g_1 = g_3 = gk\Theta = 3.4$, $g_2 = h = 0.4$, $g_4 = 0$, $g_5 = g_0 = -1.8$. The experimental results for the controller (8.8.2) correspond to the curve 3 in Figure 8.8.1. The steady-state error is absent, force decrease in the beginning of the motion is less pronounced. We explain the improvement in the regulation quality as compared to the curve 1 as follows. Since the regulator (8.8.2) only relies on the system model to estimate the object velocity $v^{(n)}$, it is less affected by the fact that the real system dynamics does not necessarily agree with the model. Besides, the feedback loop delay caused by the observer dynamics has less influence.

Oscillatory motion of the contact point.

In this case, the dynamical model of the system is given by the relationships (8.7.16), (8.7.17). Penalty weights in the LQG design of the observer and regulator are defined in the same way as for the uniformly moving contact point. The following feedback gains were obtained for the regulator:

$$g_1 = 3.49, \quad g_2 = 0.84, \quad g_3 = 2.90,$$

$$g_4 = 1.30, \quad g_5 = -1.80, \quad g_6 = 0.11 \tag{8.8.3}$$

We assume that

$$G = \Theta^{-1}\begin{bmatrix} g_1 & g_2 & g_3 & g_4 & g_5 & g_6 \end{bmatrix}$$

In the experiments, a wavy contour was created by bending a metal rod in the horizontal plane and placing it on the manipulator table. The manipulator moved along this contour and maintained a constant contact force, as in the experiments described above. The contour wavelength was $\lambda = 200$ mm. In the model, we found it useful to assume that the wave modeling the contour is decaying. The damping length was taken to be $\Delta = \lambda = 200$ mm. The parameters d and Ω in the contour model (8.7.14) are computed depending on the velocity v_y of motion along the contour.

The curves *1* shown in Figure 8.8.2 illustrate the outputs of force, velocity, and arm position sensors for the LQG controller using a full-order (six-order) extended state observer feedback. In the experiment, the manipulator first contacts the object, then, after a delay, starts moving laterally. This experiment corresponds to the stepwise increase of oscillation amplitude (zero amplitude when the manipulator is not moving along the contour and nonzero when moving). The quality of force tracking is good, even though the velocity of the motion along the contour is high, the oscillation amplitude is significant (±25 mm), and the contour wave considerably differs from a sine curve.

For comparison, the curves *2* in Figure 8.8.2 illustrate the results obtained for pure force feedback ($gk\Theta = 3$). The plot shows considerable force oscillations that coincide in frequency with contact point oscillations.

Note that by using the above dynamical regulator, the manipulator can still maintain contact with an object while moving along its contour at a velocity of up to 150 mm/s. Yet using only measurement feedback (without an observer feedback), the manipulator starts losing contact at 45 mm/s.

We have also tested a regulator of the form (8.7.10) that employs feedback with respect to both measurement and estimates of state variables describing the object motion:

$$v_d^{(n)} = -g[F(t_n) - F_d] - h\dot{x}(t_n) - g_5\hat{v}^{(n)} - g_6\hat{x}_s^{(n)} \qquad (8.8.4)$$

The feedback gains in (8.8.4) were chosen to be close to those in (8.8.3): $g_1 = g_3 = gk\Theta = 3.5$, $g_2 = h = 0.84$, $g_4 = 0$, $g_5 = -1.8$, $g_6 = 0.11$. The quality of force tracking in this case is worse than for the curves *1* in Figure 8.8.2.

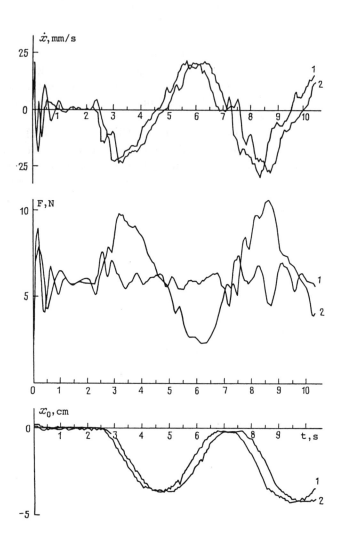

FIGURE 8.8.2
Experimental results for oscillatory motion of the contact point.
Upper plot – manipulator velocity; middle plot – contact force;
lower plot – contact coordinate

Part III.

Applications

This part of the book considers basic applications of sensing manipulators to automation of industrial operations such as assembly, surface machining, and retrieval of randomly placed workpieces. We present algorithms for these operations on the basis of control laws designed in Part II. The algorithms are experimentally tested using the research gantry manipulator UM-1.25 described in Chapter 4.

Chapter 9

Manipulator Control in Surface Machining

This chapter considers finish grinding as an example of a surface machining task. The purpose of grinding is to remove small irregularities left on the surface of a machined part. Grinding is performed by an abrasive tool that contacts a workpiece and moves along its contour. Such a motion can be considered as a motion along an imposed mechanical constraint.

The designed algorithms for controlling the motion of a manipulator performing a grinding operation follow the approaches developed in Chapters 6 and 7. Manipulator control is based on force sensor information and includes regulation of both tool tracking velocity and force of contact with a workpiece.

9.1 Requirements for Grinding Tasks

In many industrial operations, billets, workpieces, and entire units are manufactured by die casting, press work, or welding. All such parts tend to have small irregularities or excess stock material on their surfaces. For example, parts formed by die casting may have some surplus metal, or flash, along the split line of molding dies, and stock runouts around sprues and gates. Press work leaves burrs on the edges of parts, and welding produces stock runout in the form of weld beads. It is necessary to remove such excess material from part surfaces by grinding.

As mentioned above, excess stock is commonly found along a single line, such as a split line of dies, a part edge, or a weld beads. Figure 9.1.1 illustrates a widely used method of removing stock runout (burr) *1* using a peripheral section of the fast-rotating grinding wheel *2*. The stock is removed in the contact zone between the grinder and the part. The contact zone is typically comparatively small in size. In the process of grinding, the

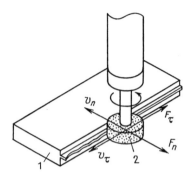

FIGURE 9.1.1
Stock removal using a grinding wheel

wheel and the contact zone travel along the part contour, i.e., along the line of burr concentration.

In modeling the grinding process, we shall assume that the wheel moves normal to its rotation axis. The motion of the grinding wheel (or any other metal-cutting tool, for that matter) is generally characterized by its longitudinal v_τ and transverse v_n velocities, or feedrates. The transverse feedrate v_n defines feed variations, while the longitudinal feedrate v_τ defines stock removal rate for a given feed (depth of the cut). The longitudinal cutting force F_τ and transverse contact force F_n describe how the grinder interacts with the part. If the part has an almost plane grinding surface, then an approximate dependence $F_\tau = f F_n$ holds true [296]. The "friction" coefficient f may take values from 0.3 to 1.0 depending on the part material, abrasive wheel type, and the wheel rotation rate [227]. Naturally, the cutting force direction does not depend on the direction of the wheel travel along the part, because the wheel rotates fast.

Herein, we consider that the contact force determines the cutting force ($F_\tau = f F_n$) and, therefore, the depth of the cut. Thus, the interaction between the abrasive wheel and the part is defined by two parameters, the feedrate v_τ and the contact force F_n. The relationship between F_n, v_τ, and the rate of material removal in grinding are discussed in [140, 296] and other papers. As far as the process is concerned, an optimal grinding regime should aim to keep a constant speed of stock removal [206]. To maintain such a regime, the tracking velocity v_τ needs to be decreased when the force F_n (or F_τ) is increased and vice versa. An alternative approach is to control the metal removal rate by controlling the power applied to the workpiece [296].

Having briefly exposed the process requirements to finish grinding, we shall turn to algorithms that enable manipulators to perform this operation.

An industrial robot can perform part grinding in one of two ways. It can either move a tool around a stationary part, or move a part to be machined around a tool mounted on the stationary base. Heavy and bulky parts are generally ground in the first way. The second way is used to grind relatively small and light parts, as well as for machining by several tools in succession. For the second case, there is little constraint on the drive weight (hence, its power) and dimensions for a stationary abrasive tool.

In the simplest case, a grinding manipulator executes a "fixed" program. Moving the tool along a preset trajectory, it cuts off the part sections (unwanted stock, burrs) that extend beyond the standard contour [94, 136, 175, 218]. It is hardly possible to maintain the required grinding regime by executing a rigid program, and the sensitivity to errors is high. To achieve an optimal grinding quality, one should control the manipulator motion to maintain the desirable process parameters.

We can control the tool motion relative to the part by regulating the contact force F_n and/or the tracking velocity v_τ.

It is useful *to regulate the contact force F_n* between an abrasive tool and a part, if the stock runout is nondense (thin) compared to the part itself. As opposed to the material of the part, the nondense stock runouts can be "easily" removed, ground off by an abrasive wheel. Such thin plates of excess stock, or flash, are common in die castings. Burrs left after press work are normally thin as well. By controlling the contact force exerted by the abrasive wheel on the part, one can remove all the unwanted stock while limiting cutting of the part surface. Such an approach is implemented in [127]. To remove a nondense stock, it may not be necessary to regulate the feed velocity, which can be constant. Grinding with a constant feed velocity can be performed for a more dense stock as well, by regulating the contact force depending on the density of the stock runout [296].

It is possible to maintain a constant contact force by mounting a tool (or a part) on an elastic suspension [2, 9, 285]. In this case, to obtain a given contact force, the commanded trajectory of the tool motion should be respectively shifted towards the part. Force information can be used for a better control of the process.

By using an active control of the contact force, one can remove even denser excess stock. It is necessary to increase the contact force where the excess material is thicker. This can be done by inspecting the part contour before grinding and comparing it with a standard contour. Motion along the contour can be realized by using the information from force sensors or other types of sensors [1, 2, 9, 10, 22, 122, 127, 182, 249, 252, 281, 282, 285, 296]. Note that grinding with force sensor feedback and preliminary scanning has little sensitivity to the wear of the abrasive tool (wheel) or deformations of manipulator links. Yet a preliminary scanning of the part contour or its individual sections may be time consuming.

For dense stock runouts such as weld beads, it is beneficial to *regulate longitudinal feedrate* v_τ of the abrasive tool. The algorithm described below is based on the assumption that the part contour should take a preset shape as a result of grinding. Any material that extends beyond the standard contour is assumed to be a stock runout. We control the manipulator motion to ensure that the grinding wheel moves along the standard contour of the part in contact with its surface. If the manipulator encounters a stock runout during grinding, then the abrasive tool applies a cutting force and a normal pressure to it. One can measure these forces by a force sensor and use this measurement to control the tracking velocity so that, for example, the volumetric rate of stock removal remains constant. Such algorithms are employed in [229, 249]. The advantage is that the algorithm does not need an additional operation to define excess stock size (scanning). The disadvantages are that the algorithm is sensitive to errors in position and shape of machined parts, abrasive wheel wear, and manipulator structure deformations.

In the subsequent sections of this chapter, we shall consider control algorithms for the manipulator motion during grinding that allow control of both the force F_n and the velocity v_τ.

9.2   Grinding of a Stationary Part

In this section, we shall consider a grinding operation where the manipulator arm holds a tool.

We assume that the part to be machined is fixed on the manipulator table. The burrs to be removed are on the lateral surface of the part and belong to the same horizontal plane. The arm of the gantry manipulator can move in the horizontal plane. Each of the two degrees of freedom moves the arm along one axis of the coordinate system OXY. A tool with an abrasive wheel of radius ρ having a vertical rotation axis is mounted on the manipulator arm. The horizontal plane passing through the burrs on the part intersects the lateral surface of the wheel. The tool is attached to the manipulator arm through a force sensor that measures two horizontal components F_x and F_y of the force $\mathbf{F}$ applied to it.

The manipulator is controlled using a velocity servo system as described in Chapter 4. The control algorithm should compute the commanded velocities V_{xd} and V_{yd} depending on measurements and, possibly, time.

To remove the burrs, the abrasive wheel should move along the part contour so as to maintain a given contact force between them. Denote by x, y the coordinates of the abrasive wheel center O_1. As the wheel moves in contact with the lateral surface of the part, the point O_1 moves along

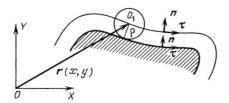

FIGURE 9.2.1
Kinematics of grinding wheel motion along the part contour

an "extended contour", i.e., along the curve each point of which is at a distance ρ from the part contour (see Figure 9.2.1). We can describe the extended contour by a parametric equation $x = x(s)$, $y = y(s)$ (see (6.1.3)). Note that the normal to the extended contour raised at the center of the abrasive wheel coincides with the normal to the part contour raised at the wheel contact point.

The motion of the wheel center along the extended contour can be controlled as described in Section 6.2. To this end, we set the commanded velocity vector $\boldsymbol{V}_d = [V_{xd}, \quad V_{yd}]^T$ in the form (6.2.1):

$$\boldsymbol{V}_d = g(F_n - F_d)\boldsymbol{n} + v_\tau \boldsymbol{\tau} \qquad (9.2.1)$$

In this equation, $\boldsymbol{n}$ and $\boldsymbol{\tau}$ are vectors of the external normal and tangent to the extended contour, v_τ is a tracking velocity, $F_n = \boldsymbol{F} \cdot \boldsymbol{n}$, and F_d is a commanded contact force. The parameters F_d and v_τ should be chosen so that the lateral surface of the part is machined with the required quality.

Chapter 6 where the control law (9.2.1) is studied assumes that there is either no friction between the manipulator and the object (workpiece), or a Coulomb friction is present. Then, the vectors $\boldsymbol{n}$ and $\boldsymbol{\tau}$ can be determined by the force sensor output (in the absence of friction, $\boldsymbol{n} = \boldsymbol{F}/F$, where $F = |\boldsymbol{F}|$). This is not true for grinding. First, the wheel interaction with the unwanted stock (burr) significantly affects the contact force between the wheel and the part. Second, due to chatter and variations in the wheel properties, and for other reasons, there is no one-to-one correspondence between the tangential cutting force F_τ and the normal contact force F_n. Figure 9.2.2 displays an experimentally obtained dependence between these forces. In this experiment, the manipulator described in Chapter 4 held a steel part against the rotating abrasive wheel. The forces F_n and F_τ measured by the force sensor were sampled and logged. As seen in Figure 9.2.2, there is no one-to-one correspondence between the forces F_n and F_τ, though a certain approximate relationship can be observed.

The commanded force F_n in the control law (9.2.1) is chosen such that

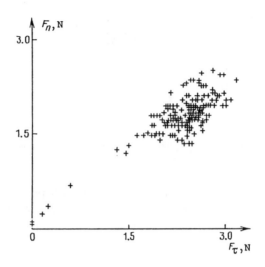

FIGURE 9.2.2
Experimentally obtained correspondence between the normal contact and cutting forces

the wear of the part is small, yet the burr is removed. Accordingly, the vectors n and τ in (9.2.1) should be vectors of normal and tangent to the contour of a finished (standard) part with the burr removed rather than to the part contour with this burr. Generally, to ensure a high quality of grinding, one should control the parameters F_d and v_τ in (9.2.1) depending on the excess stock variation Δ. For large stock runouts, it is advisable to increase F_d. Hence, to maintain a constant rate of stock removal, the velocity v_τ has to be decreased.

Thus, to implement the algorithm (9.2.1), we need to know the vectors of normal $n(s)$ and tangent $\tau(s)$ to the contour in advance, before the grinding. In some cases, one also needs to know the excess stock variation $\Delta(s)$.

If the manipulator performing grinding is part of a flexible manufacturing system, then the dependencies $n(s)$ and $\tau(s)$ can be computed from geometric parameters of parts stored in the database. In this section, we assume that the contour of a finished (standard) part needs to be programmed in the manipulator control system prior to machining a batch of parts.

A standard contour can be followed using a nonrotating abrasive wheel that is attached to the manipulator arm through force sensors. By sampling the manipulator position during the contour following motion, we obtain the contour coordinates $x(s), y(s)$ and can compute the vectors $n(s)$, $\tau(s)$

from them. By following the standard contour, one can also determine the normal vectors $n(s)$ using a force sensor. In so doing, an imitation of the abrasive wheel can be used that has a small friction coefficient against the machined part. We shall refer to the described procedure of obtaining standard contour parameters as *measuring*.

If it is necessary to know excess stock dimensions prior to grinding, one can find them by *scanning* each part before grinding. Scanning consists in measuring a part contour and comparing it with the standard contour. Similar to the contour measuring, scanning can be performed by moving a nonrotating abrasive wheel or its imitation along the contour and with the help of a force sensor.

To recapitulate, the grinding operation as considered in this chapter consists of a sequence of three procedures: measuring, scanning, and grinding proper. Related work on grinding, where measuring (contour following) and grinding procedures using force control are developed, is considered in [249, 122]. Below we consider each procedure in more detail.

Measuring.

To perform an initial measuring of the part contour, the manipulator follows the contour by using the algorithm considered in Section 6.2. If friction is negligible, then we determine the contour normal as $n = F/F$, where F is a 2×1 vector of the force sensor output, and F is the norm of F. The tangent vector τ is obtained by rotating the vector n through the angle $\pi/2$. If the Coulomb friction is significant, then we can obtain the vector n by rotating the unit vector F/F through the angle $\beta = \arctan f$, where f is a friction coefficient.

The measured information on the contour should be stored. Since the memory in the control system is limited, this information should be presented in a compact form. Let us consider how to store the measured contour information and then restore the vectors n and τ from this data.

We move the manipulator along the contour and periodically sample the position of the point O_1 of the manipulator arm using position sensors. Denote by A_i the i-th ($i = 1, \ldots, N$) sampled point (node), where $r_i = [x_i \quad y_i]^T$ is its radius-vector, $l_i = A_i A_{i+1} = r_{i+1} - r_i$. The distances l_i ($i = 1, \ldots, N$) between the nodes should be small and about the same.

Using the nodes A_i, we can find (approximately) the tangent and normal to the contour. The vector of tangent at the point A_i can be computed as

$$\tau = (l_{i-1} + l_i)/|l_{i-1} + l_i|, \tag{9.2.2}$$

where it is assumed that $l_i = 0$ for $i = 0$. To compute the vector of tangent on the interval between the nodes A_i and A_{i+1}, we shall apply linear interpolation. Let us introduce a parameter s_i for the broken line with nodes A_j from the point A_1 to a given contour point. Denote by s_i the length of the broken line up to the node A_i:

$$s_i = \sum_{j=1}^{i} l_j \qquad (9.2.3)$$

If the radius-vector of the manipulator arm is $r = [x \ \ y]^T$, then assuming that the manipulator is on the i-th segment of the contour (between the nodes A_i and A_{i+1}), the contour coordinate s_i is given by

$$s = (r - r_i) \cdot l_i / l_i + s_i \qquad (9.2.4)$$

As soon as the difference $s - s_i$ exceeds l_i $((r - r_i) \cdot l_i > l_i^2)$, we consider that the manipulator has passed from the i-th to the $i + 1$-th segment of the contour. We determine the vector of tangent inside the i-th segment by using linear interpolation as

$$\tau(s) = [(s_{i+1} - s)\tau_i + (s - s_i)\tau_{i+1}]/(s_{i+1} - s_i) \qquad (9.2.5)$$

According to the notation (9.2.3), $s_{i+1} - s_i = l_i$ in (9.2.5). The vector of normal $n(s)$ is obtained by rotating the vector $\tau(s)$ through the angle $\pi/2$. If $\tau = [\tau_x \ \ \tau_y]$, then $n = [\tau_y \ \ -\tau_x]^T$.

The described measuring procedure is applicable even if the radius ρ_1 of the imitation wheel is not equal to the radius ρ of the abrasive wheel. In this case, we need to modify the data (the coordinates of the points A_i) by substituting $r_i + (\rho - \rho_1)n(s_i)$ for the vectors $r_i = [x_i \ \ y_i]^T$.

Scanning.

As the manipulator follows the part contour performing the scanning, we can find the vectors n and τ in the control law (9.2.1) from the force sensor output. If the tangential component of the contact force between the nonrotating wheel and the burr is large, then these vectors can be found from the formulae (9.2.2) - (9.2.5). To determine the burr size, we need to compute a current value of s using (9.2.4). The burr size variation on the i-th segment of the contour is given by

$$\Delta r_i = [r'(s_i) - r_i] \cdot n_i, \qquad (9.2.6)$$

where $r'(s)$ is the contour of the part before grinding, and $n_i = n(s_i)$ is the vector of normal at the node A_i of the standard contour.

Grinding.

In grinding, the vectors n and τ are computed from (9.2.2) - (9.2.5). Assuming that the contour has already been scanned, we can change the commanded force F_d and tracking velocity v_τ depending on the burr height as follows:

$$
\begin{aligned}
F_d &= F_{d0} + h_F\Delta, \\
v_\tau &= \max(v_0,\ v_1 - h_v\Delta),
\end{aligned}
\qquad
\Delta = \begin{cases} \Delta r & \text{if } \Delta r \geq \Delta_0, \\ 0 & \text{if } \Delta r > \Delta_0 \end{cases},
\qquad (9.2.7)
$$

where Δ_0 defines the desired grinding quality; $F_{d0}, v_0, v_1, h_F, h_v$ are empirically selected constants; v_0 is the minimal, v_1, the maximal, tracking velocity.

9.3 Grinding a Part in the Manipulator Arm

Let us pass to the second grinding method described in Section 9.1. We suppose that the manipulator arm holds a part *1*, while an abrasive wheel *2* is mounted on the stationary base (see Figure 9.3.1).

We keep assuming that the machined lateral surface of the part is in the same horizontal plane as the working surface of the abrasive wheel. The wheel center is at the point O, which is the origin of the stationary coordinate system $OXYZ$. The part can be moved in the horizontal plane OXY by two translational and a rotational degrees of freedom. The latter rotates the part in the plane OXY about the point O_1. Denote by x, y the components of a radius-vector r of the point O_1, and by ψ, a rotation angle of the part relative to the initial position. Each of the controlled coordinates x, y, ψ corresponds to motion along one degree of freedom of the manipulator.

The part is held the end-effector of the manipulator arm through a two-component force sensor *3*. The sensitivity axes of the sensor are directed along the axes of the coordinate system $O_1X'Y'$ attached to the part. For the rotation angle $\psi = 0$, these axes are collinear to the axes OXY. Denote by F_x, F_y the components of the measured force vector F in the coordinate system OXY

$$
F_x = F'_x \cos\psi - F'_y \sin\psi,
$$

$$
F_y = F'_x \sin\psi + F'_y \cos\psi, \qquad (9.3.1)
$$

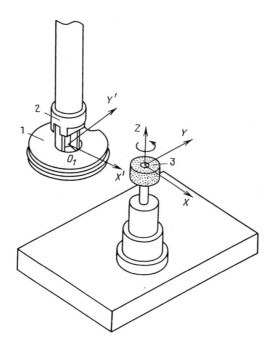

FIGURE 9.3.1
Grinding a part using a stationary tool

where F_x', F_y' are the force components measured by the force sensor in the coordinate system $O_1 X'Y'$ (the readings of the sensor channels).

Similar to Section 9.2, the manipulator motion shall be such that the part is in contact with the abrasive wheel, while the contact point follows the part contour. Once again, we shall consider an extended contour, each point of which is at a distance of the wheel radius ρ from the finished part contour. When the part is in contact with the wheel, the wheel center belongs to the extended contour (see Figure 9.3.2). To move the part in contact with the wheel, we need to set the commanded velocities V_{xd}, V_{yd}, and ω_d in the three controlled degrees of freedom such that the point O remains on the extended contour. As follows from Section 9.2, such motion can solve the grinding problem.

Let us now formulate the basic kinematic relationships we shall be using further. For the sake of simplicity, we assume that each ray drawn from the point O_1 crosses the extended contour at a single point. To describe the contour, we employ the coordinate system $O_1 X'Y'$ attached to it (see Figure 9.3.3). Let r' be the radius-vector of a contour point, σ, a counterclockwise angle the radius makes with the axis $O_1 X'$. Then the parametric

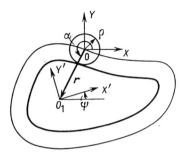

FIGURE 9.3.2
Extended contour and coordinate systems for the part motion around the grinding wheel

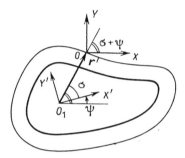

FIGURE 9.3.3
Kinematics of the wheel motion with respect to the part contour

equation for the contour has the form $r'(\sigma) = [x'(\sigma) \quad y'(\sigma)]^T$. Note that the contour is uniquely defined by the relationship $r'(\sigma) = |r'(\sigma)|$, since $x'(\sigma) = r'(\sigma)\cos\sigma$, $y'(\sigma) = r'(\sigma)\sin\sigma$. Denote by $n'(\sigma)$ and $\tau'(\sigma)$ a normal and a tangent to the contour in the coordinate system $O_1 X'Y'$.

Let us now describe the contour in the variables x, y, ψ. We introduce a matrix $\Psi(\psi)$ of rotation in the plane OXY through the angle ψ in the positive direction (counterclockwise):

$$\Psi(\psi) = \begin{bmatrix} \cos\psi & -\sin\psi \\ \sin\psi & \cos\psi \end{bmatrix} \tag{9.3.2}$$

Let the point O be on the extended contour, the angle $OO_1 X'$ be equal to σ, and the coordinates of the point O in the coordinate system $O_1 X'Y'$ be $x'(\sigma)$, $y'(\sigma)$. One can see from Figure 9.3.3 that in the coordinate system OXY, the vector $\overline{O_1 O}$ equals to $\Psi(\psi)r'(\sigma)$. The angle α measured in the positive direction between the radius-vector $r = \overline{O_1 O} = -\overline{O_1 O}$ and the axis OX is related to the angles σ and ψ by the equation $\alpha = \pi + \sigma + \psi$. The angle α is uniquely determined by the manipulator arm coordinates x, y and $\tan \alpha = y/x$. Hence, if the point O is on the extended contour of the part, then the radius-vector of the point O_1 should satisfy the relationship

$$r = -\Psi(\psi)r'(\alpha - \psi - \pi) \tag{9.3.3}$$

For $\psi = \text{const}$, the equation (9.3.3) determines a curve in the coordinate system OXY, which the point O_1 of the manipulator arm should follow so that the (stationary) point O remains on the contour. The external normal n_1 to this curve at the point O_1 is opposite to the external normal to the part contour at the point O. For the point O to follow the object contour in the positive direction, the point O_1 should move in the opposite direction. Therefore,

$$n_1 = -\Psi(\psi)n'(\alpha - \psi - \pi), \quad \tau_1 = -\Psi(\psi)\tau'(\alpha - \psi - \pi) \tag{9.3.4}$$

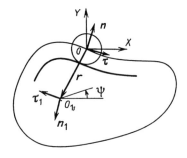

FIGURE 9.3.4
Vectors of normal and tangent to the part contour and the arm trajectory

If n and τ are vectors of normal and tangent to the part contour in the coordinate system OXY (see Figure 9.3.4), then the relationship (9.3.4) means that $n_1 = -n$ and $\tau_1 = -\tau$.

As mentioned above, the vector valued function $r'(\sigma)$ can be uniquely reconstructed from the scalar valued function $r'(\sigma)$. Hence, we can reduce (9.3.3) to a single scalar equation describing the constraint imposed on the part motion. Without deriving this equation, let us construct a vector of normal and two vectors of tangent to this constraint imposed on the three generalized part coordinates. Guided by Chapter 7, we choose the coordinates x, y, and $w = \psi L$, which have the same dimensionality, to define the position and orientation of the part. A characteristic length L will be chosen later.

To construct the vectors of tangent, let us consider which virtual displacements are compatible with the constraints. Figure 9.3.4 shows that the point O does not leave the contour during virtual translational motion of the part in the direction of the tangent τ to the contour, as well as during virtual part rotation about the point O. The tangential vectors corresponding to the virtual motions in the space (x, y, w) have the form

$$\tau_1 = \begin{bmatrix} -\tau \\ 0 \end{bmatrix}, \quad \tau_2 = \begin{bmatrix} -y/L \\ x/L \\ 1 \end{bmatrix} \tag{9.3.5}$$

Note that τ_2 is not a unit vector. A unit vector of the external normal to the constraint can be chosen as a vector orthogonal to the vectors τ_1 and τ_2

$$\bar{n} = \begin{bmatrix} -n \\ m \end{bmatrix} \cdot (1 + m^2)^{-1/2}, \quad m = (yn_x - xn_y)/L \tag{9.3.6}$$

In this expression, $|m|L = |r \times n|$ is a length of the vector product r by n considered as vectors in the three-dimensional space $OXYZ$. Note that the orientation of the vector $\bar{n}$ (the magnitude of its third component) depends on a chosen value of the characteristic length L. For example, L can be defined by the ratio of a characteristic velocity of the manipulator arm motion to a characteristic angular velocity of the end-effector rotation, or, in other words, by the radius of curvature of the contour.

According to the algorithm (7.3.5) designed in Chapter 7, to solve the general problem of motion along the constraint, we shall set the commanded velocities of motion in the coordinates x, y, and w as follows:

$$\begin{bmatrix} V_d \\ \omega_d L \end{bmatrix} = \frac{g}{1 + m^2}(F_n - F_d)\begin{bmatrix} -n \\ m \end{bmatrix} + v_\tau \begin{bmatrix} -\tau \\ 0 \end{bmatrix} + \omega L \tau_2 \tag{9.3.7}$$

In this expression, $F_n = -F \cdot n + (M_z/L)m$, where M_z is a torque applied to the arm at the point O_1, and ω_d is a commanded motion velocity in the coordinate ψ. The parameters g and F_d are a force feedback gain and a commanded normal component of the "generalized" force vector, respectively. The last two terms in (9.3.7) define an arbitrary direction of motion in the tangential space. To define this direction uniquely, one must define the relation between the parameters v_τ and ω. Note that v_τ denotes the velocity of the point O motion along the extended contour. The instant angular velocity ω has no effect on v_τ, but defines the displacement of the point of contact between the abrasive wheel and the part along the edge of the wheel (see Figure 9.3.4).

According to (9.3.7), the control of grinding a part held in the manipulator arm requires the information on three components of the generalized force vector: the forces F_x and F_y, and the torque M_z. We shall now consider two grinding algorithms which only need information on two force components, F_x and F_y.

Grinding an elongated part.

Assume that the part has an elongated shape, i.e., the contour segment to be ground is close to a straight line. Let us take as a characteristic length L an average radius of curvature of the contour segment, which is large. Then the first two components of the vector τ_2 (see (9.3.5)) and the third component of the vector $\bar{n}$ in (9.3.6) are small. We neglect these small parameters and substitute for the expression (9.3.7) an approximation

$$\begin{bmatrix} V_d \\ \omega_d \end{bmatrix} = g(F_n - F_d) \begin{bmatrix} -n \\ 0 \end{bmatrix} + v_\tau \begin{bmatrix} -\tau \\ 0 \end{bmatrix} + \omega \begin{bmatrix} 0 \\ 1 \end{bmatrix} \qquad (9.3.8)$$

In this expression, $F_n = -\boldsymbol{F} \cdot \boldsymbol{n}$ is a force component that is normal to the contour, F_d is a commanded contact force applied to the contour, and v_τ is a feed (contouring) velocity.

The first two components of the vector (9.3.8) coincide with the components of the vector (9.2.1) of the commanded motion velocities up to the sign. Since v_τ should determine the "tangential" motion to the constraint uniquely, we define ω in (9.3.8) in the form $-g_\psi(\psi - \psi_d)$. Then the equation for the third component in (9.3.8) takes the form

$$\omega_d = -g_\psi(\psi - \psi_d), \qquad (9.3.9)$$

where ψ_d is a commanded rotation angle of the part, g_ψ is a position feedback coefficient for the rotational degree of freedom. Without the loss of generality, we shall further assume that $\psi_d = 0$.

It follows from (9.3.3) that for $\psi_d = 0$, if the part contacts the wheel, the coordinates of the point 0_1 should satisfy the relation $r = r_1(\alpha)$, where

$$r_1(\alpha) = -r'(\alpha - \pi) \qquad (9.3.10)$$

The relationship (9.3.10) yields a contour that is central symmetric of the extended contour with respect to the point 0, provided the point 0_1 of the latter coincides with 0. Thus, the considered grinding problem is solved exactly as described in Section 9.2. Also in this case, by performing measurement in the same way as in Section 9.2, we directly obtain the information on the contour (9.3.10) rather than the initial contour $r'(\sigma)$.

Grinding a rounded part.

We shall assume that the extended contour of a part closely approximates a circle with the center 0_1. For an extended contour, this assumption may hold true if, for example, the radius ρ of the abrasive wheel is much larger than the part size.

The magnitude of the third component m of the vector $\bar{n}$ of normal to the constraint equals $|r \times n|/L$. Provided the contour is a circle with the center O_1, the vectors r and n are collinear, and $m = 0$. Then the control law (9.3.7) takes the following form (we have substituted the expression for τ_2 from (9.3.5)):

$$\begin{bmatrix} V_d \\ \omega_d L \end{bmatrix} = g(F_n - F_d) \begin{bmatrix} -n \\ 0 \end{bmatrix} + v_\tau \begin{bmatrix} -\tau \\ 0 \end{bmatrix} + \omega L \begin{bmatrix} -y/L \\ x/L \\ 1 \end{bmatrix} \quad (9.3.11)$$

For a contour that differs from a circle, we shall use an approximate value of normal to the constraint assuming that $m = 0$. In the right-hand side of (9.3.11), $F_n = -F \cdot n$. Therefore, the information from a two-component force sensor is sufficient for the computation.

The tracking velocity v_τ is defined from machining requirements. The angular velocity ω of rotation about the point O should be set such that the point O_1 does not move perpendicularly to the straight line O_1O. In this case, the part is always held on one side of the abrasive wheel, which is necessary if a part of the wheel is sheathed. Let us demand that the point O_1 projection of the motion velocity, which corresponds to the motion along the constraint, onto the vector $[-y \ x]^T$ normal to the vector $r = \overline{OO_1}$ be zero. To this end, we set the dot product of the sum of the two last terms in (9.3.11) and the vector $[-y \ x \ 0]^T$ to be zero and obtain $\omega = (x\tau_y - y\tau_x)v_\tau/r^2$, where $r^2 = x^2 + y^2$. Since $-y\tau_x + x\tau_y = -n_y y - xn_x = -n \cdot r$, we can rewrite the last relationship as

$$\omega = -v_\tau n \cdot r/r^2 \quad (9.3.12)$$

Note that for a circle $n \cdot r = -r$, and (9.3.12) yields that $\omega = v_\tau/r$.

The relationships (9.3.11), (9.3.12) describe a manipulator control algorithm that enables a "rounded" part to maintain a given contact force with the abrasive wheel, while their contact point moves along the part contour.

If the part contour is not a circle, then the errors will accumulate and the vector OO_1 will rotate about the point O. We can avoid the error accumulation by introducing position feedback with respect to the point O. For example, to keep the vector OO_1 parallel to the axis OX, we can modify the control law (9.3.12) as

$$\omega = -v_\tau \boldsymbol{n} \cdot \boldsymbol{r}/r^2 - g_y y, \qquad (9.3.13)$$

where g_y is a positive constant.

Our discussion, so far, leaves open the question of how to obtain the necessary information to compute the right-hand side terms in (9.3.11), (9.3.12). The components of the vector $\boldsymbol{r}$ (the arm coordinates x, y) can be obtained from position sensors. As mentioned in Section 9.2, the vectors $\boldsymbol{n}$ and $\boldsymbol{\tau}$ should be known from a priori information on the finished part contour (standard part contour). In addition, to set the tracking velocity v_τ and contact force F_d, we may need the information on burr height. Thus, generally, it is necessary to measure a standard part with burrs and to scan the part before the grinding.

The standard part can be measured using a stationary imitation of the abrasive wheel with the same radius. Provided the friction coefficient between the imitation and the part is small, we can control the part motion according to (9.3.11), (9.3.12), assuming that the vector $\boldsymbol{n} = -\boldsymbol{F}/F$, and the vector $\boldsymbol{\tau}$ is obtained by rotating $\boldsymbol{n}$ through the angle $\pi/2$. During this motion, it is sufficient to periodically sample the coordinates that define the contour of the standard part. One can calculate these coordinates using the coordinates x, y, ψ in the three degrees of freedom of the manipulator, which are defined by the part contour according to (9.3.3). We shall use the parametric equation (9.3.10) of the contour, along which the manipulator point O_1 should move if the part contacting the wheel does not rotate. It follows from (9.3.3) and (9.3.10) that

$$\boldsymbol{r}_1(\alpha - \psi) = \Psi(-\psi)\boldsymbol{r} \qquad (9.3.14)$$

Given the constant orientation of the vector OO_1, $\alpha = \text{const}$ and the argument in the left-hand side of (9.3.14) monotonically changes with rotation of the part. By periodically sampling the variables s_i and $R_i = |\boldsymbol{r}_1(s_i)|$ such that the differences $|s_i - s_{i_1}|$ are small and approximately equal, we obtain sufficiently complete information on the contour $\boldsymbol{r}_1(s)$. Then we can use this information to reconstruct an approximate normal $\boldsymbol{n}_1(s)$ and tangent $\boldsymbol{\tau}_1(s)$ to the contour $\boldsymbol{r}_1(s)$ as described in Section 9.2. Denote

$$\begin{array}{l} x_i = R_i \cos s_i, \\ y_i = R_i \sin s_i, \end{array} \qquad l_i = \begin{bmatrix} x_{i+1} - x_i \\ y_{i+1} - y_i \end{bmatrix} \qquad (9.3.15)$$

We compute the tangent $\boldsymbol{\tau}_i$ at the point s_i by the formula (9.2.2).

By using the coordinates x, y, ψ, at any time through scanning and grinding procedures, one can find the parameter $s = \alpha - \psi$, where $\sin\alpha = y/r$, $\cos\alpha = x/r$ $(r^2 = x^2 + y^2)$, and the integer i such that $s_i \geq s \geq s_{i+1}$ (or $s_i \leq s \leq s_{i+1}$). To compute the vector of tangent, we apply linear interpolation from (9.2.5), by substituting τ' for τ. Note that this expression has a somewhat different meaning in Section 9.2, since s is a length of the broken line there, and not an angle. We obtain the vector n' by rotating τ' through the angle $\pi/2$. According to (9.3.4), the vectors of normal n and tangent τ to the part contour from the control law (9.3.11), (9.3.12) equal, respectively, $n = \Psi(\psi)n'$, $\tau = \Psi(\psi)\tau'$.

For scanning, we also need to calculate the excess stock thickness. To this end, we shall use a relationship analogous to (9.2.6)

$$\Delta r_i = [x - x_i \quad y - y_i] \cdot n'_r \qquad (9.3.16)$$

The parameters F_d and v_τ in the control laws (9.2.1) and (9.3.11) have the same meaning. Hence for grinding, these parameters can be defined according to (9.2.7).

9.4 Experiments in Grinding

For the experimental study of grinding operation, we have used the robotic system described in Chapter 4.

The control algorithms computing commanded motion velocities are coded in the control computer. Voltages corresponding to these commanded velocities are applied to analog servo systems. The velocity servo systems apply the power voltages to the motors according to the expression (4.2.1).

We shall start with the grinding method where a tool is held by the manipulator arm (see Figure 9.4.1). An abrasive wheel *2* is rotated by a drive *1* which is mounted on the manipulator arm through a two-component force sensor *3* so that the wheel rotation axis is vertical. A flat steel part *4* about 3 mm thick is clamped in the vise *5* on the manipulator table. The part is in the horizontal plane. Grinding should remove a burr *6* modeled by steel plates about 0.5 mm thick attached to the part. By simulating burrs in this repeatable way, we can compare how different algorithms work under similar conditions.

When moving along the part contour during measurement, the vector of normal in the control law (9.2.1) is defined from the force sensor output. When scanning the part with the burrs, this vector is computed using the position data sampled during measurement of the standard contour, as

FIGURE 9.4.1
The grinding tool in the manipulator arm

described in Section 9.2. Figure 9.4.2 illustrates the standard part contour computed using the sampled manipulator position data with normals at some contour points, as well as the contour of the part with the excess stock shown in Figure 9.4.1. The two stock runouts are marked by shading.

The control program performs computations (9.2.1), (9.2.4), (9.2.5), and (9.2.7) very fast. Figure 9.4.3 displays plots of time dependencies of the contact force F_n and velocity v_τ registered in the experiments with the part shown in Figure 9.4.2. Figure 9.4.4 shows the experimentally obtained dependence $\Delta(t)$ (for the burr 1 in Figure 9.4.2). The variable $\Delta(t)$ is the excess stock dimension used in (9.2.7) to compute v_τ and F_d at the appropriate time instant. The numerical parameters used in the motion control algorithm (9.2.1), (9.2.7) are as follows:

$$F_{d0} = 2.5 \text{ N}, \quad h_F = 2 \text{ N/mm}, \quad h_v = 5 \text{ s}^{-1}, \quad v_1 = 7 \text{ mm/s},$$

$$v_0 = 1.7 \text{ mm/s}, \quad g = 0.35 \text{ mm/(N} \cdot \text{s)}$$

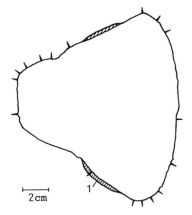

FIGURE 9.4.2
The standard part contour and the contour with simulated burrs

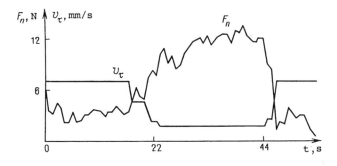

FIGURE 9.4.3
Time histories of contact force F_n and velocity v_τ in the grinding experiment

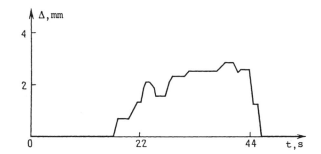

FIGURE 9.4.4
Time history of the measured burr height $\Delta(t)$

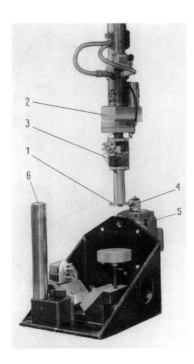

FIGURE 9.4.5
The part held in the manipulator arm

Such algorithm parameters allow the removal of the stock runout Δ up to 0.7 mm in a single pass of the abrasive wheel along the part contour. Let us now describe the experiments in which a part to be ground is held in the manipulator arm. Figure 9.4.5 displays a part *1* attached to the end-effector *2* that rotates the part about the vertical axis. This end-effector with the two-component force sensor *3* is described in Section 4.3 (see Figure 4.3.3).

For this method of grinding, a powerful stationary drive *5* is used to rotate the abrasive wheel *4*, since the drive weight is not limited by the manipulator weight lifting capacity. To measure the standard part and scan the part, they have been rotated in contact with the imitation wheel *6*. Figure 9.4.6 illustrates a photo of the part with a relatively dense simulated burr in segment *1* of the contour. The abrasive tool actuated by the powerful drive allows it to be removed successfully. Segment *2* of the part shown in Figure 9.4.6 has already been ground. We employed the grinding algorithm for an extended contour described in Section 9.2.

The plots of dependencies $F_n(t)$, $v_\tau(t)$ obtained by grinding are displayed in Figure 9.4.7. The parameters of the algorithm (9.2.1), (9.3.12) are as

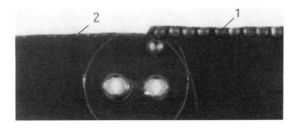

FIGURE 9.4.6
The part with the dense simulated burr 1

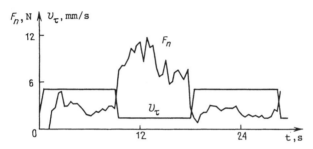

FIGURE 9.4.7
Plots of time dependencies $F_n(t)$, $v_\tau(t)$ in the grinding experiment

follows:

$$F_{d0} = 2 \text{ N}, \quad h_F = 2 \text{ N/mm}, \quad h_v = 5 \text{ s}^{-1}, \quad v_1 = 6 \text{ mm/s},$$

$$v_0 = 2 \text{ mm/s}, \quad g = 0.3 \text{ mm/(N} \cdot \text{s)}$$

Figure 9.4.8 illustrates the part contour obtained by measurement as well as the scanned stock runouts *1* and *2*. The dotted line shows the amount of stock removed by moving the part once along the abrasive wheel. The parameters of the grinding control algorithm are as follows:

$$F_{d0} = 2,5 \text{ N}, \quad h_F = 2 \text{ N/mm}, \quad h_v = 20 \text{ s}^{-1}, \quad v_1 = 6 \text{ mm/s},$$

$$v_0 = 2 \text{ mm/s}, \quad g = 0.3 \text{ mm/(N} \cdot \text{s)}$$

Figure 9.4.9 shows the plots of dependencies F_n and v_τ on time during grinding the contour section with the stock runout *1*. Figure 9.4.9 resembles

FIGURE 9.4.8
The measured part contour with scanned stock runouts

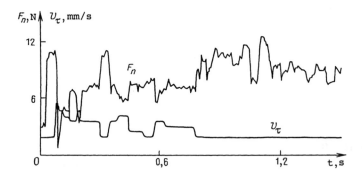

FIGURE 9.4.9
Plots of time dependencies $F_n(t)$, $v_\tau(t)$ obtained for grinding the stock runout 1

Figure 9.4.3, even though they represent different ways of grinding with distinct control algorithms. The similarity is stipulated by the common approach to grinding control, which is to control the variables F_n and v_τ simultaneously.

The experiments prove that the manipulator control algorithms for grinding designed in this chapter work well. These algorithms can be also used to control a manipulator in polishing and fettling.

Chapter 10

Assembly Operations

This chapter considers manipulator control algorithms for typical assembly operations such as mating parts with cylindrical surfaces (inserting a peg into a hole), or mating threaded parts (driving a screw down a threaded hole). These operations are performed in assembly of different parts and devices. In assembly operations, one must control mutual microdisplacements of the parts to be mated. Force sensor output reflects such relative microdisplacements with great accuracy.

In assembly, mating parts are subject to mechanical constraints. In this chapter, the manipulator performing part mating is controlled using the methods for constrained control developed in Chapters 5–7. The control design is further complicated by the fact that the character of imposed constraints may repeatedly change in the course of the part mating process. In addition, friction between parts may significantly influence the assembly operation. Despite the above issues and the simplifying assumptions made in the control design, a consistent experimental performance of the discussed algorithms is demonstrated.

10.1 Inserting a Peg into a Hole

Let us consider the problem of inserting a part shaped as a circular cylinder into a hole of the same form, i.e., the problem of inserting a peg in the base part hole. In some cases, we shall assume that the hole is chamfered. The peg is typically held in the manipulator gripper during the assembly operation. The manipulator moves the peg from the initial position till it contacts the base part. Depending on positioning accuracy, the peg may either directly enter the hole, strike the chamfer, the edge of the hole, or the base part surface outside the hole.

Assume that the peg axis remains parallel to the hole axis through the

insertion process. Then the parts can be mated by moving the peg translationally along the axis of the hole, provided that the initial misalignment
of the axes does not exceed $\Delta = (D - d)/2$, where D is the hole diameter,
and d, the peg diameter. For instance, if the difference in the diameters of
mating parts is 0.1 mm, then the initial relative positional error Δ for the
parts should not exceed 0.05 mm. In many assembly applications, such a
high positioning accuracy cannot be achieved. Note also that the mating
part positions may not be known with sufficient precision.

The peg axis may be not parallel to the hole axis. Figure 10.1.1 displays
some possible cases of mutual part positions. Figure 10.1.1a illustrates the
case where the peg edge contacts the hole edge at two points and the peg
axis makes an angle θ with the hole axis. Figure 10.1.1b shows the peg
contacting the hole edge at three points.

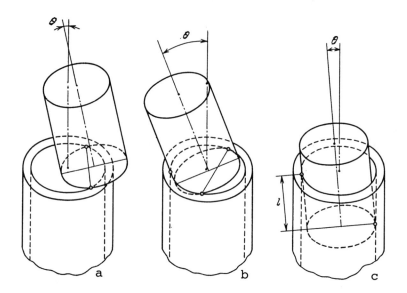

FIGURE 10.1.1
Possible cases of mutual part positions in peg-in-hole insertion:
a – two-point contact; b – three-point contact; c – two-point
internal contact

When the peg end is inserted into the hole, it can contact the inner surface
of the hole either at one point, or at two diametrically opposite points (see
Figure 10.1.1c). For a two-point contact, where the angle θ between the
axes of the peg and the hole is small, θ can be approximately expressed as
a ratio of the peg insertion depth l and the hole diameter [293]:

$$\theta \cong cD/l, \tag{10.1.1}$$

where $c = 1 - d/D$ is a relative clearance. If the peg slides inside the hole under the action of external forces and torques, then constraint forces act on the peg at the points A and B as shown in Figure 10.1.2. Tangential components T_A, T_B of these constraint forces are related to their normal components N_A, N_B through the friction coefficient f as $T_A = fN_A$, $T_B = fN_B$. The component N_A is perpendicular to the generatrix of the peg, the component N_B, to that of the hole.

The peg may stop moving inside the hole because it jams [242]. In case of jamming, the forces acting between the peg and the walls of the hole are not defined uniquely by the applied forces and torques. The system may develop large internal stress forces that obstruct the peg motion in the presence of friction. For the two-point contact, jamming is possible when the straight line connecting the contact points A and B belongs simultaneously to both friction cones (see Figure 10.1.2). In this case, the system of nonzero constraint forces, which produces a zero principal vector and torque, is physically realizable. Different solutions of the static problem can be obtained by adding this system of forces to the constraint forces that compensate for the external forces. The internal force system may arise because of unsuccessful insertion history.

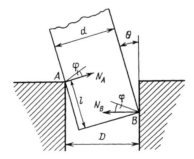

FIGURE 10.1.2
A peg jammed in the two-point contact configuration

As the peg submerges into the hole, the line AB leaves the cone of friction built about the point A. The straight line AB first leaves the friction cone through its lower boundary. The condition of being on the boundary is

$$l = d\tan\varphi = df, \qquad (10.1.2)$$

where φ is a friction angle. The jamming is impossible for $l > df$. By using (10.1.1), we can rewrite this inequality as follows:

$$\theta \leq c/f \qquad (10.1.3)$$

We shall further use (10.1.3) in Section 10.2 to design a manipulator control for a peg-in-hole insertion.

Theoretical and mechanical problems related to assembly of cylindrical parts are treated in detail in a number of papers, such as [6, 188, 242, 284, 293]. Let us consider principal methods for automated mating of parts with initial linear and angular positional errors. Note that most known control algorithms rely on manipulator compliance (passive or active).

The first automated assembly method uses *passive compliance devices.* Passive compliant elements are typically installed in the manipulator wrist immediately preceding the gripper. The elastic (spring) suspension of the peg shown in Figure 10.1.3 can be characterized by its axial, lateral, and angular stiffness, and a distance L from a compliance center C to the peg end. The compliance center is a point at which the applied force causes no angular displacement of the peg. Herein we assume that such a point exists. Papers [293, 297] demonstrate that the assembly can be facilitated by arranging the compliance center to be close to the lower peg end, i.e., be remote with respect to the elastic suspension. Remote Center Compliance devices and their applications are considered in [38, 44, 52, 98, 287, 330].

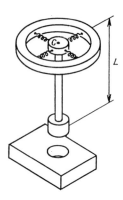

FIGURE 10.1.3
The elastic suspension of the peg

Passive compliance devices can be used in assembly subject to certain conditions. One condition for a successful assembly operation is that the peg should land on the chamfer of the hole. One more limitation is that the physical parameters of passive compliance devices must be chosen depending on the particular type and sizes of the parts to be assembled.

In many practical cases, however, the peg does not land on the chamfer, much less insert into the hole itself. Parts may not have a chamfer at all. In such cases, after establishing contact between the parts, we still have to make the peg end enter the hole. It can be achieved by the second control method which uses *information on interaction forces* between the parts. The simplest control algorithms only employ the information on an axial interaction force while using a passive compliance device. The most common algorithm of this kind works as follows. A manipulator arm, or a special assembly end-effector that incorporates compliant elements and holds a peg, performs searching motions of small amplitude around an assumed hole center. For instance, it may spiral away from the assumed hole center. The action of frictional forces and the suspension compliance make the peg tilt so that at some time instant its edge engages the hole chamfer and the peg partially enters the hole. Sharp force variation registered by a sensor indicates that the peg finds itself in the hole. This method of assembly is discussed in many papers [30, 110, 167, 189, 234, 286], but it also has its limitations.

The performance of the assembly algorithm can be significantly improved by using the information on three components of a part interaction force with a passive compliant peg suspension. A typical algorithm of this kind looks as follows. A peg attached to a compliant wrist is pressed against the part and moved towards the hole. When the peg strikes the edge of the hole, it may tilt, and its lateral surface contacts the edge of the hole. This can be detected by monitoring the lateral force components. After that the process of peg insertion begins, in which the lateral force from the hole wall acts on the peg. The manipulator moves in the direction of this force, because of the force feedback. As the peg moves laterally, the passive compliance compensates for the angular misalignment as well. Such misalignment correction is somewhat analogous to that performed by a Remote Center Compliance device. Such algorithms for the insertion of a peg into a hole are developed in [85, 112, 225, 256, 269, 273].

Algorithms that use the information on three components of the measured force for correcting the manipulator motion in assembly are applicable to a broad class of parts. To switch to a different type of assembly parts, it is sufficient to modify the software algorithm parameters, such as feedback gains, while the hardware does not need to be changed. The advantage of such algorithms is their relative simplicity.

Manipulator control algorithms that move the peg relative to the base part using the information on two lateral force components and two lateral moment components [121, 184, 191, 211, 224, 270, 291] are more versatile, though more complex. These algorithms usually use the information on axial force as well. Control is designed so that the peg can rotate about its end subject to the lateral moment, but moves without rotating, when the lateral force acts on the end. Such adjustment of the peg position is similar to the alignment produced by a Remote Center Compliance device, yet the compliance is active, not passive. The algorithms can be realized as active accommodation or artificial compliance. An accommodation or compliance matrix should be chosen diagonal in the coordinate system with the origin close to the peg end and an axis directed along the peg axis. This approach is more advanced and difficult to implement in practice. Note that a similar control algorithm can be obtained by using a concept of manipulator motion along a mechanical constraint, as pointed out in Section 1.3.

10.2   Manipulator Control in Peg-in-Hole Insertion

This section describes manipulator control algorithms for inserting a peg into a hole. Such algorithms use information on three components of the interaction force between the mating parts. The peg is moved by three translational degrees of freedom of the manipulator in the active accommodation regime. Misalignment is corrected with the help of the passive compliance of the peg suspension (in particular, force sensor compliance).

We shall perform the insertion of the peg into a hole in four consecutive stages. In the first stage, the peg moves till it contacts the base part surface. This stage can be considered as a transition from a free motion to a mechanically constrained one. In the second stage, a search for the hole is conducted. In the search, the peg end or its chamfer moves along the surface of the base part till it contacts the edge of the hole. The third assembly stage is aligning the peg with the hole. The alignment can also be considered as a transition to a mechanically constrained motion, since the character of the constraint changes at this stage. Finally, in the fourth assembly stage, the peg moves translationally inside the hole to the prescribed depth of insertion. This motion can be considered as a motion along the constraint imposed at the third stage.

The development and algorithms of the first three assembly stages depend on whether the ends of mating parts have chamfers. Subsections 10.2.1 and 10.2.2 consider the following two cases. In the first case, we suppose that the hole has a chamfer and the peg end lands on it immediately, so there is no hole searching stage. In the second case, neither the hole nor the peg are chamfered, and searching motions are necessary to insert the peg in the hole.

We shall assume that the peg is held in the arm of a gantry manipulator with three degrees of freedom. To describe the control algorithms, we use a stationary orthogonal frame $OXYZ$. The manipulator degrees of freedom move the arm along the axes of the frame. The axis OZ is vertical, the axes OX and OY are horizontal. For both cases in hand, we assume that the force sensor sensitivity axes are aligned with the frame axes.

The base part with a hole of diameter D is placed on the manipulator base so that the hole axis is vertical, i.e., parallel to the axis OZ. We assume that in the absence of contact with the part, the peg axis is nearly vertical. The insertion of the peg in the hole is performed in a downward direction. The peg diameter d is a little smaller than the hole diameter D.

The manipulator control algorithms presented below are defined by setting commanded velocities of motion along each translational degree of freedom (V_{xd}, V_{yd}, V_{zd}) depending on time and readings of force sensors and position sensors. The sensor signals are fed to the control computer that calculates commanded velocities. A control voltage applied to each drive of the manipulator is computed by an analog servo system according to the expression (4.2.1).

10.2.1 Insertion into a chamfered hole

Let us start with the case where the base part has a chamfered hole. We suppose that the hole coordinates are not known exactly a priori. We shall set the commanded velocities of motion for the manipulator establishing contact with the part as follows:

$$V_{xd} = g[x_d(t) - x], V_{yd} = g[y_d(t) - y], V_{zd} = g[z_d(t) - z], \quad (10.2.1)$$

where g is a feedback gain, x, y, and z are measured manipulator arm coordinates, and $x_d(t)$, $y_d(t)$, $z_d(t)$ are commanded values for these coordinates. We assume that such motion will land the peg on the hole chamfer. The position-controlled motion (10.2.1) stops when the vertical force component exceeds a threshold value. Then the next assembly stage begins, which combines alignment of the peg and the hole and insertion of the peg into the hole.

During the insertion, the peg should first move along the conical surface of the chamfer, then along the cylindrical lateral surface of the hole. If the peg axis is vertical, then in contact with the surface each point of the peg, including its center, moves first along the conical surface and then along the cylindrical surface of the radius $D - d$. Figure 10.2.1 displays the surface where the peg center should belong during the insertion. This surface defines the constraint imposed on the peg center coordinates as the peg moves in contact with the inner surface of the hole.

To set a commanded velocity of the manipulator motion along this constraint, we shall use the formula (7.3.7). In the projections on the axes OX, OY, OZ, the commanded velocity vector has the form:

$$V_{xd} = gF_x + v\tau_x, \quad V_{yd} = gF_y + v\tau_y, \quad V_{zd} = gF_z + v\tau_z, \quad (10.2.2)$$

where F_x, F_y, F_z are components of the external force vector $\boldsymbol{F}$ acting on the peg, g is a force feedback gain, which is the same for all degrees of

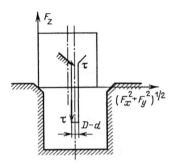

FIGURE 10.2.1
Constraint imposed on a peg position during the insertion

freedom, τ_x, τ_y, τ_z are components of the unit vector τ of tangent to the surface directed along the desired motion, $v = \text{const} > 0$ is a commanded velocity of the tangential motion. Since the peg should move downwards, the vector τ at each point of the surface can be determined uniquely. From all the vectors in the tangential plane, we should choose a vector that belongs to the vertical plane. At points of the conical surface $\tau_z < 0$, and the vector $(\tau_x, \tau_y, 0)$ is directed towards the cone axis. At points of the cylindrical surface (see Figure 10.2.1)

$$\tau = [0 \ \ 0 \ \ -1]^T \tag{10.2.3}$$

By substituting the vector (10.2.3) into (10.2.2), we obtain the commanded velocities for the peg motion in the hole

$$V_{xd} = gF_x, \quad V_{yd} = gF_y, \quad V_{zd} = gF_z - v \tag{10.2.4}$$

The two vectors τ constructed for the conical and cylindrical surfaces respectively are at a sharp angle (see Figure 10.2.1). Hence, we can use the formula (10.2.4) to compute commanded values of the velocity vector for the peg motion along the chamfer as well. In this way, the control of the entire motion is designed based on the description of the constraint imposed at the last stage of the motion.

The horizontal component of the commanded velocity vector (10.2.4) coincides in direction with the force exerted on the peg by the lateral surface of the hole and is zero only in the absence of this force. By setting the commanded velocity in this way, we ensure that a zero interaction force between the peg and the lateral hole surface is eventually established.

We have described the algorithm for computing the commanded velocity (10.2.4) under the assumption that both the peg and hole axes are vertical.

In practice, a misalignment may develop during the insertion. However, the experiments demonstrate that if the misalignment is not too large, it can be compensated by angular compliance of the force sensor.

The last expression in (10.2.4) can be written in the form

$$V_{zd} = g(F_z - F_d), \qquad (10.2.5)$$

where $F_d = v/g$ is a permissible value of the vertical force F_z. By setting the velocity V_{zd} in this way, the manipulator downward motion stops as soon as the force F_z exceeds F_d.

Let us discuss the choice of the force F_d. We assume that the peg is attached to the manipulator through an elastic suspension. The compliance of this suspension is very important for the insertion. It is the compliance that corrects angular misalignments between the axes of the peg and hole.

As mentioned in Section 10.1, the assembly is facilitated if the compliance center of the suspension is close to the lower end of the peg. Let us consider a more difficult case, where the compliance center C is much higher than the lower peg end, at a distance L from it. The point C is supposed to be on the peg axis. We also assume that the angular stiffness k_θ of the suspension is small compared to its lateral or axial stiffness.

For a small deviation of the peg axis from the vertical line, Coulomb friction force F_f between the hole surface and the edge of the downward moving peg should satisfy the inequality

$$|F_f| < F_d \qquad (10.2.6)$$

For a one-point contact between the peg and the wall of the hole during the vertical peg motion, the condition

$$|F_f| \cong f|F| \qquad (10.2.7)$$

should be satisfied, where F is a horizontal component of the force the peg exerts on the wall of the hole. One can estimate this force by assuming that it is caused by an angular deformation of the elastic suspension. For a small angular deviation θ of the peg axis from the vertical, the force is given by

$$|F| \cong k_\theta \theta / L \qquad (10.2.8)$$

It follows from (10.2.6)–(10.2.8) that the peg stalls and cannot move downwards for

$$F_d < f k_\theta \theta / L \qquad (10.2.9)$$

Now let us require that the peg inserted in the hole does not move downwards if a misalignment angle θ exceeds the critical angle at which the jamming is possible (see (10.1.3)). From (10.2.9) and (10.1.3), we obtain

$$F_d < k_\theta c/L \qquad (10.2.10)$$

If (10.2.10) holds, then the peg can start moving down and establish a two-point contact only at an angle smaller than the critical angle.

For the peg moving inside the hole, the initial stage of the insertion is the most important. If the peg is inserted further than the critical depth (see (10.1.2)), there is no risk of jamming. Therefore, it is important to monitor the depth of the peg insertion. The peg is considered to be fully inserted when the depth of insertion reaches a prescribed value.

Experiments have been done using the robotic system described in Chapter 4. The peg is held in the gripper fingers which are used as three-component force sensors (see Figure 10.2.2). The gripper is attached to the manipulator arm through the wrist (see Figure 4.3.2) and has a fixed orientation relative to the arm. The peg is moved by the three translational degrees of freedom of the manipulator. To provide sufficient angular compliance, the gripper fingers hold the peg through rubber spacers. The compliance center of such elastic suspension is inside the part of the peg held by the fingers. This helps to correct misalignment during the insertion.

The diameter of the peg used in the experiments is 20 mm, the hole diameter is 20.02 mm, the hole chamfer diameter is 24 mm. In the control law (10.2.4), the force feedback gain is $g = 10$ mm/(N·s), the commanded value of the force F_d is chosen 0.3 N at the alignment stage, and 5 N at the insertion stage ($v = 50$ mm/s). The time required to insert the peg by 15 mm into the hole does not exceed 0.5 s. In the assembly, jamming does not occur even when angular misalignments between the axes of the peg and hole are up to 0.1 rad in the beginning of the motion.

10.2.2 Peg-in-hole insertion in the absence of chamfers

Let us now proceed with the second case, where both *the hole and the peg have no chamfers*, and it is necessary to search for the hole in order to mate the parts. For the algorithm we are going to describe, as for the algorithm of Subsection 10.2.1, the compliance of the peg suspension is of great importance. Such compliant suspension can be provided, for instance, by an elastic element of the force sensor through which the peg is connected to the manipulator (see Figure 10.1.3). We shall consider the elastic suspension of the peg as a lumped elastic rotational element with an angular stiffness k_θ.

Let us consider the algorithms for each assembly stage.

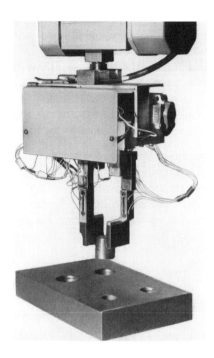

FIGURE 10.2.2
The peg held in the force sensing fingers

Establishing contact.

Similar to Subsection 10.2.1, we shall control the manipulator motion at
the first assembly stage according to (10.2.1). We assume that as a result
of this motion, the peg end contacts the base part surface outside the hole.

Searching for the hole.

We start with moving the peg along the base part surface towards the
assumed location of the hole center. Suppose that the commanded motion
is along the axis OX. To ensure that the peg moves as desired at a constant
velocity and simultaneously applies a given contact force to the base part,
we shall set the commanded velocities of the manipulator degrees of freedom
as:

$$V_{xd} = v_x + g_x F_x, \quad V_{yd} = g_y F_y, \quad V_{zd} = g_z(F_z - F_{d1}), \quad (10.2.11)$$

where v_x is a desired (commanded) velocity of motion along the axis OX;
g_x, g_y, g_z are feedback gains ($g_y > 0$, $g_z > 0$, and the gain g_x may be zero);

F_x, F_y, F_z are measured force components, F_{d1} is a commanded contact force between the peg and the surface.

For the peg motion outside the hole, the lateral force $F_y = 0$, so setting the velocity V_{yd} in the form (10.2.11) does not make the peg deviate from the straight path. The searching is considered to be completed, when the lateral surface of the peg contacts the edge of the hole (see Figure 10.2.3a). The contact is assumed to be established if the inequality

$$|F_x| + |F_y| > F_0 \qquad (10.2.12)$$

is satisfied, where F_0 is an empirically chosen threshold value. When the hole is found, we proceed with the next assembly stage.

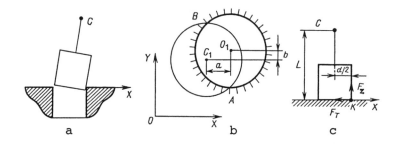

FIGURE 10.2.3
Searching for the hole with the peg: a – lateral surface of the peg contacting the hole edge; b – schematics of the peg moving over the hole; c – peg tilting caused by the friction force

Let us consider conditions for the moving peg to get into the hole. Denote by C_1 a projection of the compliance center C on the surface of the base part (the plane XYZ). Denote by b a distance from the hole center O_1 to the trajectory of the point C_1, which is parallel to the axis OX (see Figure 10.2.3b). Clearly, for $b < (D - d)/2$, the peg will be eventually inserted into the hole.

If the trajectory of the point C_1 passes at a distance $b > (D - d)/2$ from the center O_1, then the lateral surface of the peg can still establish contact with the edge of the hole. Suppose that because of the angular compliance of the suspension, the friction force tilts the peg and its edge contacts the edge of the hole. Then in addition to frictional and vertical forces, a force directed towards the center of the hole is applied at each point of contact between the peg and the hole. Under the action of this force, the control (10.2.11) will move the peg towards the center of the hole and can insert it.

If the friction between the peg and the base part is small, the peg may rotate about the straight line that connects the two points A and B of contact between its edge and the edge of the hole (see Figure 10.2.3b). The peg will tilt if the vertical force creates a tilting moment relative to the straight line AB. For this to happen, the center C_1 of the peg end and the hole center O_1 should be on the same side from the line AB, i.e., the distance $C_1 O_1$ should be less than $(D^2/4 - d^2/4)^{1/2}$. Assuming that $c \ll 1$, we can write this condition in the form

$$b < D\sqrt{c/L} \qquad (10.2.13)$$

The condition (10.2.13) is much weaker than the condition $b < (D-d)/2$ or $b < Dc/2$ for a peg falling directly into the hole, since $c \ll 1$.

The friction between the peg end and the surface of the base part may cause the peg to tilt even though the condition (10.2.13) does not hold. In some case, the peg may tilt even being outside the hole. Let us consider this case assuming that a tilting angle is infinitely small, i.e., peg only contacts the surface at a point K (see Figure 10.2.3c). The frictional force F_f applied to the peg at the point K produces a moment $F_f L$ about the center C of the compliant suspension, where L is a distance from the suspension center C to the peg end. The vertical force F_z creates an opposite in direction moment $F_z d/2$ about the point C. If $F_z d/2 < F_f L$, then the peg will tilt. By taking into account that $F_f = f F_z$, where f is a coefficient of friction, we rewrite the condition for the peg tilting in the form

$$f L > d/2 \qquad (10.2.14)$$

Assume now that the condition (10.2.14) is satisfied and the peg tilts at a small angle about the horizontal axis OY. If the point C_1 gets inside the hole during the motion of the tilted peg, then the peg edge establishes contact with the hole edge. This will happen for

$$b < D/2 \qquad (10.2.15)$$

Even if the condition (10.2.14) for the peg tilting is not satisfied, the peg can still rotate about the axis normal to the direction of motion as in the case described above. Suppose the point C_1 is inside the hole (see Figure 10.2.3b). Denote by A that of the two intersection points of the peg end and the hole edges, which has a larger x-coordinate. Denote by a a distance between the projections of the points C_1 and A along the axis OX. The peg can turn about the straight line passing through the point A and parallel to the axis OY, provided the condition $f L > a$ similar to (10.2.14) is satisfied. Let us demonstrate that a is minimal when the point A lies on the straight line parallel to the axis OY and passing through the hole center O_1. Denote this minimum by a_m. One can readily see that

$$a_m = \sqrt{(d/2)^2 - (D/2 - b)^2}$$

Thus, if $fL > a_m$, then the peg will tilt in motion. From this, we obtain a condition on a deviation b of the peg center trajectory from the hole center for which the peg is tilted

$$b < D/2 - \sqrt{(d/2)^2 - f^2L^2} \qquad (10.2.16)$$

The moving peg can also tilt about the straight line AB connecting the points of intersection of the peg end and the hole edges (see Figure 10.2.3b). Such rotation can occur easier than the above considered tilt about the straight line parallel to the axis OY. The tilt about the axis AB can happen for a friction coefficient smaller than the one needed for the inequality (10.2.16) to hold. Without a detailed study of conditions under which the peg will turn about the straight line AB, we can estimate a permissible deviation of the trajectory in the case of small friction coefficients using the inequality (10.2.13).

To recapitulate, if the trajectory of the peg motion passes sufficiently close to the point O_1, then the hole will be detected, and we may proceed with the next assembly stage – initial alignment.

Alignment.

When the lateral surface of the peg contacts the edge of the hole (see Figure 10.2.3a), a force acts from the inner surface of the hole on the peg. To align the hole and peg axes, we move the compliance center C along the horizontal component of this force. Simultaneously with the peg motion in the horizontal plane, we shall move it along the hole axis. Let us set the commanded velocities of the motion in the form analogous to (10.2.4), (10.2.5)

$$V_{xd} = gF_x, \quad V_{yd} = gF_y, \quad V_{zd} = g_z(F_z - F_d) \qquad (10.2.17)$$

To avoid jamming during the alignment, the force F_d should satisfy the inequality (10.2.10). The alignment stage is considered complete when the depth of the peg insertion exceeds a critical value determined by (10.1.2).

Peg motion in the hole.

In the last assembly stage, we continue to compute the commanded velocities using (10.2.17), which is analogous to (10.2.4), and varying only the parameters. For instance, the restriction (10.2.10) on the commanded value of the vertical force aimed to prevent jamming is irrelevant at this stage. The insertion is deemed complete when the peg is inserted in the hole to a given depth, or when the vertical interaction force between the parts exceeds a threshold value. These conditions are monitored by the readings of position sensors or a force sensor.

Experiments have been performed using the robotic system described in Chapter 4. For inserting the peg into the hole, three translational degrees of freedom of the manipulator are used. A three-component force sensor, such as the probe shown in Figure 2.2.7, is used to measure the forces. Figure 10.2.4 illustrates a step of the experiment. The peg is being inserted into a hole in the base part placed on the manipulator table, *1* denotes the force sensor module measuring the axial force F_z, *2* denotes the force sensor module measuring the lateral components F_x and F_y, *3* denotes the peg attached to the sensor. The three-component force sensor can be considered as an elastic suspension. The compliance center C of this suspension is located inside the module *1* at a distance $L \cong 12$ cm from the lower end of the peg. The angular stiffness coefficient of the suspension is $k_\theta = 15$ N·m/rad. The diameter of the peg d is 20 mm, the hole diameter is 20.02 mm. The base part with the hole and the peg are both made of steel and have no chamfer. The steel-on-steel friction coefficient f is 0.1 to 0.2. The condition (10.2.14) is satisfied for the given parameters L, d, and f.

FIGURE 10.2.4
Experiment in peg-in-hole insertion

The experiments demonstrate that the assembly is performed success-fully, if the trajectory of the peg end center passes at a distance of less than 4 mm from the hole center. If the hole position is known with a greater error, then a scanning motion is necessary. The peg can be successfully inserted in the hole when the initial angular misalignment is less than 0.05 rad. The feedback gains have been experimentally chosen for each stage of assembly. In the alignment stage, the gains in the algorithm (10.2.17) are $g = 5$ mm/(N·s), $g_z = 3$ mm/(N·s). The commanded value F_d of the force F_z chosen in accordance with the inequality (10.2.10) should not exceed 0.12 N. In the experiments, the force F_d has been set to 0.1 N.

Figure 10.2.5 displays plots of forces F_x, F_y, and F_z obtained in the experiment. The numbers $I - IV$ denote the assembly stages: I is the stage of establishing contact between the peg and the base part, II is searching for a hole, III is the alignment, IV is the peg motion in the hole. The increase of the force F_z at the end of Stage I indicates that the contact between the parts has been established. In the second stage, the force F_z is maintained constant, the force F_x is nonzero due to friction. The lateral force components somewhat increase at the end of this stage as the peg surface establishes contact with the hole edge. At the end of the alignment stage, all force components are close to zero. At the last stage, i.e., peg motion in the hole, the vertical component of the force F_z grows due to the friction of the lateral peg surface against the inside surface of the hole, but remains small.

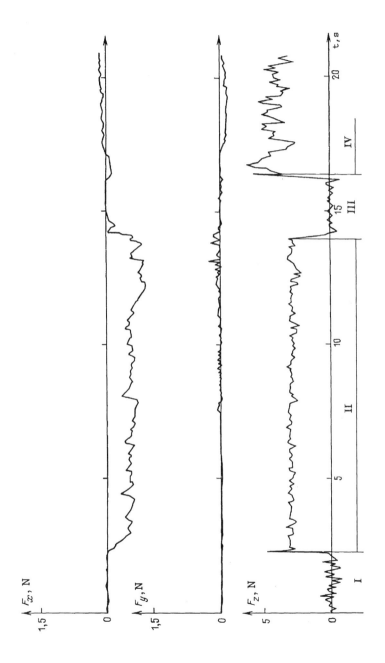

FIGURE 10.2.5
Force history in the-peg-in-hole insertion experiment

10.3   Threaded Joint Assembly

A threaded joint is a connection between a part with an external thread, such as a screw or a bolt, and a part with a threaded hole, such as a nut. A threaded joint is very common in assembly and can be found in various units and devices. The operations of assembling a threaded joint and inserting a peg into a hole have some common features.

When parts are being screwed together, a screw moves translationally in the threaded hole as it rotates about its longitudinal axis. Through the rotation, the threads of the screw move along the grooves of the threaded hole. Clearly, an assembly can be performed provided that the thread diameters of the screw and the hole correspond to each other, and their thread angle and pitch h are the same. Linear and angular misalignment of parts should be small, otherwise the screw can be jammed and the thread damaged in the assembly attempt.

For the automated assembly of threaded joints, special assembly end-effectors, such as a powered screwdriver, can be used. These devices employ various orientation and passive compliance fixtures to compensate for part positioning errors. Such devices are typically designed to handle parts of specific type and size. For a greater flexibility and reliability, assembly operations need to be partially or completely controlled. Relatively few methods of controlling a threaded joint assembly using force sensors have been published [170, 283, 304, 346]. Among them, we set apart methods that use information on an axial component of the part interaction force. This information makes the hole search much easier and allows control of the initial assembly stage, that of alignment.

The alignment can be performed as follows [346]. An elastically suspended screw is brought to the threaded hole so that its end lands on the chamfer of the hole. Then the screw is rotated about the vertical axis (the axes of the screw and the hole are supposed to be vertical) in the direction opposite to that required for driving the screw in. Before the initial portions of the thread can mate, the screw "skips" down by the thread pitch and the parts mate along the threads. This instant can be registered by a microdisplacement sensor – a force sensor. After mating the initial portions of the thread, the screw can be driven in.

If the screw does not land on the chamfer of the hole because of a positional error, or if it strikes the chamfer but the parts to be assembled have a fine thread, the information on the axial force alone may not suffice. In order to search for holes, align the axes of the screw and hole, and control the fastening operation, it is desirable to have information on three components of the interaction force and axial torque. Algorithms using this information are considered below. These algorithms, similar to those for peg insertion described above, allow correction of small angular positional errors by using a passive compliant suspension of the screw.

10.4 Manipulator Control in Threaded Joint Assembly

The considered algorithm for controlling a threaded joint assembly employs information on three components of the force vector and an axial moment of forces of interaction between the screw and the threaded hole. The manipulator arm can be moved by three translational degrees of freedom along the axes of the orthogonal coordinate system $OXYZ$. The end-effector mounted on the arm can rotate the screw about the vertical axis OZ. The end-effector has a four-component force sensor measuring three force components and an axial torque. The upper flange of the sensor is attached to the end-effector, and the screw is rigidly attached to the lower flange. Translational displacements of the arm and a rotation angle ψ of the upper flange of the sensor are measured by the respective position sensors.

If the rotation angle of the upper flange of the sensor is zero, then the sensor sensitivity axes are directed along the corresponding axes OX, OY, and OZ. If the sensor rotates, one can transform its measurements into the stationary coordinate system as (see the expressions (7.5.5), (7.5.7)).

$$F_x = F'_x \cos \psi - F'_y \sin \psi,$$

$$F_y = F'_x \sin \psi + F'_y \cos \psi,$$

$$F_z = F'_z, \qquad M_z = M'_z \qquad (10.4.1)$$

In the above expressions, F'_x, F'_y, F'_z, and M'_z are force sensor readings in the coordinate system attached to the upper flange, ψ is a rotation angle of this flange about the vertical axis given by the rotational degree of freedom.

The part with the threaded hole is placed on the manipulator base so that the hole axis is vertical. The screw and the threaded hole have standard chamfers to facilitate the assembly. The screw is driven in the downward direction.

The manipulator control algorithm should set commanded velocities V_{xd}, V_{yd}, and V_{zd} of the translational degrees of freedom and a commanded velocity ω_d of the screw rotation about the vertical axis depending on sensor readings and time. These commanded velocities are applied to the velocity servo systems.

We shall subdivide the operation of the threaded joint assembly into five consecutive stages. At the first two stages, i.e., establishing contact and searching for the hole, the manipulator with the screw performs the same motion as when inserting a peg into the hole (see Section 10.2).

When *establishing contact*, we control the manipulator motion in accordance with the expression (10.2.1). During the motion, the screw moves vertically downwards till it contacts the part with the threaded hole. The instant of establishing contact can be defined by an increase of the axial force F_z.

When *searching for the hole*, the manipulator presses the screw against the part and moves in the direction of the supposed location of the hole center. The commanded velocities of this motion are given by (10.2.11). The stage is completed when the lateral force components are sufficiently large to satisfy the condition of the form (10.2.12). The force sensor orientation at the first two stages is constant.

The following, third stage is *alignment*. As in assembly of cylindrical parts, this is the most complicated and important stage. It is necessary to align the screw so that the initial portion of thread at the end of the screw mates that of the hole. To this end, we shall employ the method presented in [346]. The manipulator presses the screw down to the hole with a small force and slowly rotates it in the direction opposite to driving the screw in. We set the commanded velocities of the motion as follows:

$$V_{xd} = gF_x, \quad V_{yd} = gF_y, \quad V_{zd} = g_z(F_z - F_d), \quad \omega_d = \omega_1 \quad (10.4.2)$$

In this equation, F_x, F_y, and F_z are components of the force (10.4.1) measured by the force sensor, g and g_z are feedback gains, F_d is a commanded contact force between the screw and the hole, ω_1 is a constant angular velocity. For a right-hand thread, one should set $\omega_1 > 0$. Note that the first three relations in (10.4.2) coincide with the expressions (10.2.4) or (10.2.17). This control method allows alignment of the screw axis with that of the hole by rotating the screw and moving it along the chamfer towards the center of the hole. Angular misalignment is corrected due to mechanical compliance of the screw suspension. When rotating the screw, the vertical contact force with the hole is kept constant. At some instant, the initial thread portions mate and the screw "skips" down by the pitch of the thread, while the vertical component F_z of the force sharply decreases. This instant can be detected by the condition $F_z < F_1$, where F_1 is an empirically chosen threshold force. Once the condition $F_z < F_1$ is satisfied, the screw stops rotating in the opposite direction, and the next stage may begin.

The fourth stage is *driving the screw in*. At this stage, the screw moves along the screw constraint. Commanded velocities of the manipulator arm motion can be defined by (7.5.14). In the experiment, we replaced the last two expressions in (7.5.14) by the simpler expressions

$$V_{xd} = gF_x, \quad V_{yd} = gF_y, \quad V_{zd} = g_z F_z - v_z, \quad \omega_d = g_\psi M_z - \omega_2, \quad (10.4.3)$$

where v_z and ω_2 are positive commanded values of the axial, translational, and angular velocities. These velocities are related as $v_z = \omega_2 h/(2\pi)$, where h is a screw thread pitch which describes the motion along the screw constraint. If $F_z < 0$, then $V_{zd} < -v_z$, and the screw is driven (downwards) into the hole. If $F_z > 0$, then $V_{zd} > -v_z$, and the screw is driven (upwards) out of the hole. The negative sign at ω_2 means that the parts have a right-hand thread. The stage is considered to be completed, if either of the two conditions hold:

$$|z - z_0| \geq l_0, \quad |M_z - \omega_2/g_\psi| < M_1, \quad (10.4.4)$$

where l_0 is a depth by which the screw needs to be inserted into the hole, z_0 is an arm coordinate at the initial moment of assembly, M_1 is a threshold value that defines the torque of preliminary tightening of the threaded joint.

The last, fifth stage produces the *final tightening* with a given torque M_d. At this stage, the screw experiences practically no linear displacement, it is just rotated through a small angle. The commanded velocities are set in the following way:

$$V_{xd} = gF_x, \quad V_{yd} = gF_y, \quad V_{zd} = g_z(F - F_d'), \quad \omega_d = -\omega_3$$

The tightening stops when the torque M_z exceeds the given torque M_d.

Experiments have been conducted on the robotic system described in Chapter 4. The assembly end-effector with a drive for rotation about the vertical axis shown in Figure 4.3.3 is mounted on the manipulator arm. The screw is attached to the end-effector through a four-component force sensor (see Figure 4.3.3). The commanded velocities are calculated by the control computer, the power voltages are formed by the velocity servo systems in accordance with (4.2.1).

Screws with M 8×1, M 12×1, M 16×1,5, and M 16×1 metric threads were used in the experiments. These are so-called machine screws that are typically used in manufacturing to provide for an enhanced strength of threaded joints.

Figure 10.4.1 displays plots of time histories for the force components F_x, F_y, and F_z at the alignment stage (the screw M 16×1 is rotated in the direction opposite to driving the screw in). The plot of the dependence $F_z(t)$ (see Figure 10.4.1a) shows periodical sharp force drops. At times of these drops, the lateral components F_x, F_y also change sharply (see Figures 10.4.1b and 10.4.1c). The instants of abrupt changes of the axial force F_z correspond to mating of initial thread portions. For the alignment, it suffices to detect the first change of the force F_z. To determine this instant more precisely, the velocity of the screw rotation during the alignment should be small compared to that of driving the screw in.

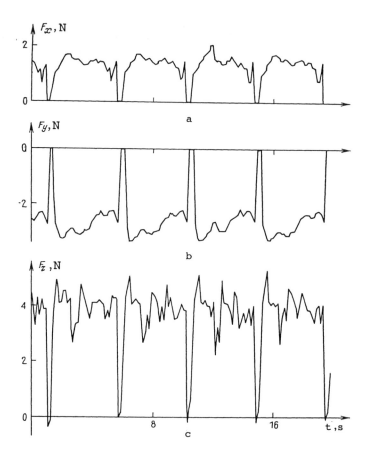

FIGURE 10.4.1
Forces logged in the threaded joint alignment experiment during reverse rotation of the screw

The parameters of the algorithm (10.4.2) in the experiment were as follows:

$$F_d = 4 \text{ N}, \quad \omega_1 = 1.3 \text{ s}^{-1}, \quad g = 0.5 \text{ mm/(N} \cdot \text{s)}, \quad g_z = 1 \text{ mm/(N} \cdot \text{s)}$$

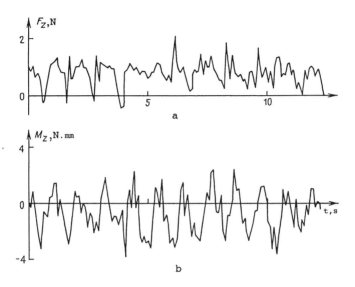

FIGURE 10.4.2
Forces logged in the threaded joint assembly experiment when driving the screw in

Figure 10.4.2 illustrates plots of time histories for the axial force F_z (see Figure 10.4.2a) and the axial torque M_z (see Figure 10.4.2b) when driving the screw in for the following parameters of the algorithm (10.4.3):

$$\omega_2 = 6 \text{ s}^{-1}, \quad g = 0.75 \text{ mm/(N} \cdot \text{s)}, \quad g_z = 1 \text{ mm/(N} \cdot \text{s)},$$

$$g_\psi = 5 \cdot 10^{-3} (\text{N} \cdot \text{s} \cdot \text{mm})^{-1}$$

When driving the screw in, a small force F_z acts on the screw "pulling" it out of the hole. The torque M_z also varies slightly, which proves that the screw driving goes normally. The time-history of the torque M_z shows a periodic component corresponding to the screw rotation.

Chapter 11

Searching for Parts and Grasping

This chapter considers tasks of grasping randomly positioned parts. In manufacturing, such tasks are typically performed prior to assembly and machining, in particular, prior to loading machines with billets for machining. Randomly positioned workpieces typically come to the work station in bins or tubs, or on a conveyor belt. It is necessary to pick up and orient such parts one by one. Manipulators with force sensors can be used for such tasks.

11.1   Part Pickup Tasks

Picking up randomly stored parts from a bin or a work station is a complicated task. A number of ways to perform this task are known. Parts can be picked up one by one manually and positioned in a definite sequence on an intermediate work station. Parts can also be separated by using various mechanical devices, such as mechanical part feeders. Though simple and reliable, these devices are not always flexible. They have to be changed if the type, shape, or weight of parts vary significantly.

The pick-up operation can be automated by using sensor-based robotic manipulators. We shall further subdivide the part pickup process into the following separate stages: *separating a part among all other parts, defining its position* (includes orientation), *grasping, transferring and placing* the part with a desired position and orientation. These stages can be performed in different sequence. In some cases, a part is first separated from the others by using an electromagnet or a vacuum cup. In other cases, to separate a part means to determine its position by inspection prior to the physical separation of the part.

A number of algorithms for pickup of separated workpieces are known. Papers proposing algorithms for picking up parts from a random heap are not as numerous. Some of them employ visual sensor systems that allow one to determine the position and orientation of parts (inspection and isolation), as well as monitor grasping [12, 18, 129]. To ensure reliable grasping of parts, the gripper may be instrumented with tactile or force sensors.

In this book, we consider part pickup by a manipulator using force sensors. In addition to force sensors, proximity sensors (ultrasound, optical, inductive, etc.) or tactile sensors may be used. Below, we consider three methods for part pickup using force sensors.

Parts can be picked up one by one without previously determining their position and orientation. A part picked in this way can be randomly oriented in the gripper. Hence, additional handling is required to orient the part properly. Parts made from ferromagnetic materials can be picked up by electromagnetic grippers [60, 87, 349, 352, 353, 315].

Force sensors can be used to monitor attraction of parts to the electro magnet (see, for example, [60]). A method and device for picking up ferromagnetic parts is described later in this chapter. The method employs a force sensor for measuring the weight of the parts attracted to the electro magnet and controlling the power supply to the magnet. This method allows to pick up parts of essentially any weight and shape stored in a bin random-scramble.

To place a picked part into a desired position, it is necessary to determine the part position (including orientation). As mentioned above, it is relatively simple to determine the position of a separated part. This task can be performed in many ways, by using devices such as sensory tables with optical, tactile, or ultrasound sensors, visual sensor systems, or scanners [13, 97, 226, 345].

The second method of part pickup is based on collecting preliminary information on the part surface using a force sensor contacting the surface. One can use this information in algorithms for grasping, transferring, and orienting parts. To collect the contact information, sensory probes and gripper fingers with integrated tactile or force sensors can be used [88, 90, 187, 221].

Finally, the third pickup method employs a gripper with sensing fingers. When the fingers move in contact with a part, it is possible to determine the shape of the part and simultaneously grasp it. This method is considered in Section 11.3.

11.2 One-by-one Pickup of Parts Using Electromagnetic Gripper

Let ferromagnetic parts be stored random-scramble in a tub that is placed within the working space of a manipulator with an electromagnet pickup device. A force sensor installed between the manipulator arm and the electromagnet measures a vertical force component. The parts can be extracted from the tub one by one using the following method. To grasp a part, an electric current certainly greater than required to pickup a single part is supplied to the electromagnet. The weight P of one part is known, so a force sensor allows one to determine the number of picked up parts by their weight. Then by gradually reducing the current, all the parts except one are dropped from the electromagnet.

Let us consider a sequence of actions needed to pick up *a single part* in more detail. First, the manipulator moves the electromagnet to the position above the tub with the parts. Then the arm with the gripper starts moving down. The downward motion stops as the force sensor registers a force directed upwards. This means that the electromagnet contacts either the parts or the bottom of the tub. At that time, the electromagnet is switched on, and a large current passes through its coil. As a result, a few parts can be attracted to the magnet. Then the pickup device lifts. Note that to pick up a single part, it may be necessary to pull up a force exceeding the weight of several parts. This may happen, for instance, if the picked up part is covered by other parts. As the electromagnet moves up, the measured force is used to count the number of the picked parts. For example, if this force signal is between 0.5 P and 1.5 P, we can safely assume that a single part is held by the electromagnet. The signal less than 0.5 P means that no parts are on the magnet (either none was picked up, or all fell down in the lifting). If the signal is greater than 1.5 P, we assume that more than one part is extracted. In order to leave only one part on the electromagnet, its power is gradually reduced. In so doing, the parts are dropped from the magnet one by one, till the force sensor shows that only one is left. At that time the power is sharply increased, so that the remaining part is held reliably. If all the parts are dropped while reducing the power, then the pickup process has to be repeated. This can happen, for instance, if the lifted parts are interlocked with each other.

Figure 11.2.1a,b shows bin-picking of washers. Each part has a mass of about 100 g. In the experiments, three-to-five parts tend to be extracted simultaneously at first. It takes about 5 s to complete the cycle of picking a single part. No more than 5% of the experiments end up with a failure to extract a single part.

Several electromagnet pickup devices for parts ranging from 10–20 g to

FIGURE 11.2.1
Extracting a single part from a bin using the electromagnet gripper: a – several parts hanging from the magnet; b – all parts but
one dropped

five–seven kg were designed using the described principles.

11.3 Part Pickup Using Force Sensing Fingers

This section considers a manipulator control algorithm that allows recog-
nition of a part position simultaneously with grasping. The algorithm is
based on the motion of force sensing gripper fingers along the surface of
the part. According to the approach developed in Chapters 6 and 7, we
design control of the grasping motion as a superposition of motions along
the normal and tangent to the constraints imposed on the finger positions.

We shall use two translational degrees of freedom of a gantry manipulator to move the manipulator arm in the horizontal plane OXY, and a
rotational degree of freedom to turn the gripper about the vertical axis.
Each of the two gripper fingers is independently actuated by its own drive.
The motion of the manipulator with the gripper is controlled by velocity
servo systems that track the commanded velocities. The control algorithm

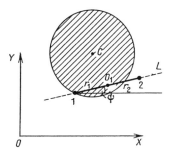

FIGURE 11.3.1
Schematics of the grasping process with only one finger contacting the part

described below consists in setting these commanded velocities.

We shall consider the task of grasping a part represented by a right circular cylinder of diameter D. The base of the cylinder is on the manipulator table. Assume that by scanning the table, one gripper finger has established contact with the part lateral surface, while the other finger does not contact it. This position is schematically illustrated in Figure 11.3.1, where *1* is the first finger, *2* is the second finger, and C is the center of the part. The figure does not take into consideration the size of the fingers. After contacting the object with one finger, we start rotating the gripper about the vertical axis, until the second finger makes contact with the part. In so doing, we try to maintain contact of the first finger with the part.

Denote by ψ a rotation angle of the gripper counted counterclockwise from the axis OX (see Figure 11.3.1), by O_1, the intersection point of the gripper rotation axis with the straight line L connecting the fingers; by r_1 and r_2, distances from the point O_1 to the respective fingers *1* and *2*. Since we do not move the finger *1*, the distance r_1 is constant. We shall control the coordinates x, y, ψ, and r_2, which correspond to motion along the respective degree of freedom each.

The condition that the finger *1* maintains the contact with the part constrains the coordinates x, y, and ψ

$$(x - r_1 \cos \psi - x_c)^2 + (y - r_1 \sin \psi - y_c)^2 = (D/2)^2, \qquad (11.3.1)$$

where x_c, y_c are the coordinates of the part center C. By differentiating (11.3.1) with respect to the time, we obtain:

$$(x - r_1 \cos \psi - x_c)\dot{x} + (y - r_1 \sin \psi - y_c)\dot{y}$$

$$+[(x - r_1 \cos \psi - x_c) \sin \psi - (y - r_1 \sin \psi - y_c) \cos \psi] r_1 \dot{\psi} = 0 \quad (11.3.2)$$

For the motion at a tangent to the constraint (11.3.1) (see Chapter 7), the velocities $\dot{x}$, $\dot{y}$, and $\dot{\psi}$ should satisfy (11.3.2). Among the variety of possible motions along the constraint, we choose the motions for which the finger with the coordinates $x - r_1 \cos \psi$, $y - r_1 \sin \psi$ remains stationary. For these motions,

$$\dot{x} = -r_1 \omega \sin \psi, \quad \dot{y} = r_1 \omega \cos \psi, \quad \dot{\psi} = \omega \quad (11.3.3)$$

The relationships (11.3.3) will be further used to determine the last term $v\tau$ in the expression (7.3.5) for the commanded velocity vector.

The vector of normal to the surface (11.3.1) is directed along the respective function gradient. The normal vector components are proportional to the weights at $\dot{x}$, $\dot{y}$, and $\dot{\psi}$ in (11.3.2). If the angle γ between the straight line L and the segment connecting the part center C with the finger 1 is small, then the expression in the square brackets in (11.3.2) is close to zero. In this case, the first two components of the vector of normal to the constraint differ little from the components n_x, n_y of the normal n to the lateral surface of the part at its contact with the finger 1, and the third component of the normal is close to zero. Then the vector of normal to the constraint closely approaches the vector

$$\begin{bmatrix} \frac{2}{D}(x - r_1 \cos \psi - x_c) \\ \frac{2}{D}(y - r_1 \sin y - y_c) \\ 0 \end{bmatrix} = \begin{bmatrix} n_x \\ n_y \\ 0 \end{bmatrix} \quad (11.3.4)$$

The vector n can be found using the force sensor in the finger 1 as $n = F/F$, where F is the force vector acting on this finger from the part.

The vector (11.3.4) is exactly the normal to the constraint for $\gamma = 0$, i.e., in the end of grasping, when the gripper fingers are at the diametrically opposite points of the part. We shall use the vector (11.3.4) instead of the vector of normal to the constraint for all γ.

By using (11.3.3) and (11.3.4) in view of (7.3.5), we obtain the components V_{xd}, V_{yd} of the commanded velocity for the translational motion of the manipulator arm (the rates of the coordinates x, y) and the commanded velocity ω_d for the rotation of the gripper (the rate of the angle ψ):

$$V_{xd} = g(F_n - F_d)n_x - r_1 \omega \sin \psi,$$

$$V_{yd} = g(F_n - F_d)n_y + r_1 \omega \cos \psi,$$

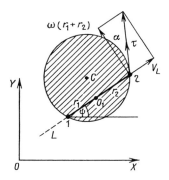

FIGURE 11.3.2
Schematics of the grasping process with one finger contacting the
part and the other sliding along its surface

$$\omega_d = \omega \qquad\qquad\qquad\qquad\qquad (11.3.5)$$

In (11.3.5), g is a constant positive gain, $F_n = n \cdot F$ is a projection to the
normal n of the force vector acting on the finger 1 from the the part, and
F_d is a commanded force. The angular velocity ω of the gripper rotation
can be any desired, for example, it can be constant.

If the gripper rotates with the velocity ω, and the finger 1 does not
move, then the finger 2 rotates about it with the velocity $\omega(r_1 + r_2)$ (see
Figures 11.3.1, 11.3.2). If the distance between the fingers is $r_1 + r_2 < D$,
then the finger 2 will contact the part lateral surface at some time instant.
After that, we keep rotating the gripper. In order to maintain the contact
of the finger 1 with the part, we also continue to move the manipulator arm
translationally, using V_{xd}, V_{yd} in (11.3.1) as the commanded velocities. In
addition, we demand the second finger to slide along the lateral surface of
the part (with τ being the tangent vector).

We consider the motion of the second finger, assuming that the first finger
remains stationary. Then the motion of the second finger is defined by its
displacement along the straight line L (the variation of the coordinate r_2)
and by the gripper rotation.

In accordance with the algorithms of Chapters 6 and 7, the commanded
velocity of the second finger should be designed as a sum of vectors directed
at a tangent and normal to the part surface (to the constraint). The velocity
component $\omega(r_1 + r_2)$ of the finger 2, caused by the gripper rotation, is
perpendicular to the straight line L. By varying the coordinate r_2, the
second finger can be moved along the straight line L. Therefore, for the
commanded velocity vector to be directed at a tangent to the constraint,

we should set the commanded value V_L of the velocity $\dot{r}_2$ as

$$V_L = \omega(r_1 + r_2)\tan\alpha$$

In this equation, α is an angle between the straight line L and the normal to the lateral part surface at the point of contact with the finger 2 (see Figure 11.3.2). We can find the normal vector from the force measured by the finger 2. Hence the angle α can be found in the course of the manipulator motion.

The normal to the lateral part surface at the point of contact with the finger 2 is close to the line L, if the distance $r_1 + r_2$ between the fingers is close to the part diameter D, i.e., the angle γ is small. When both fingers contact the part surface, the angle $\gamma < \pi/2$. By setting the commanded velocity vector according to (7.3.5), we shall substitute a unit vector directed along the line L for the vector of normal, regardless of the angle γ. Thus, in order to move the second finger along the part surface in contact with it, the following commanded velocity is set in the velocity servo system that controls the second finger position r_2:

$$V_{Ld} = g_1(F_L - F_{Ld}) + \omega(r_1 + r_2)\tan\alpha \tag{11.3.6}$$

In (11.3.6), g_1 is a constant positive gain to be chosen, F_L is a projection onto L of the force measured by the second finger, and F_{Ld} is a constant commanded force. If $F_L > F_{Ld}$, then the finger 2 velocity $g_1(F_L - F_{Ld})$ is directed away from the finger 1.

When grasping a round part, the second finger slides along its lateral surface, until the distance $r_1 + r_2$ ceases to increase. This happens when the fingers are at the diametrically opposite points of the lateral surface of the part. At this point, on the corresponding command, the fingers start closing and clamp the part with a given force. Then the manipulator lifts the part and transfers it to the work station.

The above described control algorithm was implemented as a computer program. The commanded velocities are computed in accordance with (11.3.5), (11.3.6) using signals of the force sensors in both fingers, measurements of the gripper angular position, and finger positions. The experiments were performed with the robotic system described in Chapter 4. The system allows rotation of the gripper about the vertical axis and move the force sensing fingers independently (see Figure 4.3.2). The manipulator DC motor voltages are controlled by velocity servo systems according to (4.2.1).

In the experiments, round parts of 90 mm diameter were grasped and picked up. Figure 11.3.3 shows one stationary finger contacting the part, while the other finger slides along its lateral surface. The time of grasping one part is about 2 s. The described control algorithm can also be used to grasp parts other than circular in shape, both stationary and slowly moving.

FIGURE 11.3.3
Experiment in grasping a round part with two independently actuated force sensing fingers

References

[1] E. Abele, D. Boley, and W. Sturz. Interactive programming of industrial robots for deburring. In *14th International Symposium on Industrial Robots*, pages 505-516, 1984.

[2] E. Abele and W. Sturz. Sensors for the adaptive control of fettling tasks with industrial robots. In *2nd International Conference on Robot Vision and Sensory Control*, pages 133-145, 1982.

[3] L.D. Akulenko and N.N. Bolotnik. Synthesis of optimal control of transport motions of manipulation robot. *Mechanics of Solids*, 21(4):18-26, 1986.

[4] C.H. An and J.M. Hollerbach. The role of dynamic models in Cartesian force control of manipulators. *International Journal of Robotics Research*, 8(4):51-71, 1989.

[5] R.J. Anderson and M.W. Spong. Hybrid impedance control of robotic manipulators. *IEEE Journal of Robotics and Automation*, 4(5):549-556, 1988.

[6] M. Aner. Fügemechanismen mit Sensorik in der automatischen Montage. *Robotersysteme*, 8:101-106, 1992.

[7] V.V. Avetisyan, L.D. Akulenko, and N.N. Bolotnik. Modeling and optimization of transport motion for an industrial robot. *Soviet Journal of Computer and Systems Sciences*, 24(6):107-114, 1986.

[8] H. Asada and N. Goldfine. End effector design for robotic grinding. In *Preprints of the 10th World IFAC Congress on Automatic Control*, volume 4, pages 198-203, 1987.

[9] H. Asada and N. Goldfine. Optimal compliance design for grinding robot tool holders. In *IEEE International Conference on Robotics and Automation*, pages 316-322, 1985.

[10] H. Asada and Y. Sawada. Design of an adaptable tool guide for grinding robots. *Robotics and Computer Integrated Manufacturing*, 2(1):49-54, 1985.

[11] H. Asada and J.J. Slotine. *Robot Analysis and Control*. John Wiley and Sons, New York, 1986.

[12] J. Back et al. Bin-picking using a 3-D sensor and a special gripper. In *5th International Conference on Robot Vision and Sensory Control*, pages 529-540, 1985.

[13] D.J. Balek and R.B.Kelley. Using gripper mounted infrared proximity sensors for robot feedback control. In *IEEE International Conference on Robotics and Automation*, pages 282-287, 1985.

[14] E. Baumann. *Electrische Kraftmesstechnik*. VEB Verlag Technik, Berlin, 1976.

[15] J.S. Bendat and A.G. Piersol. *Engineering Application of Correlation and Spectral Analysis*. John Wiley and Sons, New York, 1980.

[16] E.A. Barbashin. *Introduction to the Theory of Stability*. Wolters–Noordhoff, Groningen, 1970.

[17] A. K. Bejczy and R. S. Dotson. A force-torque sensing and display system for large robot arms. In *IEEE Southeast Conference*, pages 458-465, 1982.

[18] J. Birk. A computation for robots to orient and position hand-held workpieces. *IEEE Transactions on Systems, Man and Cybernatics*, 6(10):665-671, 1976.

[19] K.B. Biggers, S.C. Jacobsen, and G.E. Gerpheide. Low-level control of the Utah/MIT dextrous hand. In *IEEE International Conference on Robotics and Automation*, volume 1, pages 61-66, 1986.

[20] M. Blauer and P.R. Belager. State and parameter estimation for robotic manipulators using force measurements. *IEEE Transactions on Automatic Control*, 32(12):1055-1066, 1987.

[21] V.G Boltyanskii. *Mathematical Methods of Optimal Control*. Holt, Rinehart and Winston, new York, 1971. Translation from the Russian.

[22] G.M. Bone, M.A. Elbestawi, R. Lingarkar, and L. Liu. Force control for robotic deburring. *Transactions of ASME. Journal of Dynamic Systems, Measurement and Control*, 113(9):395-401, 1991.

[23] A. H. Bottomley. Progress with the British myoelectric hand. In *Symposium on Automatic Control in Prosthetics*, pages 114-124, 1962.

[24] A. Bray, G. Barbato, and R. Levi. *Thery and Practice of Force Measurements*, Academic Press, New York, 1990.

[25] H. Bruyninckx and J. De Schutter. Specification of force-controlled actions in the "Task frame formalism" – a synthesis. *IEEE Transactions on Robotics and Automation*, 12(4):581-589, 1996.

[26] L. Cai and A.A. Goldenberg. An approach to force and position control of robot manipulators. *IEEE International Conference on Robotics and Automation*, pages 86-91, 1989.

[27] R.H. Cannon and E. Shmitz. Initial experiments on the end-point control of a flexible one-link robot. *International Journal of Robotic Research*, 3(3):62-75, 1985.

[28] F. Chernousko, N. Bolotnik, and V. Gradetsky. *Manipulation Robots. Dynamics, Control and Optimization*. CRC Press, Boca Raton, 1994. Translation from the Russian.

[29] N.G. Chetaev. *The Stability of Motion*. Pergamon Press, 1961. Translation from the Russian.

[30] H.S. Cho, H.J. Warnecke, and D.G. Gweon. Robotic assembly: a synthetic overview. *Robotica*, 5(2):153-165, 1987.

[31] S. Chiaverini and L. Sciavicco. The parallel approach to force/position control of robotic manipulators. *IEEE Transactions on Robotics and Automation*, 9(4):361-373, 1993.

[32] S. Chiaverini, B. Siciliano, and L. Villani. Force/position regulation of compliant robot manipulators. *IEEE Transactions on Automatic Control*, 39(3):647-652, 1994.

[33] P. Coiffet. *Interaction with the Environment*. Kogan Page, London, 1983. Translation from the French.

[34] P. Coiffet et al. *Robot Technology*. McGraw-Hill, New York, 1983. Translation from the French, *Les Robots*.

[35] A. Cornut and T. Schultz. Multicompontent force/torque transducer for industrial robot. In *Weighting and Force Measurement in Trade and Industry: 9th Conference of TC3 on Measurement of Force and Mass*, pages 1-9, 1983.

[36] J.J. Craig. *Introduction to Robotics: Mechanics and Control*. Addison Wesley, Reading, Mass., 1989.

[37] M.R. Cutkosky. *Robotics: Mechanic Grasping and Fine Manipulation*. Addison Wesley, Reading, Mass., 1989.

[38] M.R. Cutkosky and P.K. Wright. Position sensing wrists for industrial manipulators. In *12th International Symposium on Industrial Robots*, pages 427-438, 1982.

[39] L. Daneshmend, V. Hayward, and M. Pelletier. Adaptation to environment stiffness in the control of manipulators. In: *Experimental Robotics I*, editors V. Hayward and O. Khatib, Springer-Verlag, 1990.

[40] P. Dario and G. Buttazzo. An anthropomorphic robot finger for investigating artificial tactile perception. *International Journal of Robotic Research*, 6(3):25-48, 1987.

[41] P. Dauchez, P. Coiffet, and A. Fournier. Cooperation of two manipulators in assembly tasks. *Digital Systems for Industrial Automation*, 2(2):179-200, 1984.

[42] A. D'Auria and M. Salmon. Sigma – an integrated general purpose system for automatic manipulation. In *5th International Symposium on Industrial Robots*, pages 185-201, 1975.

[43] D.M. Dawson, F.L. Lewis, and J.F. Dorsey. Robust force control of a robot manipulator. *International Journal of Robotics Research*, 11(4):313-319, 1992.

[44] T. L. De Fazio. Displacement state monitoring for the remote center compliance (RCC): Realizations and applications. In *1st International Conference on Assembly Automation*, pages 33-42, 1980.

[45] A. De Luca and C. Manes. Modeling of robots in contact with a dynamic environment. *IEEE Transactions on Robotics and Automation*, 10(4):542-548, 1994.

[46] J. De Schutter. *Compliant Robot Motion: Task Formulation and Control.* PhD Thesis, Katholieke Universiteit Leuven, 1986.

[47] J. De Schutter and H. Van Brussel. Compliant robot motion, I. A formalism for specifying compliant motion tasks, II. A control approach based on external control loops. *International Journal of Robotics Research*, 7(4):3-33, 1988.

[48] E.A. Devjanin, V.S. Gurfinkel, E.V. Gurfinkel, E.V. Kartashev, A.V. Lensky, A.Yu. Schneider, and L.G. Stilman. The six-legged walking robot capable of terrain adaptation. *Mechanism and Machine Theory*, 18(4):257-260, 1983.

[49] R. Dillman and B. Faller. Kraft-Momenten Servosystem für Industrieroboter. *Electronik*, pages 89-95, 1982.

[50] J. Dorsey. *Semiconductor Strain Gage Handbook.* BLH Electronics Inc., Waltham, Mass., 1994.

[51] J.C. Doyle and J.S Stein. Multivariable feedback design: Concepts for a classical/modern synthesis. *IEEE Transactions on Automatic Control*, 26(1):4-16, 1981.

[52] S.H. Drake, P.C. Watson, and S.N. Simunovic. High speed robot assembly of precision parts using compliance instead of sensory feedback. In *7th International Symposium on Industrial Robots*, pages 878-98, 1977.

[53] H.A. Elmaraghy and S. Payandeh. Contact prediction and reasoning for compliant robot motions. *IEEE Transactions on Robotics and Automation*, 5(4):533-538, 1989.

[54] P. Elosegui, R.W. Daniel, and P.M. Sharkey. Joint servoing for robust manipulator force control. In *IEEE International Conference on Robotics and Automation*, pages 246-251, 1990.

[55] S.D. Eppinger and W.P. Seering. On dynamic models of robot force control. In *IEEE International Conference on Robotics and Automation*, volume 1, pages 29-34, 1986.

[56] S.D. Eppinger and W.P. Seering. Three dynamic problems in robot force control. *IEEE Transactions on Robotics and Automation*, 8(6):751-758, 1992.

[57] H. A. Ernst. Computer-operated mechanical hand. In *Proceedings of the Spring Joint Conference*, volume 21, pages 39-51, 1962.

[58] B. Espiau and J.V. Catros. Use of optical reflectance sensors in robotics application. *IEEE Transactions on Systems, Man and Cybernetics*, 10(12):904-912, 1980.

[59] B. Espiau and A. Gruy. Sequential algorithms related to optical proximity sensors. In *First International Conference on Robot Vision and Sensory Control*, pages 223-232, 1981.

[60] A. Ferloni et al. Ordinatore: a dedicated robot that orients objects in a predetermined direction. In *10th International Symposium on Industrial Robots*, pages 471-475, 1980.

[61] G. Ferretti, G. Magnani, and P. Rocco. On the stability of integral force control in case of contact with stiff surfaces. *Transactions of ASME, Journal of Dynamic Systems, Measurement and Control*, 117(4):547-553, 1995.

[62] A.F. Filippov. Differential equations with discontinuous right-hand sides. *Transactions of the American Mathematical Society*, 62, 1964. Translation from the Russian.

[63] A.F. Filippov. Differential Equations with Discontinuous Right-hand Sides. Nauka, Moscow, 1985. In Russian.

[64] W.D. Fisher and M.S. Mujtaba. Hybrid position/force control: A correct formulation. *International Journal of Robotics Research*, 11(4):300-311, 1992.

[65] A.E. Fitzgerald, C. Kingsley, Jr., and A. Umans. *Electric Machinery*, 4th edition, McGraw-Hill, New York, 1983.

[66] K. Fock et al. Multicomponent force measurement for computer-aided control of intelligent robots. In *2nd International Conference on Arti-*

ficial Intelligence and Information – Control Systems of Robots, pages 249-251, 1980.

[67] A.M. Formal'skii. Controllability and Stability of Constrained Systems. Nauka, Moscow, 1974. In Russian.

[68] K.S. Fu, R.C. Gonzalez, and C.S.G. Lee. *Robotics: Control, Sensing, Vision and Intelligence*, McGraw-Hill, New York, 1987.

[69] A. Gaillet and C. Reboulet. An isostatic six-component force and torque sensor. In *13th International Symposium on Industrial Robots*, pages 102-111, 1983.

[70] I.M. Gelfand, V.S. Gurfinkel, M.L. Tsetlin, and M.L. Shik. *Models of the Structural-Functional Organization of Certain Biological Systems*, MIT Press, Cambridge, Mass., 1971.

[71] A.H. Gelig, G.A. Leonov, and V.A. Yakubovich. *Stability of Nonlinear Systems with Nonunique Equilibrium State*, Nauka, Moscow, 1978. In Russian.

[72] C.C. Geschke. A system for programming and controlling sensor-based robot manipulators. *IEEE Transactions on Pattern Analysis and Machine Intelligence*, 5(1):1-7, 1983.

[73] R.E. Goddard, Y.F. Zheng, and H. Hemami. Dynamic hybrid velocity/force control of robot compliant motion over globally unknown objects. *IEEE Transactions on Robotics and Automation*, 8(1):132-138, 1992.

[74] R.C. Goertz and F. Bevilaqua. A force-reflecting positional servomechanism. *Nucleonics*, 10(11):43-45, 1952.

[75] A.A. Goldenberg. Analysis of force control based on linear models. In *IEEE International Conference on Robotics and Automation*, pages 1348-53, 1992.

[76] H. Goldstein. *Classical Mechanics*. Addison-Wesley, Cambridge, Mass., 1981.

[77] J.J. Gonzalez and G.R. Widmann. A force commanded impedance control scheme for robots with hard nonlinearities. *IEEE Transactions on Control Systems and Technology*, 3(4):398-408, 1995.

[78] D.M. Gorinevsky and A.M. Formal'sky. Stability of an elastic manipulator with force feedback. *Mechanics of Solids*, 20(3):48-56, 1985.

[79] D.M. Gorinevsky and A.B. Lisunov. Controlling a manipulator with a force sensor and an allowance for elasticity of the manipulator parts. *Automation and Remote Control*, 3:21-23, 1989.

[80] D.M. Gorinevsky, A.V. Lensky, and A.Yu. Schneider. Dependence of the natural frequences of a force sensor on the design parameters. *Moscow University Mechanics Bulletin*, 41(3):10-16, 1986.

[81] Gorinevsky D.M. and Schneider A.Yu. Dynamics of small motions of a walking robot when there is feedback of the support reactions. *Mechanics of Solids*, 22(6):37-46, 1987.

[82] Gorinevsky D.M. and Schneider A.Yu. Force control in locomotion of legged vehicles over rigid and soft surfaces. *International Journal of Robotics Research*, 9(2):4-23, 1990.

[83] S.M. Gorlin and I.I. Slezinger. *Wind Tunnels and Their Instrumentation*. Izrael Program for Scientific Translations, 1966. Translation from the Russian.

[84] T. Goto, T. Inoyama, and K. Takeyasu. Precise insert operation by tactile controlled robot "HI-T-HAND EXPERT 2". In *4th International Symposium on Industrial Robots*, pages 209-217, 1974.

[85] T. Goto, T. Inoyama, and K. Takeyasu. Precise insert operation by tactile controlled robot. *Industrial Robot*, 1(5):225-228, 1974.

[86] T. Goto, K. Takeyasu, and T. Inoyama. Control algorithm for precision insert operation robots. *IEEE Transactions on Systems, Man and Cybernatics*, 10(1):19-25, 1980.

[87] A.A. Grishin, E.V. Gurfinkel, S.B. Mozhzhevelov, E.V. Efremov, A.Yu. Schneider, A.V. Lensky, and G.P. Nikitin. RPG-060 – robot for extracting parts stored 'ramble-scramble'. *Pribory i Sistemy Upravlenija (Control Devices and Systems)*, (1):35–37, 1985. In Russian.

[88] V.S. Gurfinkel, E.A. Devjanin, S.A. Anischenko, A.V. Lensky, S.B. Mozhzhevelov, and A.Yu. Schneider. Adaptive manipulator with force sensors. In *4th CISM-IFToMM Symposium on Theory and Practice of Robots and Manipulators, RoManSy*, pages 327-336, 1981.

[89] V.S. Gurfinkel, E.A. Devjanin, S.A. Anischenko, A.V. Lensky, S.B. Mozhzhevelov, and A.Yu. Schneider. An adaptive manipulator with force sensors. *Mechanism and Machine Theory*, 18(4):267-270, 1983.

[90] V.S. Gurfinkel, E.A. Devjanin, A.V. Lensky, S.B. Mozhzhevelov, A.M. Formalsky, and A.Yu. Schneider. Controlling a manipulator using sensory motor interaction. *Robotica*, 2(2):155-159, 1984.

[91] V.S. Gurfinkel, E.A. Devjanin, A.V. Lensky, S.B. Mozhzhevelov, A.M. Formalsky, and A.Yu. Schneider. Force feedback in the manipulator control system. *Mekhanika Tverdogo Tela*, 6:56-64, 1984. In Russian.

[92] V.S. Gurfinkel, D.M. Gorinevsky, A.B. Lizunov, A.Yu. Schneider, E.A. Devjanin, A.M. Formalsky, and A.V. Lensky. Control of manipulator with force sensing capability. *Information Control Problems in Manufacturing Technology, Proceedings of the 5th IFAC Symposium*, pages 325-330, 1986.

[93] V.S. Gurfinkel et al. Control of adaptive walking robot. In *Proceedings of the 10th IFAC World Congress*, pages 218-225, 1987.

[94] L. Gustafsson. Cleaning of castings – a typical job for an industrial robot. In *13th International Symposium on Industrial Robots*, volume 1, pages 24-30, 1983.

[95] H. Hanafusa and H. Asada. A robot hand with elastic fingers and its application to assembly process. In M. Brady et al., editors, *Robot Motion: Planning and Control*, pages 337-359, MIT Press, Cambridge, Mass., 1982.

[96] L.D. Harmon. A sense of touch begins to gather movement. *Sensor Review*, 18(4):267-270, 1981.

[97] L.D. Harmon. Automated tactile sensing. *International Journal of Robotics Research*, 1(2):3-32, 1982.

[98] S. Havlic. A new elastic structure for a compliant robot wrist. *Robotica*, 1(1):95-102, 1983.

[99] S. Hayati. Hybrid position/force control of multi-arm cooperating robots. *IEEE International Control Conference on Robotics and Automation*, volume 1, pages 82-89, 1986.

[100] D. Helms. Entwicklung eines Sechs-Komponenten Kraft-Moment-Aufnehmers. *Feinwerktechnik und Messtechnik*, 85(2):75-81, 1977.

[101] S. Hirose and K. Yoneda. Development of optical 6-axial force sensor and its signal calibration considering non-linear interference. In *IEEE International Conference on Robotics and Automation*, pages 46-53, 1990.

[102] G. Hirzinger. Sensory feedback in robotics state-of-the-art in research and industry. In *Preprints of the 10th World IFAC Congress on Automatic Control*, volume 4, pages 204-210, 1987.

[103] G. Hirzinger and M. Brunet. Fast and self-improving compliance using digital force-torque control. In *4th International Conference on Assembly Automation*, pages 268-281, 1983.

[104] G. Hirzinger and J. Dietrich. Multisensory robots and sensor-based path generation. In *IEEE International Conference on Robotics and Automation*, 1986.

[105] G. Hirzinger and K. Landzettel. Sensory feedback structures for robots with supervised learning. In *IEEE International Conference on Robotics and Automation*, pages 627-635, 1985.

[106] N. Hogan. Adaptive control of mechanical impedance by coactivation of antagonist muscles. *IEEE Transactions on Automatic Control*, 29(8):681-690, 1984.

[107] N. Hogan. Impedance control – an approach to manipulation, Part 1 – Theory, Part II – Inplementation, Part III – Applications. *Trans-*

actions of ASME. Journal of Dynamic Systems, Measurement and Control, 107(1):1-24, 1985.

[108] N. Hogan. Mechanical impedance control in assistive devices and manipulators. In M. Brady et al., editors, *Robot Motion: Planning and Control*, pages 362-371, MIT Press, 1983.

[109] Industrial evolution. *Robotics World*, 4(10):1, 1986.

[110] H. Inoue. An artificial hand controlled by a logical machine. *Journal of the Society of Instrument and Control Engineers*, 7(12):839-844, 1968.

[111] H. Inoue. Computer controlled bilateral manipulator. *Bulletin of JSME*, 14(69):199-207, 1971.

[112] H. Inoue. *Force feedback in precise assembly tasks*. Memo 308, MIT AI Laboratory, 1974.

[113] An intelligent, self-adaptive robot by IBM. *Epsilonics*, 3(3):3-4, 1983.

[114] S. C. Jacobsen et al. Design of the Utah/MIT dexterous hand. In *IEEE International Conference on Robotics and Automation*, pages 1520-1532, 1986.

[115] *JR3 Multi-Axis Load Cell Technologies. Sensor Overview*. Marketing material, 1996.

[116] K.P. Jankowski and H.A. Elmaraghy. Constraint formulation for invariant hybrid position/force control of robots. *Transactions of ASME. Journal of Dynamic Systems, Measurement and Control*, 118(6):290-299, 1996.

[117] D. Jeon and M. Tomizuka. Learning hybrid force and position control of robotic manipulators. *IEEE Transactions on Robotics and Automation*, 9(4):423-431, 1993.

[118] A.R. Johnston. *Optical proximity sensors for manipulators*. Technical Memorandum 33-112, Jet Propulsion Laboratory, Pasadena, 1973.

[119] S. Kaczanowski and A. Aderek. Adaptive robot performed assembly. In *4th CISM-IFToMM Symposium on Theory and Practice of Robots and Manipulators, RoManSy*, pages 278-288, 1981.

[120] R.K. Kankaanranta and H.N. Koivo. Dynamics and simulation of compliant motion of a manipulator. *IEEE Journal of Robotics and Automation*, 4:163-173, 1988.

[121] M. Kasai and K. Takeyasu. Trainable assembly system with an active sensory table possessing six axes. In *11th International Symposium on Industrial Robots*, pages 393-404, 1981.

[122] K. Kashiwagi, K. Ono, E. Izusi, T. Kurenuma, and K. Yamada. Force controlled robot for grinding. In *International Intelligent Robots and Systems Symposium*, pages 1001-1007, 1990.

[123] I. Kato, E. Okazaki, and H. Nakamura. The electrically controlled hand prosthesis using command disk and/or EMG. *Journal of the Society of Instrument and Control Engineers,* 6(4):12-17, 1967.

[124] S. Kawamura, E. Miyazaki, and S. Arimoto. Hybrid position/force control of robot manipulators based on learning method. In *International Conference on Advanced Robotics,* pages 235-242, 1985.

[125] P. Kazanzides, N.S. Bradley, and W.A. Wolovich. Dual-drive force/volocity control: Implementation and experimental results. In *IEEE International Conference on Robotics and Automation,* pages 92-97, 1989.

[126] H. Kazerooni. Contact instability of the direct drive robot when constrained by a rigid environment. *IEEE Transactions on Automatic Control,* 35(6):711-714, 1990.

[127] H. Kazerooni, J.J. Bausch, and B.M. Kramer. An approach to automated deburring by robot manipulators. *Transactions of ASME, Journal of Dynamic Systems, Measurement and Control,* 108(12):354-359, 1986.

[128] H. Kazerooni, P.K. Houpt, and T.B. Sheridan. Robust compliant motion for manipulators, Part 1: The fundamental concepts of compliant motion, Part 2: Design method. *IEEE Journal of Robotics and Automation,* 2(2):83-105, 1986.

[129] R.B. Kelly et al. Three vision algorithms for acquiring workpieces from bins. *Proceedings IEEE,* 71(7):803-820, 1983.

[130] J. Kerr and B. Roth. Analysis of multifingered hands. *International Journal of Robotics Research,* 4(4):3-17, 1986.

[131] O. Khatib. A unified approach for motion and force control of robot manipulators: The operational space formulation. *IEEE Journal of Robotics and Automation,* 3(1):43-53, 1987.

[132] O. Khatib and J. Burdick. Motion and force control of robot manipulators. In *IEEE International Conference on Robotics and Automation,* volume 3, pages 1381-1386, 1986.

[133] Ch.A. Klein and R.L. Briggs. Use of active compliance in control of legged vehicles. *IEEE Transactions on Systems, Man and Cybernetics,* 10(7):393-400, 1980.

[134] A.J. Koivo. *Fundamentals for Control of Robotic Manipulators.* John Wiley and Sons, New York, 1989.

[135] G.A. Korn and T.M Korn. *Mathematical Handbook for Scientists and Engineers.* McGraw-Hill, New York, 2nd edition, 1968.

[136] C. Kossin. Roboter gegen Grate. *Moderne Fertigung,* 10:70-72, 1986.

[137] F.M. Kulakov. *Supervisory Control of Robot Manipulators*. Nauka, Moscow, 1980. In Russian.

[138] H.B. Kuntze. Closed-loop algorithms for industrial robots – status and recent trends. In *Robot Control (Syroco'85), Proceedings of the First IFAC Symposium*, pages 437-443, 1986.

[139] H.B Kuntze, A. Jacubasch, and E. Brodbeck. Force and position control of industrial robots for manufacturing purposes. In *Robot Control (Syroco'85), Proceedings of the First IFAC Symposium*, pages 379-385, 1986.

[140] T.R. Kurfess, D.E. Whitney, and M.L. Brown. Verification of a dynamic grinding model. *Transactions of ASME, Journal of Dynamic Systems, Measurement and Control*, 110(12):403-409, 1988.

[141] H. Kwakernaak and R. Sivan. *Linear Optimal Control Systems*, John Wiley and Sons, New York, 1972.

[142] J.C. Latombe et al. The LM robot programming system. In H. Hanafasa and H. Inoue, editors, *The Robotics Research: Proceedings of the 2nd International Symposium*, volume 2, pages 377-391, 1985.

[143] N.A. Lehtomaki, N.R. Sandell, and M. Atans. Robustness results in linear-quadratic gaussian based multivariable feedback systems. *IEEE Transactions on Automatic Control*, 26(1):75-92, 1981.

[144] E.K. Lavrovsky and A.M. Formal'sky. Eigenvalues of systems with a digital computer in the control contour. *Soviet Journal of Computer and Systems Sciences*, 23(4):82-90, 1985.

[145] E.K. Lavrovsky and A.M. Formal'sky. On the theory of controlled systems with delays. *Moscow University Mechanics Bulletin*, 41(5):15-20, 1986.

[146] A.V. Lensky, A.B. Lizunov, A.M. Formalsky, and A.Yu. Schneider. Control of motion of a manipulator along a constraint. *Mechanics of Solids*, 22(5):37-45, 1987.

[147] A.V. Lensky, A.B. Lizunov, A.M. Formalsky, and A.Yu. Schneider. Manipulator motion along a constraint. *Robotica*, 4(4):247-253, 1986.

[148] R. Levi. Multicomponent calibration of machine-tool dynamometers. *ASME Transactions, Journal of Engineering for Industry*, 94(4):1067-1072, 1972.

[149] A. Liegeois, E. Dombre, and P. Borrel. Learning and control for computer-controlled manipulator. *IEEE Transactions on Automatic Control*, 25(6):1097-1102, 1980.

[150] H. Lipkin and J. Duffy. Hybrid twist and wrench control for a robotic manipulator. *Transactions of ASME, Journal of Mechanisms, Transmissions, and Automation in Design*, 110:138-144, 1988.

[151] R. Little. Force/torque sensing in robotic manufacturing. *Sensors. The Journal of Machine Perception*, 9(11), 1992.

[152] A.B. Lizunov, A.M. Formal'sky, and A.Yu. Schneider Controlling motion of manipulator with proximity sensor along an object contour. *Avtomatika i Telemekhanika (Automation and Remote Control*, (5):48-57, 1987. In Russian.

[153] G.J. Liu and A.A. Goldenberg. Robust hybrid impedance control of robot manipulators. In *IEEE International Conference on Robotics and Automation*, pages 287-292, 1991.

[154] T. Lozano-Perez. Robot programming. *Proceedings IEEE*, 71(7):821-841, 1983.

[155] T. Lozano-Perez. Compliance in robot manipulator. *Artificial Intelligence*, 25(1):5-12, 1985.

[156] T. Lozano-Perez, M.T. Mason, and R.H. Taylor. Automatic synthesis of fine motion strategies for robots. *International Journal of Robotics Research*, 3(1):3-24, 1984.

[157] Z. Lu and A.A. Goldenberg. Robust impedance control and force regulation: Theory and experiments. *International Journal of Robotics Research*, 14(3):225-254, 1995.

[158] J.Y.S. Luh, W.D. Fisher, and R.P.C. Paul. Joint torque control by a direct feedback for industrial robots. *IEEE Transactions on Automatic Control*, 25(2):153-161, 1983.

[159] J.Y.S. Luh and Y.F. Zheng. Constraint relation between two coordinated industrial robots for motion control. *International Journal of Robotics Research*, 6(3):60-70, 1987.

[160] Z. Luo and M. Ito. Control design of robot for compliant manipulation on dynamic environments. *IEEE Transactions on Robotics and Automation*, 9(3):286-296, 1993.

[161] H. Makino and N. Furuya. The SCARA robot and its family. In K. Rathmill, editor, *Robotic Assembly*, pages 13-26, 1985.

[162] J.A. Maples and J.J. Becker. Experiments im force control of robotic manipulators. In *IEEE International Conference on Robotics and Automation*, volume 2, pages 695-702, 1986.

[163] M.T. Mason. Compliance and force control of computer controlled manipulator. *IEEE Transactions on Systems, Man and Cybernetics*, 11(6):418-432, 1981.

[164] M. T. Mason. Compliant motion. In M. Brady et al., editors, *Robot Motion: Planning and Control*, pages 305-322, MIT Press, 1983.

[165] M. T. Mason and J. K. Salisbury. *Robot Hands and the Mechanics of Manipulation*. MIT Press, Cambridge, Mass., 1985.

[166] R. Masuda and K. Hasegawa. Total sensory system for robot control and its design approach. In *11th International Symposium on Industrial Robots*, pages 159-166, 1981.

[167] H. MacCallion and P.C. Wong. Some thoughts on the automatic assembly. *Industrial Robot*, 2(12):141-148, 1975.

[168] N.H. McClamroch and D. Wang. Feedback stabilization and tracking of constrained robots. *IEEE Transactions on Automatic Control*, 33(5):419-426, 1988.

[169] *Micro Electric Motors for Automatic Systems: Technical Handbook.* E.A. Lodochnikov and Yu.M. Yuferov, editors, Energia, Moscow, 1969. In Russian.

[170] J. Millberg and C. Maier. Contribution to the automation of screw-driving with the aid of industrial robots. *Annals of the CIRP*, 34(1):49-52, 1985.

[171] J.K. Mills and A.A. Goldenberg. Force and position control of manipulators during constrained motion task transition. *IEEE Transactions on Robotics and Automation*, 5(1):30-46, 1989.

[172] J.K. Mills and D.M. Lokhorst. Stability and control of robotic manipulators during contact/noncontact task transition. *IEEE Transactions on Robotics and Automation*, 9(3):148-163, 1993.

[173] E.E. Mitchell and J. Vranish. Magnetoelastic force feedback sensors for robots and machine tools: An update. In *Robots 9 Conference*, volume 1, pages 11-28, 1985.

[174] N.N. Moiseev. *Asymptotical Methods in Nonlinear Mechanics.* Nauka, Moscow, 1981. In Russian.

[175] A. Mortenson. Automatic grinding. In *13th International Symposium on Industrial Robots*, pages 1-11, 1983.

[176] R.M. Murray, Z. Li, and S.S. Sastry. *A Mathematical Introduction to Robotic Manipulation.* CRC Press, Boca Raton, 1994.

[177] Y. Nakamura. *Advanced Robotics: Redundancy and Optimization.* Addison-Wesley, Reading, Mass., 1991.

[178] Y. Nakamura and H. Hanafusa. A new optical proximity sensor for three dimensional autonomous trajectory control of robot manipulators. In *International Conference on Advanced Robotics*, pages 179-186, 1983.

[179] Y. Nakamura, T. Yoshikawa, and I. Futamata. Design and signal processing of six-axis force sensor. In *4th International Symposium on Robotics Research*, pages 75-81, 1987.

[180] E. Nakano, S. Ozaki, T. Ishida, and I. Kato. Cooperational control of the anthropomorphous manipulator "Melarm". In *4th International Symposium on Industrial Robots*, pages 251-260, 1974.

[181] Yu.I. Neimark. *Dynamical systems and control processes* Nauka, Moscow, 1978. In Russian.

[182] J.L. Nevins. Information – control aspects of sensor systems for intelligent robots. In *Robot Control (Syroco'85), Proceedings of the First IFAC Symposium*, pages 1-6, 1986.

[183] J.L. Nevins and D.E. Whitney. Computer-controlled assembly. *Scientific American*, 238(2):62-74, 1978.

[184] J.L. Nevins and D.E. Whitney. The force vector assembler concept. In *On the Theory and Practice of Robots and Manipulators, Proceedings of the First CISM-IFToMM Symposium*, 1973.

[185] A new assembly robot: the DEA PRAGMA 3000. *Industrial Robot*, 7(2):106-116, 1980.

[186] V. Nicolo et al. Industrial robots with sensory feedback applications to continuous arc welding. In *10th International Symposium on Industrial Robots*, pages 15-22, 1980.

[187] Y. Nishida, Y. Hirata, and A. Watanabe. A method of grasping unoriented objects using force sensing. In *Robotics and Factories of the Future: Proceedings of the First Conference*, pages 329-336, 1984.

[188] M.S. Ohwovoriole, J.W. Hill, and B. Roth. On the theory of single and multiple insertions in industrial assemblies. In *10th International Symposium on Industrial Robots*, pages 545-557, 1980.

[189] D.E. Okhotsimskii. Sensorization of robots for the adaptation and control of automatic assembly. *Soviet Journal of Computer and Systems Sciences*, 25(6):46-52, 1987.

[190] D.E. Okhotsimskii and Yu.F. Golubev. *Mechanics and Motion Control of Walking Machine*. Nauka, Moscow, 1984. In Russian.

[191] N. Onda, K. Asakawa, and T. Kamada. Precise robotic assembly using active compliance control. In *16th International Symposium on Industrial Robots*, pages 445-455, 1986.

[192] *Operational Amplifiers. Design and Applications*. J. G. Graeme, G. E. Tobey, and L. P. Huelsman, editors, New York, 1971.

[193] D.E. Orin and S.Y. Oh. Control of force distribution in robotics mechanisms containing closed kinematic chains. *Transactions of ASME, Journal of Dynamic Systems, Measurement and Control*, 103(2):134-141, 1981.

[194] S.N. Osipov and A.M. Formal'sky. On the theory of manipulation systems with force sensitization. *Mechanics of Solids*, 22(2):62-70, 1987.

[195] R. Palm. Design of a fuzzy controller for a second guided robot manipulator. In *Proceedings of 2nd International Symposium of the Foundation of Kinematics, Dynamics and Control*, part 3, pages 1-12, 1987.

[196] L.A. Pars. *A Treatise on Analytical Dynamics*. John Wiley and Sons, New York, 1965.

[197] R.P. Paul. *Robot Manipulators: Mathematics, Programming and Control*. MIT Press, Cambridge, Mass., 1981.

[198] R.P. Paul. WAVE – a model based language for manipulator control. *Industrial Robot*, 4(1):10-17, 1977.

[199] R.P. Paul and B. Shimano. Compliance and control. In *Joint Automatic Control Conference*, pages 694-699, 1976.

[200] S. Payandeh and A.A. Goldenberg. A robust force controller. In *IEEE International Conference on Robotics and Automation*, pages 36-41, 1991.

[201] D. Pessen. Tactile gripper system for robotic manipulators. In *12th International Symposium on Industrial Robots*, pages 411-416, 1982.

[202] F. Pfeiffer, K. Richter, and H. Wapenhans. Elastic robot trajectory planning with force control. In: *Proceedings of the IFIP TC 7 Conference on Modelling the Innovation: Communications, Automation and Information Systems*, pages 201-212, 1990.

[203] *Piezo-Instrumentation KISTLER, an Outstanding Tool in Biomechanics: Quartz measuring platform: 63.201e*, 1983.

[204] G. Piller. A compact six-degree-of-freedom force sensor for assembly robots. In *12th International Symposium on Industrial Robots*, pages 121-130, 1982.

[205] L.E. Pfeffer and O. Khatib. Joint torque sensory feedback in the control of a PUMA manipulator. *IEEE Transactions on Robotics and Automation*, 5(4):418-425, 1989.

[206] G. Plank and G. Hirzinger. Controlling a robot's motion speed by a force-torque sensor for deburring problems. In *Information Control Problems in Manufacturing Technology, Proceedings of the 4th IFAC Symposium*, pages 97-102, 1983.

[207] L.S. Pontryagin, V.G. Boltyanskii, R.V. Gamkrelidze, and E.F. Mishchenko. *The Mathematical Theory of Optimal Processes*. John Wiley and Sons, New York, 1962.

[208] E.P. Popov, F.A. Vereschagin, and S.L. Zenkevich. *The Manipulation Robots: Dynamics and Algorithms*. Nauka, Moscow, 1978. In Russian.

[209] J.A. Purbrick. A multiaxis force sensing finger. In *Proceedings of the Second International Computer Engineering Conference*, pages 53-58, 1982.

[210] H.P. Qian and J. De Schutter. The role of damping and low pass filtering in the stability of discrete time implemented robot force control. In *IEEE International Conference on Robotics and Automation*, pages 1368-73, 1992.

[211] M.H. Raibert and J.J. Craig. Hybrid position/force control of maniuplators. *Transactions of ASME, Journal of Dynamic Systems, Measurement and Control*, 103(2):126-133, 1981.

[212] K. Richter. *Kraftregelung Elastischer Roboter*. Dr.-Ing. Dissertation, Technische Univerität München, Fortschritt-berichte, Reihe 8: Mess-, Steuerungs- und Regelungstechnik, Nr. 259, VDI Verlag, 1991.

[213] P.K. Roberts, R.P. Paul, and B.M. Hillberry. Effect of wrist force sensor stiffness on the control of robot manipulators. In *IEEE International Conference on Robotics and Automation*, pages 269-274, 1985.

[214] *Robot Motion: Planning and Control*, M. Brady et al., editors, MIT Press, Cambridge, Mass., 1983.

[215] *Robotic Assembly*, K. Rathmill, editor, IFS, Kempston, Bedford, UK, 1985.

[216] A.D. Rodic and M.K. Vucobratovic. Regulator of minimal variance in hybrid control strategy of manipulation robots. In *IEEE International Conference on Robotics and Automation*, pages 1232-37, 1992.

[217] Ya.N. Roitenberg. *Automatic control*. 3d edition. Nauka, Moscow, 1992. In Russian.

[218] B.W. Rooks. The fettling of castings – a job for industrial robot. In *8th International Symposium on Industrial Robots*, pages 92-105, 1978.

[219] C.A. Rosen and D. Nitzan. Development in programmable automation. *Manufacturing Engineering*, 75(3):26-30, 1975.

[220] C.A. Rosen and D. Nitzan. Use of sensors in programmable automation. *Computer*, 10(12):12-23, 1977.

[221] T. Sakata. An experimental bin-picking robot system. In *3d International Conference on Assembly Automation*, pages 615-626, 1982.

[222] J.K. Salisbury. Active stiffness control of a manipulator in Cartesian coordinates. In *19th Conference on Decision and Control*, pages 95-100, 1980.

[223] J.K. Salisbury. Interpretation of contact geometries from force measurement. In *International Conference on Robotics*, pages 240-247, 1984.

[224] J.K. Salisbury and J.J. Craig. Articulated hands: force control and kinematic issues. *International Journal of Robotics Research*, 1(1):4-17, 1982.

[225] M. Salman. Assembly by robots. *Industrial Robot*, 4(2):81-85, 1977.

[226] N. Salman. New part orienting device. In *7th International Symposium on Industrial Robots*, pages 185-194, 1977.

[227] I.I. Sankov and A.I. Zibenberg. *Mechanization and Automation of Casting Abrasive Grinding*, Mashinostroenie, Moscow, 1972. In Russian.

[228] V.D. Scheinman. *Design of a Computer Controlled Manipulator*. Stanford AI project memo, AIM-92, Stanford University, 1969.

[229] D. Schmid and F. Sachsenmaier. Automatisiertes Entfernen von Angussen und Gussgraten. *Industrie–Anzeiger*, 108(46):76-77, 1986.

[230] L. Schmieder. Kraft-Momenten Fühler. *Fertigung*, 9(6):161-164, 1978.

[231] A. Schneider, U. Schmucker, T. Ihme, E. Devjanin, and K. Savitsky. Force control in locomotion of legged vehicle and body movement for mounting operations. In: *Proceedings of the 9th World Congress on Theory of Machines and Mechanisms*, volume 3, pages 2363-67, 1995.

[232] A.J. Schneider. Steuerung von Robotern mit Kraftrückkopplung. *Maschinenbautechnik*, 36:160-163, 1987.

[233] S.A. Schneider and R.H. Cannon, Jr. Impedance control for cooperative manipulation: Theory and experimental results. *IEEE Transactions on Robotics and Automation*, 8(3):383-394, 1992.

[234] M. Schweizer. *Tactile Sensoren für Programmierbare Handhabungsgeräte*. Dr.-Ing. Dissertation, Univerität Stuttgart, 1978.

[235] M. Schweizer and E. Abele. Sensor-Gesteuerte Industrieroboter für Bearbeitungs- Aufgaben. In *8th International Symposium on Industrial Robots*, pages 904-919, 1978.

[236] B. Shariat, P. Coiffet, and A. Fournier. A strategy to achieve an assembly by means of an inacurate, flexible robot. *Digital Systems for Industrial Automation*, 2(2):153-178, 1984.

[237] A. Sharon, N. Hogan, and D.E. Hardt. Controller design in the physical domain (Application to robot impedance control). In *IEEE International Conference on Robotics and Automation*, pages 552-559, 1989.

[238] B. Shimano and B. Roth. On force sensing information and its use in controlling manipulators. In *Information – Control Problems in Manufacturing Technology: Proceedings of IFAC International Symposium*, pages 119-126, 1978.

[239] K.B. Shimoga and A.A. Goldenberg. Grasp admittance center: A concept and its implications. In *IEEE International Conference on Robotics and Automation*, pages 293-298, 1991.

[240] J. Simons and H. Van Brussel. Force control schemes for robot assembly. In *Robotic Assembly*, K. Rathmill, editor, pages 253-265, 1985.

[241] J. Simons, H. Van Brussel, J. De Shutter, and J. Verhaert. A self learning automation with variable resolution for high precision assembly by industrial robots. *IEEE Transactions on Automatic Control*, 27(5):1109-1113, 1982.

[242] S. Simunovic. Force information in assembly processes. In *5th International Symposium on Industrial Robots*, pages 415-431, 1975.

[243] P. Sinha and A.A. Goldenberg. A unified theory for hybrid control of manipulators. In *IEEE International Conference on Robotics and Automation*, pages 343-348, 1993.

[244] J.J. Slotine and W. Li. *Applied Nonlinear Control*. Prentice-Hall, Englewood Cliffs, N.J., 1991.

[245] Ch.H. Spaulding. A three-axis force sensing system for industrial robots. In *The Third International Conference on Assembly Automation*, pages 565-567, 1982.

[246] M.W. Spong. On the force control problem for flexible joint manipulators. *IEEE Transactions on Automatic Control*, 34(1):107-111, 1989.

[247] M.W. Spong and M. Vidyasagar. *Dynamics and Control of Robot Manipulators*. John Wiley and Sons, New York, 1989.

[248] G.P. Starr. Edge-following with a PUMA manipulator using VAL-II. In *IEEE International Conference on Robotics and Automation*, pages 379-383, 1986.

[249] T.M. Stepien, L.M. Sweet, M.C. Good, and M. Tomizuka. Control of tool/workpiece contact force with application to robotic deburring. *IEEE Journal on Robotics and Automation*, 3(1):7-18, 1987.

[250] V. Strejc. *State Space Theory of Discrete Linear Control*. John Wiley and Sons, New York, 1981.

[251] D. Stokic, M. Vukobratovic, and D. Hristic. Implementation of force feedback in manipulator robots. *International Journal of Robotics Research*, 5(1):66-76, 1986.

[252] G. Stute and H. Erne. The control design of an industrial robot with advanced tactile sensitivity. In *9th International Symposium on Industrial Robots*, pages 518-528, 1979.

[253] M.M. Svinin and M. Uhciyama. Analytical models for designing force sensors. In *IEEE International Conference on Robotics and Automation*, pages 1778-83, 1994.

[254] M. Takahashi et al. Assembly robot system with force sensor and 3-D vision. In *Robotic Assembly*, K. Rathmill, editor, pages 267-274, 1985.

[255] K. Takeyasu et al. An approach to the integrated intelligent robot with multiple sensory feedback: Construction and control functions. In *7th International Symposium on Industrial Robots*, pages 523-527, 1977.

[256] K. Takeyasu, T. Goto, and T. Inoyama. Precision insertion control robot and its application. *Transactions of ASME, Journal of Dynamic Systems, Measurement and Control*, 98(11):1313-1318, 1976.

[257] T.J. Tarn, Y. Wu, and A. Isidori. Force regulation and contact transition control. *IEEE, Control Systems*, 16(1):32-40, 1996.

[258] R.H. Taylor and D.D. Grossman. An integrated robot system architecture. *Proceedings IEEE*, 71(7):842-855, 1983.

[259] A.N. Tikhonov. Systems of differential equations containing small parameters at derivatives. *Matematicheskij Sbornik*, 31(3):575-586 1952. In Russian.

[260] S. Timoshenko. *Elements of Strength of Materials*. Van Nostrand, New York, 1986.

[261] S.P. Timoshenko and D.H. Young. *Theory of Structures*. McGraw-Hill, New York, 1965.

[262] T. Tsujimura and T. Yabuta. Object detection by tactile sensing method employing force/torque information. *IEEE Transactions on Robotics and Automation*, 5(4):444-450, 1989.

[263] Ia.Z. Tsypkin. *Relay Control Systems*. Cambridge University Press, Cambridge, New York, 1984.

[264] T.L. Turner, J.J. Craig, and W.A. Gruver. A microprocessor architecture for advanced robot control. In *14th International Symposium on Industrial Robots*, pages 407-416, 1984.

[265] M. Uchiyama, E. Bayo, and E. Palma-Villalon. A systematic design procedure to minimize a performance index for robot force sensors. *Transactions of ASME. Journal of Dynamic Systems, Measurement and Control*, 113(9):388-394, 1991.

[266] V.I. Utkin. Variable structure systems with sliding mode: A survey. *IEEE Transactions on Automatic Control*, 22(2), 1977.

[267] V.I. Utkin. *Sliding Modes in Control and Optimization*. Springer-Verlag, Berlin, 1992.

[268] H. Van Brussel, H. Belien, and H. Thielemans. Force sensing for advanced robotic control. *Robotics*, 2(2):139-148, 1986.

[269] H. Van Brussel and J. Simons. Automatic assembly by active force feedback accomodation. In *8th International Symposium on Industrial Robots*, pages 181-193, 1978.

[270] H. Van Brussel, H. Thielemans, and J. Simons. Further developments of the active adaptive compliant wrist (AACW) for robot assembly. In *11th International Symposium on Industrial Robots*, pages 377-384, 1981.

[271] H. Van Brussel and L. Vastmas. A compensation method for controlling time varying robot configuration. In *14th International Symposium on Industrial Robots*, pages 323-333, 1984.

[272] D. Vischer and O. Khatib. Design and development of high-performance torque-controlled joints. *IEEE Transactions on Robotics and Automation*, 11(4):537-544, 1995.

[273] J. Volmer et al. Montage mit Sensorgesteuertem Greifer. *Maschinenbautechnik*, 4:165-169, 1980.

[274] R. Volpe and P. Khosla. Experimental verification of a strategy for impact control. In *IEEE International Conference on Robotics and Automation*, pages 1854-60, 1991.

[275] R. Volpe and P. Khosla. Analysis and experimental verification of a fourth order plant model for manipulator force control. *IEEE Robotics and Automation Society. Robotics and Automation Magazine*, 1(2):4-13, 1994. pages 1854-60, 1991.

[276] M. Vukobratovic and D. Stokic. *Applied Control of Manipulation Robots: Analysis, Synthesis and Exercises*. Springer Verlag, 1989.

[277] B.J. Waibel and H. Kazerooni. Theory and experiments on the stability of robot compliance control. *IEEE Transactions on Robotics and Automation*, 7(1):95-104, 1991.

[278] K.J. Waldron. Force and motion management in legged locomotion. *IEEE Journal on Robotics and Automation*, 2(4):214-220, 1986.

[279] C. Wampler. Multiprocessor control of a telemanipulator with optical proximity sensors. *International Journal of Robotics Research*, 3(1):40-50, 1984.

[280] S.S.S. Wang and P.M. Will. Sensors for computer controlled mechanical assembly. *Industrial Robot*, 5(1):9-18, 1978.

[281] H.J. Warnecke and E. Abele. Fettling of castings with industrial robots. *Annals of the CIRP*, 32(1):405-409, 1983.

[282] H.J. Warnecke, E. Abele and R.D. Schraft. Pilot installation for fettling of castings with industrial robots – basic equipment, strategies, experiments and results. In *11th International Symposium on Industrial Robots*, pages 713-722, 1981.

[283] H.J. Warnecke, E. Abele, J. Walther, and G. Fisher. Investigations of the screwdriving process with sensor-controlled industrial robots. *Annals of the CIRP*, 34(1):41-44, 1985.

[284] H.J. Warnecke and R.D Schraft. *Handbuch Handhabungs-, Montage- und Industrierobotertechnik*, Verlag Moderne Industrie, 1993.

[285] H.J. Warnecke, M. Schweizer, and E. Abele. Cleaning of castings with sensor-controlled industrial robot. In *10th International Symposium on Industrial Robots*, pages 535-544, 1980.

[286] H.J. Warnecke, M. Schweizer, and D. Haaf. An adaptable programmable assembly using compliance system and visual feedback. In *10th International Symposium on Industrial Robots*, pages 481-490, 1980.

[287] P.C. Watson. The RCC concept and its application to assembly system. In *Information Control Problems in Manufacturing Technology, Proceedings of IFAC Symposium*, 1977.

[288] P.C. Watson and S.H. Drake. Pedestal and wrist sensors for automatic assembly. In *5th International Symposium on Industrial Robots*, pages 501-511, 1975.

[289] H. West and H. Asada. A method for the design of hybrid position/force control for manipulators constrained by contact with the environment. In *IEEE International Conference on Robotics and Automation*, pages 251-259, 1985.

[290] J.T. Wen and S. Murphy. Stability analysis of position and force control for robot arms. *IEEE Transactions on Automatic Control*, 36(3):365-371, 1991.

[291] D.E. Whitney. Force feedback control of manipulator fine motions. *Transactions of ASME, Journal of Dynamic Systems, Measurement and Control*, 99(2):91-97, 1977.

[292] D.E. Whitney. Historical perspective and state of the art in robot force feedback control. *International Journal of Robotics Research*, 6(1):3-14, 1987.

[293] D.E. Whitney. Part mating in assembly. In *Handbook on Industrial Robots*, Pergamon Press, pages 1084-1116, 1985.

[294] D.E. Whitney. Resolved motion rate control of manipulators and human prostheses. *IEEE Transactions on Man-Machine Systems*, 10(2):47-53, 1969.

[295] D.E. Whitney and A.C. Esdall. Modelling robot contour process. In H. Hanafusa and H. Inoue, editors, *The Robotics Research: 2nd International Symposium*, volume 2, pages 163-170, 1985.

[296] D.E. Whitney, A.C. Esdall, A.B. Todtenkopf, T.R. Kurfess, and A.R. Tate. Development and control of an automated robotic weld bead grinding system. *Transactions of ASME, Journal of Dynamic Systems, Measurement and Control*, 112(6):166-176, 1990.

[297] D.E. Whitney and J.L. Nevins. What is Remote Center Compliance (RCC) and what can it do? In *9th International Symposium on Industrial Robots*, pages 135-152, 1979.

[298] D. E. Whitney, P. C. Watson, S. H. Drake, and S. N. Simunovic. Robot and manipulator control by extroceptive sensors. In *Joint Automatic Control Conference*, pages 155-163, 1977.

[299] L.S. Wilfinger, J.T. Wen, and S.H. Murphy. Integral force control with robustness enhancement. *IEEE, Control Systems*, 14(1):31-40, 1994.

[300] P.M. Will and D.D. Grossman. An experimental system for computer controlled mechanical assembly. *IEEE Transactions on Computers*, 24(9):879-888, 1975.

[301] C. Wu. Compliance control of a robot manipulator based on joint torque servo. *International Journal of Robotics Research*, 4(3):55-71, 1985.

[302] T. Yabuta. Nonlinear basic stability concept of the hybrid position/force control scheme for robotic manipulators. *IEEE Transactions on Robotics and Automation*, 8(5):663-670, 1992.

[303] T. Yabuta, T. Yamada, T Tsujimura, and H. Sakata. Force control of servomechanism using adaptive control. *IEEE Journal of Robotics and Automation*, 4(2):223-228, 1988.

[304] V.A. Yakhimovich, V.E. Golovashenko, and I.Ya. Kulinich. *Automation of assembly of threaded joints*. Visha-Shkola, Lviv, 1982. In Russian.

[305] T. Yoshikawa. Dynamic hybrid position/force control of robot manipulators: Description of hand constraints and calculation of joint driving force. *IEEE Journal of Robotics and Automation*, 3(5):386-392, 1987.

[306] T. Yoshikawa, T. Sugie, and M. Tanaka. Dynamic hybrid position/force control of robot manipulators - controller design and experiment. *IEEE Journal of Robotics and Automation*, 4(6):699-705, 1988.

[307] T. Yoshikawa, K. Harada, and A. Matsumoto. Hybrid position/force control of flexible-macro/rigid-micro manipulators systems. *IEEE Journal of Robotics and Automation*, 12(5):633-639, 1996.

[308] T. Yoshikawa and T. Miyazaki. A six-axis force sensor with three-dimensional cross-shape structure. In *IEEE International Conference on Robotics and Automation*, pages 249-255, 1989.

[309] T. Yoshikawa and A. Sudou. Dynamic hybrid position/force control of robot manipulators – on-line estimation of unknown constraint. *IEEE Transactions on Robotics and Automation*, 9(2):220-226, 1993.

[310] K. Youcef-Toumi and D.A. Gutz. Impact and force control. In *IEEE International Conference on Robotics and Automation*, pages 410-416, 1989.

[311] H. Zhang and R.P. Paul. Hybrid control of robot manipulators. In *IEEE International Conference on Robotics and Automation*, pages 602-607, 1985.

[312] Y.F. Zheng and Y. Fan. Robot force sensor interacting with environments. *IEEE Transactions on Robotics and Automation*, 7(1):156-164, 1991.

[313] Anmeldung BDR 2132012, ICI G01L 1/22. Kraftmesseinrichtung, H. Eder, 1973.

[314] Patent BDR 2313953, ICI G01L 7/05. Kraftmessumformer, M. Parker, 1973.

[315] Patent GDR 209419, ICI B65G 47/12. Handhabeeinrichtung zum Vereinzelln und Ordnen von Ferromagnetischen Werkstucken, T. Ungethuem, A. Hubner, and W. Titz, 1984.

[316] U.S. Patent 2865200, NCI 73-147. Wind tunnel roll moment balance, L. P. Gieseler, 1958.

[317] U.S. Patent 2866059, NCI 339-5. Slotter type multiple bending beam, E. Laiming, 1959.

[318] U.S. Patent 3004231, NCI 338-5. Parallelogram beam type load cell, E. Laiming, 1961.

[319] U.S. Patent 3159027, NCI 73-147. Multicomponent internal strain gauge balance, T. M. Curry, 1964.

[320] U.S. Patent 3201977, NCI 73-885. Strain gauge/A. I. Kutsau, 1965.

[321] U.S. Patent 3616690, NCI 73-172. Weight measuring apparatus, C. M. Harden, 1971.

[322] U.S. Patent 3640130, NCI 73-133. Force and moment arrangement, G. Spescha et all, 1972.

[323] U.S. Patent 3780573, NCI 73-146. Uniformity test machines, W. Reus, 1973.

[324] U.S. Patent 3921445, NCI 73-133. Force and torque sensing method and means for manipulators and the like, J. W. Hill and A. J. Sword, 1975.

[325] U.S. Patent 3939704, NCI 73-133. Multiaxis load cell, R. B. Zipin, 1976.

[326] U.S. Patent 3948093, NCI 73-133. Six-degree-of-freedom force transducer for a manipulator system, G. A. Folchi, G. L. Shelton, and S. S. Wong, 1976.

[327] U.S. Patent 4076131, NCI 214-1. Industrial robot, L. Dahlstrom and B. Nilsson, 1978.

[328] U.S. Patent 4092854, NCI 73-133. Multiaxis load cell, J. L. Henry and C. F. Ru-off, 1978.

[329] U.S. Patent 4094192, NCI 73-133. Method and apparatus for six-degree-of-freedom force sensing, P. C. Watson and S. H. Drake, 1978.

[330] U.S. Patent 4098001, NCI 33-163. Remote center compliance system, P. C. Watson, 1978.

[331] U.S. Patent 4099409, NCI 73-133. Multiaxis load cell with arcuate flextures, J. Jr. Edmond, 1978.

[332] U.S. Patent 4107985, NCI 73-141A. Load cell, T. R. Sommer, 1978.

[333] U.S. Patent 4132118, NCI 214-1. Asymmetric six-degree-of-freedom force transducer system for a computer-controlled manipulator system, S. S. M. Wang, M. Lake, M. A. Wesley, and P. M. Will, 1979.

[334] U.S. Patent 4156835, NCI 318-561. Servo-controlled mobility device, D. E. Witney and J. L. Nevins, 1974.

[335] U.S. Patent 4178799, NCI 73-141. Force and bending moment sensing arrangement and structure, L. Schmieder, A. Vilgertshofer, and F. Mettin, 1979.

[336] U.S. Patent 4216467, NCI 341-365. Hand controller, J. E. Colston, 1980.

[337] U.S. Patent 4259863, NCI 73-862.04. Multiaxis load cell, G. C. Rieck and A. J. Malarz, 1981.

[338] U.S. Patent 4398429, NCI 73-862.04. Force platform construction and method of operating same, N. H. Cook and F. J. Caringan, 1983.

[339] U.S. Patent 4448083, NCI 73-862.04. Device for measuring components of force and moment in plural directions, J. Hayashi, 1983.

[340] U.S. Patent 4478089, NCI 73-862.04. Three-axial force transducer for a manipulator gripper, H. F. Aviles et al., 1984.

[341] U.S. Patent 4552028, NCI 73-862.04. Device for measuring force, C. W. Burckhardt, P. Stauber, and G. Piller, 1985.

[342] U.S. Patent 4577513, NCI 73-862.04. Strain sensing arrangement, A. R. Hardwood and J. A. Ward, 1986.

[343] U.S. Patent 4621533, NCI 73-862.04. Tactile load sensing transducer, S. S. Gindy, 1986.

[344] U.S. Patent 4672855, NCI 73-862.04. Device for measuring forces and torques in different directions, L. Schmieder, 1987.

[345] USSR Patent 474435, ICI W25J 9/00. Sensing table, A.A. Kobrinskii, A.I. Korendyasev, and B.L. Salamandra, 1975, Bul. 23.

[346] USSR Patent 654382, ICI W23 R19/06. Method for assembly of threaded joints, A.G. Gerasimov and Yu. M. Budnikov, 1979, Bul. 12.

[347] USSR Patent 766854, ICI W25J 9/00. Manipulator, Yu.G. Kozyrev, V.S. Gurfinkel, V.B. Velikovich, E.M. Kanaev, S.V.Zhitomirskii, B.L. Samorodskikh, A.Yu. Schneider, E.V. Gurfinkel, 1980, Bul. 26.

[348] USSR Patent 795940, ICI B25J 15/00. Manipulator wrist sensor, E.V. Pismennaya, V.E. Novakovskii, S.G. Vekshin, and G.V. Pismennyi, 1981, Bul. 2.

[349] USSR Patent 918090. Electromagnetic gripper, S.A. Anischenko, E.V. Gurfinkel, V.S. Gurfinkel, E.A. Devjanin, S.A. Anischenko, A.V. Lensky, S.B. Mozhzhevelov, and A.Yu. Schneider, 1982, Bul. 13.

[350] USSR Patent 928375, ICI G06G 7/60. A device for modelling a whisker mechanoreceptor, D.V. Krivetz, 1982, Bul. 18.

[351] USSR Patent 974155, ICI G01L 1/22. Force and torque sensors, A.Sh. Koliskor, 1982, Bul. 42.

[352] USSR Patent 975393, ICI W25J 15/06. Manipulator for grasping ferromagnetic materials, L.S. Yampol'skii, S.A. Resnikov, and S.L. Yampol'skii, 1982, Bul. 43.

[353] USSR Patent 994225, ICI W25J 15/00. Device for grasping workpieces, V.G. Ostapchuk, V.V. Kirsanov, E.M. Kanaev, Yu.G. Kozyrev, and A.A. Nikolaev, 1983, Bul. 5.

[354] USSR Patent 1016713, ICI G01L 5/16. A six-component sensor, E.V. Grigalyuk, V.I Kuzmin, and A.A. Mikhailov, 1983, Bul. 17. A.A. Nikolaev, 1983, Bul. 5.

[355] USSR Patent 1082337, ICI G01L 5/16. A six-degrees-of-freedom sensor, D. Venturello and O. Salvatore (Italy), 1984, Bul. 14. A.A. Nikolaev, 1983, Bul. 5.

[356] USSR Patent 1216680, ICI G01L 1/22, 5/16. A force-torque sensor, V.G. Zapuskalov, A.K. Legkobyt, and R.H. Chernyshev, 1986, Bul. 9. A.A. Nikolaev, 1983, Bul. 5.

[357] USSR Patent 1281938, ICI G01L 5/16. An instrumented platform for measuring multicomponent forces and torques, M.A. Trakhimovich, V.V. Bagreev, and N.Yu. Struchkova, 1987, Bul. 1. A.A. Nikolaev, 1983, Bul. 5.

[358] USSR Patent 1308466, ICI B25J 19/02. Manipulator wrist sensor, V.A. Godzikovskii, D.M. Gorinevsky, A.V. Lensky, and A.Yu. Schneider, 1987, Bul. 17.

[359] USSR Patent 1308467, ICI B25J 19/02. Manipulator wrist sensor, V.A. Godzikovskii, D.M. Gorinevsky, A.V. Lensky, and A.Yu. Schneider, 1987, Bul. 17.

[360] USSR Patent 1308468, ICI B25J 19/02. Manipulator wrist sensor, V.A. Godzikovskii, D.M. Gorinevsky, A.V. Lensky, and A.Yu. Schneider, 1987, Bul. 17.

[361] USSR Patent 1421535, ICI B15J 15/00. A sensing device, V.A. Godzikovskii, Yu.Ja. Krashennikov, and A.A. Tsyvin, 1988, Bul. 33.

[362] USSR Patent 1460630, ICI 601B. A platform for determining a part position, D.M. Gorinevsky, 1989, Bul. 7.

Index